Beta-blockers –
present status and future prospects

Beta-blockers—present status and future prospects

Edited by W. Schweizer, Basle

Hans Huber Publishers
Bern Stuttgart Vienna

An international symposium
Juan-les-Pins, 27th–29th May 1974

Chairman: Prof. W. Schweizer

Manuscripts coordinated and translated under the supervision of:

Dr. C. Adams
Mr. I. C. W. Bigland, M.A.
Prof. W. Grüter
Dr. W. Jochum
Dr. H. Karrer
Dr. R. Ludwig
Mr. H. D. Philps, M.A.

ISBN 3-456-80203-X

Contents

Address of chief coordinator:
Dr. C. Adams, c/o ciba-geigy limited, Klybeckstrasse 141,
CH–4002 Basle, Switzerland

List of speakers

Prof. C. Bartorelli, Direttore dell'Istituto di Ricerche Cardiovascolari dell'Università di Milano, Via F. Sforza 35, I–20122 Milan, Italy

Prof. H. J. Bein, Meisenstrasse 11, CH–4104 Oberwil, Switzerland

Prof. K. D. Bock, Leiter der Abteilung für Nieren- und Hochdruckkranke, Medizinische Klinik and Poliklinik, Universitätsklinikum der Gesamthochschule Essen, Hufelandstrasse 55, D–4300 Essen, German Federal Republic

Prof. H. Bricaud, Hôpital du Tondu, F–33000 Bordeaux, France

Prof. J. Brod, Dr. Sc., F.R.C.P. (Lond.), Abteilung für Nephrologie, Medizinische Klinik, Medizinische Hochschule Hannover, Karl-Wiechert-Allee 9, D–3000 Hannover-Kleefeld, German Federal Republic

Prof. H. Brunner, Research Department (Pharmacology), Pharmaceuticals Division, ciba-geigy limited, Klybeckstrasse 141, CH–4002 Basle, Switzerland

Dr. F. R. Bühler, Leiter des Hypertonie-Service, Departement für Innere Medizin der Universität, Kantonsspital, Spitalstrasse 21, CH–4004 Basle, Switzerland

Dr. D. M. Burley, M.B., B.S., M.R.C.S. (Eng.), L.R.C.P. (Lond.), ciba laboratories limited, Wimblehurst Road, Horsham, Sussex RH12 4AB, England

Dr. M. E. Carruthers, M.D., M.R.C.G.P., Senior Lecturer in Chemical Pathology, Department of Chemical Pathology (Research), St. Mary's Hospital Medical School, London W2 1PG, England

Prof. E. K. Chung, Director of the Heart Station, Division of Cardiology, Department of Medicine, Thomas Jefferson University, Philadelphia 19107, Pennsylvania, U.S.A.

Prof. P. Cottier, Chefarzt, Bezirksspital, Weissenaustrasse 27, CH–3800 Interlaken, Switzerland

Dr. M. D. Day, Ph.D., Department of Pharmacy, The University of Aston in Birmingham, Gosta Green, Birmingham B4 7ET, England

Prof. U. C. Dubach, Direktor der Medizinischen Poliklinik, Hebelstrasse 1, CH–4056 Basle, Switzerland

Prof. A. Eisalo, University Medical Clinic, Helsinki, Finland

Prof. Ag. A. Froment, Médecin des Hôpitaux, Hôpital Cardio-Vasculaire et Pneumologique, 59 boulevard Pinel, F–6900 Lyons, France

Priv.-Doz. Dr. B. Garnier, Hôpital Daler, Route de Bertigny, CH–1700 Fribourg, Switzerland

Dr. D.W. Gau, M.A., M.R.C.P., "Upper Burgess", 11 Stratton Road, Beaconsfield, Buckinghamshire HP9 1HR, England

Prof. F. Gross, Direktor des Pharmakologischen Institutes der Universität Heidelberg, Hauptstrasse 47, D–6900 Heidelberg, German Federal Republic

Prof. W. Grüter, Head of the Medical and Pharmaceutical Information Service, CIBA-GEIGY LIMITED, Klybeckstrasse 141, CH–4002 Basle, Switzerland

Doz. Dr. G. Hitzenberger, Leiter der Abteilung für klinische Pharmakologie, I. Medizinische Universitätsklinik, Spitalgasse 23, A–1097 Vienna, Austria

Prof. J. Hodler, Leiter der Abteilung für klinische Pathologie, Medizinische Poliklinik der Universität Bern, Freiburgstrasse 3, CH–3008 Berne, Switzerland

Prof. M. Ikeda, M.D., The Third Department of Internal Medicine, Faculty of Medicine, University of Tokyo, 7-3-1 Hongo, Bunkyo-ku, Tokyo 113, Japan

Dr. P. R. Imhof, Research Department (Human Pharmacology), Pharmaceuticals Division, CIBA-GEIGY LIMITED, Klybeckstrasse 141, CH–4002 Basle, Switzerland

Prof. C. R. B. Joyce, Medical Department, CIBA-GEIGY LIMITED, Klybeckstrasse 141, CH–4002 Basle, Switzerland

Dr. P. Karhunen, Central Hospital, 13100 Haemeenlinna, Finland

Prof. P. Kielholz, Direktor der Psychiatrischen Universitätsklinik, Wilhelm-Klein-Strasse 27, CH–4056 Basle, Switzerland

Dr. J. G. Lewis, M.D., F.R.C.P., Consultant Physician, Edgware General Hospital, Edgware, Middlesex HA8 0AD, England

Dr. P. E. Lucchelli, CIBA-GEIGY s.p.a., Servizio Studi e Ricerche Cliniche, Via Piranesi 44, I–20137 Milan, Italy

Prof. Ag. J. Ménard, Hôpital Broussais, 96 rue Didot, F–75014 Paris, France

Prof. P. Milliez, Clinique Médicale Propédeutique, Hôpital Broussais, 96 rue Didot, F–75014 Paris, France

Dr. M. Motolese, CIBA-GEIGY s.p.a., Servizio Studi e Ricerche Cliniche, Viale di Villa Massimo 29, I–00161 Rome, Italy

Prof. G. Muiesan, Direttore dell'Istituto di Semeiotica Medica dell'Università di Perugia, Policlinico Monteluce, I–06100 Perugia, Italy

Prof. R. Oberholzer, Head of the Medical Department, CIBA-GEIGY LIMITED, Klybeckstrasse 141, CH–4002 Basle, Switzerland

Dr. B. N. C. Prichard, M.B., M.Sc., M.R.C.P., Senior Lecturer and Consultant Physician, University College Hospital Medical School, University Street, London WC1E 6JJ, England

Dr. W. Riess, Research Department (Pharmacological Chemistry), Pharmaceuticals Division, CIBA-GEIGY LIMITED, Klybeckstrasse 141, CH–4002 Basle, Switzerland

Prof. J.-L. Rivier, Médecin-Chef, Département de cardiologie des Services de médecine et de chirurgie, Hôpital Cantonal Universitaire, rue du Bugnon 17, CH–1011 Lausanne, Switzerland

Ass. Prof. R. Sannerstedt, Medicinska kliniken I, Sahlgrenska sjukhuset, University of Gothenburg, S–41345 Gothenburg, Sweden

Prof. J. Schwartz, Institut de Pharmacologie et de Médecine expérimentale, Faculté de Médecine, Université Louis Pasteur, 11 rue Humann, F–67000 Strasbourg, France

Prof. W. Schweizer, Leiter der Kardiologischen Abteilung, Departement für innere Medizin der Universität, Kantonsspital, Spitalstrasse 21, CH–4004 Basle, Switzerland

Dr. Jane Somerville, M.B., B.S., M.D., M.R.C.P., M.R.C.S., Senior Lecturer, Institute of Cardiology and Child Health, University of London, 2 Beaumont Street, London W1N 2DX, England

Dr. W. Somerville, M.D., F.R.C.P., Department of Cardiology, The Middlesex Hospital, Cleveland Street, London W1P 7PN, England

Dr. E. Sowton, M.D., F.R.C.P., Consultant Cardiologist, Cardiac Department, Guy's Hospital, London Bridge, London SE1 9RT, England

Dr. P. Taggart, M.B., B.S., M.R.C.P., Cardiac Department, The Radcliffe Infirmary, Oxford OX2 6HE, England

Dr. R. C. Tarazi, M.D., Research Division, Cleveland Clinic Foundation and Cleveland Clinic Educational Foundation, 9500 Euclid Avenue, Cleveland, Ohio 44106, U.S.A.

Dr. S. H. Taylor, B.Sc., M.B., Ch.B., F.R.C.P. (Edin.), Consultant Physician, Department of Medicine, The Martin Wing, The General Infirmary, Leeds LS1 3EX, England

Prof. L. Werkö, M.D., Medicinska kliniken I, Sahlgrenska sjukhuset, University of Gothenburg, S–41345 Gothenburg, Sweden

Priv.-Doz. Dr. K. H. Winterhalter, Friedrich Miescher-Institut, Klingelbergstrasse 70, CH–4056 Basle, Switzerland

Prof. A. Zanchetti, Direttore dell'Istituto di Semeiotica Medica I, Università di Milano, Via F. Sforza 35, I–20122 Milan, Italy

Opening address

by P. Milliez*

Ladies and Gentlemen,

It strikes me as being somewhat curious that – through a quirk of fate or, rather, an accident of birth – I, who hail from the north, should be welcoming you here on behalf of my compatriots to the southernmost part of France. I feel in fact that this brief opening address could more appropriately have been given by my friend Cesare Bartorelli, who is much closer to the Mediterranean than I.

Since the title of our symposium, "Beta-blockers – present status and future prospects", indicates quite clearly what we are gathered together here to discuss, I do not propose to anticipate any of the specific points that are likely to arise in the course of our deliberations. Instead, I should like to couch these few remarks of mine in a more general vein.

At a time when the Western world is faced with a perilous situation and seems to be about to plunge into chaos, I think we would be well advised to reconsider the whole problem as to the extent to which the scientific and financial efforts we are now making are indeed justified. Not long ago, following a general strike on the part of members of the medical profession in one of the Mediterranean countries, it was found that the death rate had diminished by 25%. I am sure you are also aware of the fact that there are philosophers and sociologists who claim to have demonstrated that medical science has thus far contrived to improve only the quantity, but not the quality, of life on this earth. According to this argument, all the successes achieved during recent years have in the last analysis merely served to prolong the sufferings of the privileged few.

You will no doubt all have read the article by Franz Gross which he published under the title "Future drug research – drugs of the future" [Clin. Pharmacol. Ther. *14*, 1 (1973)] and in which he raises the question as to whether it will still be possible to make progress in the treatment of those diseases that are now of chief concern to us.

Despite misgivings of the kind to which I have just referred, I hope that the review of the beta-blockers upon which we are about to embark will nevertheless contribute to improving the lot of suffering humanity, particularly since studies such as those of Prichard and of Schröder and Werkö have already furnished clinical proof to justify further high hopes of treatment with these drugs.

---

* Clinique Médicale Propédeutique, Hôpital Broussais, Paris, France.

Introductory remarks on the use of beta-blockers in hypertension

by C. Bartorelli*

By way of introduction to this opening session on hypertension, I should like to begin by emphasising those points on which all of us are likely to agree and then to anticipate briefly some of the points on which we may well disagree and which I hope will, in the course of the proceedings, prove the subject of active and fruitful discussion.

The first point on which I am sure we are unanimously agreed is that established hypertension requires to be treated. The second point is that all of us would like to have at our disposal a really ideal antihypertensive drug.

Thanks to my age and to my long and keen interest in the problem of hypertension – which I think may have been the main reasons why I was given the honour of introducing this session – I am in a position to view the treatment of hypertension both in retrospect as well as in perspective. I have personal memories of the sad times when we were still powerless to provide effective help for patients suffering from severe hypertension, and I also witnessed what might be termed the "antihypertensive gold rush" which set in when, at long last, it became clear that malignant hypertension could be reversed and that it was no longer an inevitably fatal disease. I look back, too, with gratification on the years during which the greatest improvements in our antihypertensive armamentarium were made, i.e. to the time when the advent of the oral diuretics rendered treatment for hypertension not only feasible on a far wider scale but also much easier for the patient to tolerate. And now that we seem to have reached the flatter portion of the parabola, I am following current developments with close attention, because, though the improvements being achieved at present may not be quite as dramatic as in past decades, they are nevertheless still encouraging. Meanwhile, however, the quest for an ideal antihypertensive drug continues. This ideal drug would, I suggest, have to display all of the following properties: it should be capable of lowering elevated blood pressure gradually and progressively, and it should be equally effective in both the lying and the standing position; it should be therapeutically active by the oral route, and its action should last at least 12 hours and preferably 24 hours, thus permitting a dosage schedule of, at the most, two doses daily; it should produce little or no negative inotropic action on the heart; on the other hand, it should avoid eliciting tachycardia; it should not produce tolerance; it should not give rise to any troublesome side effects when administered either on a short-term or on a long-term basis; and, finally, it should be effective in all cases of essential hypertension.

* Istituto di Ricerche Cardiovascolari, Università di Milano, Milan, Italy.

I deliberately mention this latter requirement last of all, because this is one point on which I foresee some disagreement and discussion. I realise that, if there are different mechanisms responsible for different types of essential hypertension, then we cannot expect to be able to find any one single drug that would be ideally suitable for use in each and every patient suffering from essential hypertension. But I shall not pursue this line of thought any further now, as I am sure we shall have quite a lot to hear on the subject later in our discussions. If I may be permitted at this point to express my own personal view, I can only say that I remain firmly convinced that oral diuretics still constitute the back-bone of antihypertensive treatment in practically all hypertensive patients.

The extensive armamentarium of antihypertensive drugs currently available certainly allows us quite a lot of flexibility in our approach to the treatment of hypertension and enables us to achieve a large measure of success in our attempts at controlling high blood pressure. But the use that is made of these various drugs does not seem to depend solely on physiological, pharmacological, and clinical considerations. Statistics show, for example, that the combinations of reserpine plus a diuretic are employed in Italy and the German Federal Republic to treat 56% and 79%, respectively, of all hypertensive patients in receipt of drug therapy, whereas in Great Britain the corresponding figure is only 7%! On the other hand, methyldopa is being prescribed in a mere 6% of cases in the German Federal Republic, as compared with 50% of cases in Great Britain. This, of course, does not mean that there is any difference between British and, say, German or Italian hypertensive patients, but is largely a reflection of differing marketing policies and of their effectiveness in the countries concerned.

When questioned by pharmaceutical companies as well as by patients – to both of whom we are in the very delicate position of acting as consultants – about the future of a new drug, we must therefore analyse the respective pros and cons critically and objectively, i.e. disregarding fashionable trends and bearing in mind only the advantages and the disadvantages that our patients are likely to enjoy or to endure as a result of its use. It is in this light that today we have to consider those drugs which act by blocking the beta-adrenergic receptors. These drugs have been widely employed in the treatment of various cardiovascular disorders, and attempts have now been in progress for almost a decade to find a place for them in the management of arterial hypertension. Before we try to answer questions as to their usefulness, we ought first to consider certain aspects of the kind which Dr. BRUNNER will be discussing in his paper. We have to determine the potency and the reliability of the antihypertensive action of the beta-blockers; we have to compare these beta-blockers with other antihypertensive drugs in order to see whether and to what extent one type of drug is superior to another in terms of efficacy and of the quantity or quality of side effects; and, finally, we must endeavour to arrive at an understanding of the mechanisms underlying the antihypertensive action of the beta-blockers, since this would undoubtedly help in providing a clearer guide for their clinical use. Though this last question is, of course, a very difficult one, it is of critical importance, and we shall no doubt be hearing a good deal on this topic in the course of our discussions.

The only hypertensive state in which the effect of beta-blockers in lowering the blood pressure appears quite obviously to be directly attributable to blockade of the beta-adrenergic receptors is hypertension associated with primary hyperkinetic heart syndrome. The pathophysiological basis of this disorder is believed to be an increase in the reactivity of the beta-adrenergic receptors of the heart to adrenergic stimuli. This in turn would cause an increase in cardiac output and stroke volume, leading in most of these cases to a rise in arterial blood pressure. Allow me at this point to present some data from a study in which the antihypertensive effect of propranolol was investigated in 12 patients suffering from primary hyperkinetic heart syndrome. The average systolic and diastolic pressures recorded before and during long-term treatment with doses averaging 120 mg. daily are indicated in Figure 1. The open bars show the readings based on auscultation, and the full bars the pressures obtained by direct intravascular measurement. The time at which the measurements were made is indicated at the bottom. Both in the earlier and in later phases of treatment, propranolol induced a significant decrease in systolic and diastolic pressure to normal or near-normal levels. The haemodynamic mech-

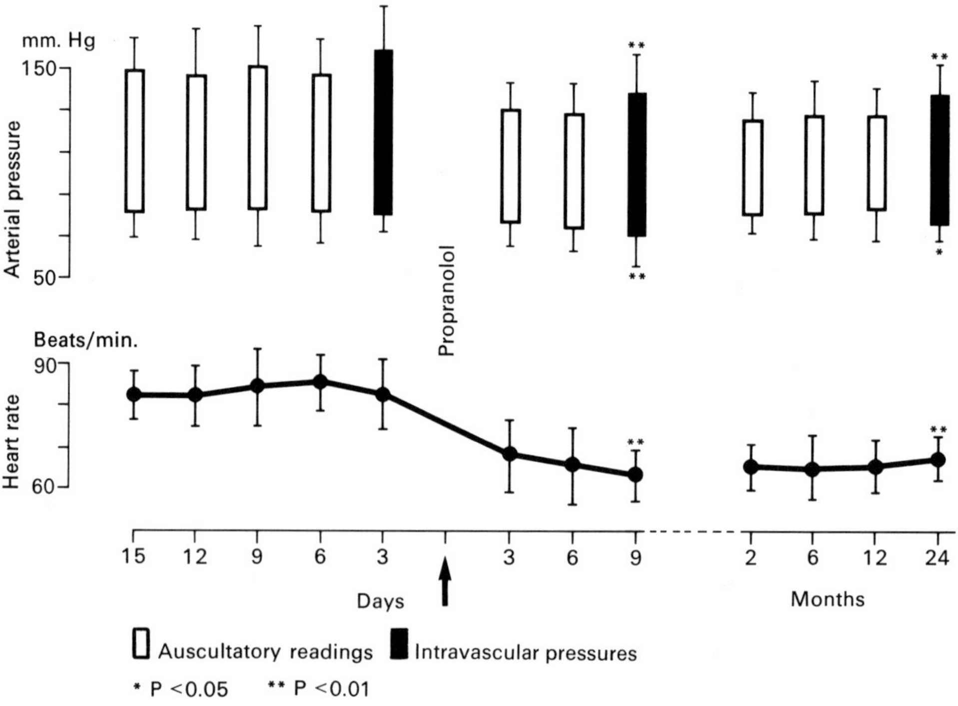

Fig. 1. Systolic and diastolic blood pressure levels, as well as heart rate, before and during long-term treatment with propranolol (average daily dose: 120 mg.) in 12 patients suffering from primary hyperkinetic heart syndrome.

anism responsible for this antihypertensive effect is outlined in Figure 2. Here, the reduction in blood pressure is shown to have been induced via a decrease in cardiac output which, after two years of treatment, was just as pronounced as during the early phase of propranolol administration.

The pathophysiological basis of primary hyperkinetic heart syndrome, the dosage in which beta-blockade therapy proves effective in this condition, and, lastly, the mechanism of the antihypertensive response in such patients provide strong support for the conclusion that in this disorder it is the beta-blocking effect of the drug as such that is directly responsible for producing the decrease in elevated blood pressure. But these observations – relating as they do to a very peculiar form of hypertension – are of little help in elucidating the anti-hypertensive effect of beta-blockers in the incomparably greater population of primary hypertensive subjects.

To conclude this brief introduction, I will now venture to outline very briefly my own current notions on the clinical use of beta-blockers in hypertension, i. e. to indicate some of the thoughts and doubts which are in my mind at the start of this symposium and which I hope the papers to be presented and the discussions following them will serve to clarify. The first thought is that the combination of a beta-blocker with a diuretic may perhaps well prove an

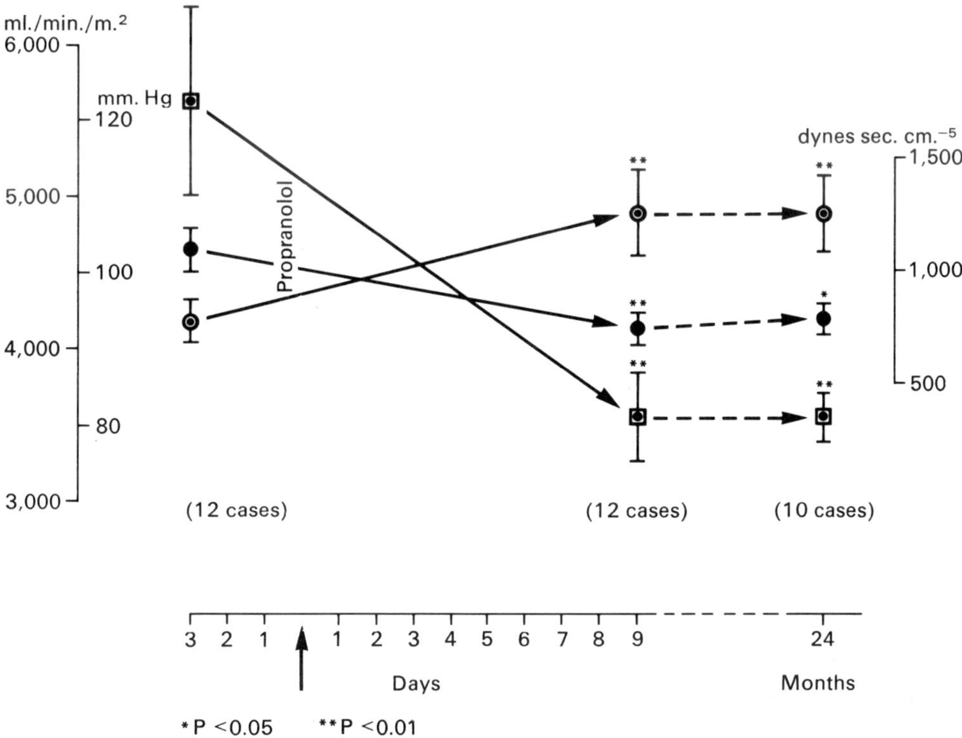

Fig. 2. Mean arterial pressure (●), cardiac index (▣), and total peripheral resistance (◉) before and during long-term treatment with propranolol in patients with primary hyperkinetic heart syndrome. The reduction in blood pressure is shown to have been induced via a decrease in cardiac output.

effective form of treatment for a good number of mild and moderate hypertensives. The first doubt is whether this type of combination has more advantages to offer, or at least fewer disadvantages, than the combinations of diuretics with reserpine, methyldopa, or clonidine that are at present available. For instance, will the fact that beta-blockers display milder side effects always outweigh the convenience of treatment with chlortalidone plus reserpine – a combination which has a long-lasting effect and enables the medication to be given only once a day or once every two days? I doubt whether there is any single answer to this question; instead, we shall probably have to continue the present practice of selecting the best-acceptable combination for each individual patient by a process of trial and error. My second thought is that beta-blockers are certainly advantageous when added to vasodilator drugs such as the hydralazines, because they counteract the tachycardia which such drugs provoke and also potentiate their antihypertensive action. But this brings me to my second doubt – a doubt which concerns the role that combinations of hydralazines, beta-blockers, and diuretics really deserve to play in the treatment of hypertension. Here, too, I suspect that we shall not be able to predict precisely which patients will benefit most from this kind of regimen, and that we shall have to find out which is the most successful and acceptable drug combination by resorting once again to trial and error. Finally, what about plasma renin activity as a predictor of the effectiveness of beta-blockers? This is a question on which I have no opinions, but only doubts. By the end of this symposium, however, I hope to have fewer doubts and more opinions! I therefore eagerly await the pleasure of hearing all that the participants here will have to say in the course of our proceedings.

The antihypertensive effect of beta-blockers

by H. BRUNNER*

When the antihypertensive effect of the beta-blockers was first described some ten years ago by PRICHARD[54, 55] and by SCHRÖDER and WERKÖ[68], this finding came as a surprise, because it could not have been foreseen either on the basis of knowledge about the functioning of the sympathetic nervous system or about the pathophysiology of hypertension or in the light of results yielded by pharmacological experiments. Meanwhile, however, this antihypertensive effect has been confirmed by so many investigators and in such a large number of patients that it is now universally accepted as proven. But, since the anti-hypertensive activity of the beta-blockers continues to command keen interest as one of the most intriguing properties of these drugs, I shall attempt here to define its characteristics in somewhat greater detail by reference to the following pertinent questions:

- How strong and how reliable is the antihypertensive effect of the beta-blockers?
- What advantages do the beta-blockers offer in the treatment of hypertension, including long-term treatment in particular, as compared with the well-established antihypertensives?
- What are the contra-indications and side effects of beta-blockers that may limit the scope of their use in hypertension?
- What is known about the mechanism by which they lower elevated blood pressure?

This last question is not quite such a theoretical one as it might appear, because it also has a very practical bearing on the problem of which other antihypertensive agents should preferably be prescribed together with a beta-blocker.

Potency of the antihypertensive effect of beta-blockers

A perusal of the literature on the antihypertensive activity of the beta-blockers reveals that all compounds belonging to this drug category display a number of features in common:
The first point to emerge is that their antihypertensive effect does not evidently exhibit any clear-cut dose-response relationship. The responses, and hence the

---

* Research Department (Pharmacology), Pharmaceuticals Division, CIBA-GEIGY LIMITED, Basle, Switzerland.

therapeutic doses required, vary very considerably from patient to patient[15, 57, 70]. At present, none of the test procedures available – such as that based on isoprenaline antagonism, for example – enables one to predict which patients will react strongly to beta-blockade and which patients will show hardly any response[31, 70]. When examining the antihypertensive effect exerted by a beta-blocker in a large group of patients, one gains the impression that the dose-response curve must have an extremely flat pattern, i.e. flatter than the corresponding curves which can be plotted for other effects of the beta-blockers, such as inhibition of tachycardia induced by isoprenaline, by emotional tension, or by physical effort. What is more, the antihypertensive effect of the beta-blockers seems to attain a certain maximum, which cannot be exceeded by resorting to further dosage increments[21, 42, 70].

Differences between the various beta-blockers arise chiefly from the positioning of their dose-response curves within the system of coordinates employed, a positioning which in turn is strictly dependent upon their inhibitory action on the peripheral beta-adrenergic receptors. In the case of pindolol, which exerts a very strong beta-blocking action per unit weight, the dose-response curve – also in respect of the drug's antihypertensive activity – lies further to the left, i.e. in the lower dosage range, than the curves for propranolol or oxprenolol (®Trasicor), whereas the curve for practolol, a compound displaying only a relatively weak beta-blocking effect, is situated further to the right, i.e. in the higher dosage range. As shown in Table 1, WAAL-MANNING[82], after having compared the antihypertensive activity of the aforementioned beta-blockers, concluded that a daily dosage of 12.5 mg. pindolol was roughly equivalent to 160 mg. propranolol, 200 mg. oxprenolol, or 650 mg. practolol. WAAL-MANNING's figure for pindolol, however, is probably too low, since according to the results of a study by KINCAID-SMITH[38], which admittedly involved cases of severe hypertension, 58 mg. pindolol is equivalent in effect to 289 mg. propranolol daily. In this respect, alprenolol would appear to lie roughly midway

Table 1. Comparison between the beta-blocking activity of five beta-blockers in the cat and the respective doses reportedly required to produce an antihypertensive effect in man.

Beta-blockers	Beta-blocking activity (inhibition of isoprenaline-induced tachycardia in the anaesthetised cat)	Doses required to produce an anti-hypertensive effect in man		
	ED_{50} (mg./kg. i.v.)	(mg./day)		
Pindolol	0.004	12.5	58	
Propranolol	0.03	160	289	180
Oxprenolol	0.03	200		
Alprenolol	0.02			450
Practolol	1.0	650		
References		82	38	5

between propranolol and practolol, a daily dosage of 450 mg. alprenolol having been found to exert the same effect as one of 180 mg. propranolol[5]. Apart from the differences in the positions which their dose-response curves occupy in the system of coordinates, the very few comparative studies that have thus far been undertaken with beta-blockers fail to reveal any significant differences in the maximum effects attainable with them – with the possible exception of practolol, which seems to have a weaker effect than the others[69, 78, 86].

When making comparisons of this kind, it must always be borne in mind that, provided a beta-blocker is given in an adequate dosage, it will already begin to exert an antihypertensive action within the first few days of treatment, but it will take two to three weeks to attain its full effect. The fall in blood pressure is usually steepest during the first week, after which it gradually flattens out[15, 57, 70]. This time-dependent pattern of the response considerably adds to the difficulty of performing comparative studies.

When employed in the treatment of hypertension, all beta-blockers have to be prescribed in doses higher than those required simply to block the peripheral beta-receptors. In the case of oxprenolol, for instance, a dose of 20–40 mg. is sufficient to suppress emotionally induced tachycardia, and one of 80 mg. is capable of inhibiting an isoprenaline-induced increase in heart rate in healthy subjects for a period of 8–12 hours. For the treatment of hypertension, by contrast, daily doses of oxprenolol amounting to 160–480 mg., and sometimes even more, are needed; on the other hand, extremely high doses of 1,000 mg. or more daily are employed only in exceptional cases. From this it might be concluded that it is not beta-blockade as such but some other property or properties of the molecule which are essentially responsible for its antihypertensive effect. Incompatible with this conclusion, however, is the strict correlation existing between beta-blocking and antihypertensive activity, to which reference has already been made. Another counterargument is the finding that D-propranolol, which displays all the properties of the optically inactive racemic mixture with the exception of a beta-blocking effect, also exerts no antihypertensive action[83]. Despite this, on the basis of comparative studies undertaken with oxprenolol and propranolol[61], as well as with the dextrorotatory and laevorotatory isomers of N-isopropyl-p-nitrophenylethanolamine (administered, however, in only one single dose), certain authors nevertheless believe that the antihypertensive effect occurs independently of beta-blockade[28].

As already mentioned, the extent of the fall in blood pressure produced by beta-blockers varies considerably from patient to patient. In relatively large groups of patients, average decreases in systolic pressure of 15–20%, and in diastolic pressure of 10–15%, have been reported[1, 8, 32, 50, 67, 75, 81]. From the quantitative aspect, the antihypertensive effect of the beta-blockers can thus perhaps best be compared with that of the thiazide diuretics[20, 26, 52, 67, 77]. Like these diuretics, beta-blockers too elicit a clinically satisfactory reduction in blood pressure in only 20–40% of cases when prescribed alone[15, 40, 45, 50, 57, 75, 81]. This in itself is sufficient to justify the conclusion that beta-blockers are seldom suitable as monotherapy for hypertension. On the contrary, like diuretics, they

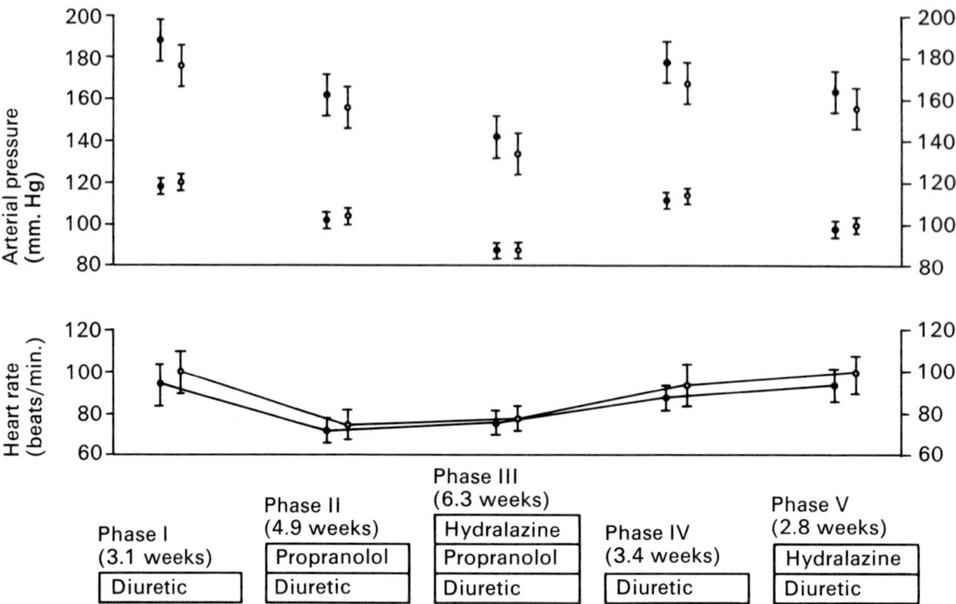

Fig. 1. Mean systolic and diastolic pressures *(above)* and mean heart rates *(below)* in the supine (•) and standing (◦) positions recorded in 23 patients with moderate or severe essential hypertension during five phases of treatment, in which propranolol and hydralazine were successively added to diuretic therapy (Phases I–III) and the patients then treated with a diuretic alone (Phase IV) followed by a diuretic plus hydralazine (Phase V). The mean duration of each phase of treatment is indicated in brackets. (Adapted from: ZACEST et al.[87])

should preferably be employed in combined regimens; in other words – depending on the patient's condition and response – they must as a rule be supplemented by treatment with other antihypertensives.

The commonest type of combination is that of a beta-blocker plus a thiazide diuretic. This generally produces an additive antihypertensive effect, thanks to which the proportion of cases showing an adequate drop in blood pressure increases to some 70–88%[22, 48, 51, 63, 70, 87, 88]. Here again, however, there may be exceptions to the rule: SAFAR et al.[64], for example, report that in their hands a combination of pindolol and clopamide did not prove significantly more effective than pindolol alone; and DORPH and BINDER[20] obtained similar findings with oxprenolol and hydrochlorothiazide.

Another frequently employed type of combination consists of a beta-blocker together with a vasodilator[33, 36, 38, 65, 66, 79]. An important advantage of such treatment lies not only in the mutual reinforcement of the antihypertensive activity of the two drugs, but also in the fact that at the same time their side effects are attenuated – a point to which further reference will be made later. A third possibility is to build up step by step to a suitable regimen, e.g. by prescribing only a beta-blocker plus a diuretic to begin with and, if the patient fails to respond adequately, adding a vasodilator afterwards[39, 70]. Such combinations of three differently acting antihypertensives are reported to be extremely effective[38, 70, 87]; shown in Figure 1 is an example illustrating the response to treatment in which a threefold combination of a diuretic, a beta-blocker, and a vasodilator was built up progressively. Combinations with other antihypertensive preparations, e.g. methyldopa or guanethidine, have also been tried[46, 53], and here, too, the beta-blocker has been found to potentiate their effect[45, 57, 75].

Advantages of beta-blockers as compared with conventional antihypertensives

The main reason why more and more doctors are now resorting to beta-blockers in the treatment of hypertension is their exceptionally good tolerability as compared with other antihypertensive agents[13, 15, 22, 57]. Beta-blockers hardly ever give rise to headache or palpitation, such as the vasodilators may produce, or to postural hypotension and disturbances of potency in the male, which can occur as side effects of the post-ganglionic sympathetic inhibitors; nor, in contrast to methyldopa and reserpine, are they liable to provoke depression. As exemplified by Figure 2, the decrease in blood pressure induced by beta-blockers is virtually the same in both the lying and the standing position, and even physical work does not markedly reduce the blood pressure any further[15, 32, 55, 58, 59]. Beta-blockers also have the advantage of not causing such handicapping sedation as methyldopa or clonidine. On the contrary, in extensive trials carried out with oxprenolol, 20–30% of the patients actually reported an enhanced sensation of well-being[22] – a positive effect which we believe to be of very great importance and which may well play a decisive role in the management of young hypertensives suffering from few if any subjective symp-

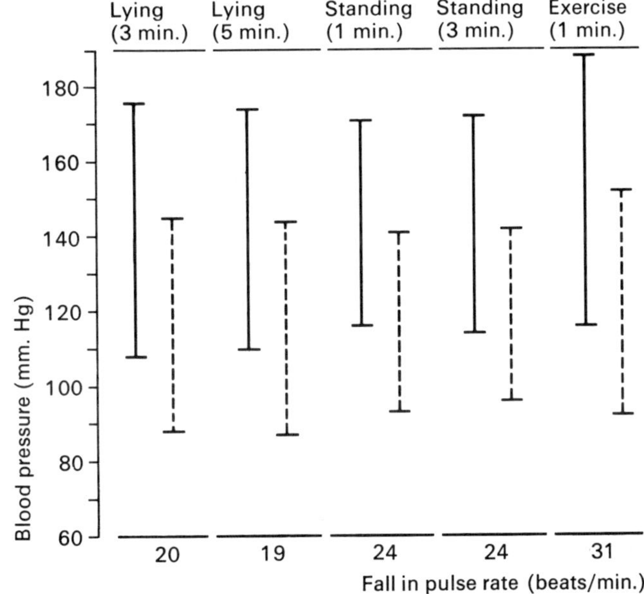

Fig. 2. Average blood pressures – recorded in the lying and standing positions, as well as after one minute of exercise – in 17 hypertensive patients before (unbroken vertical lines) and after (interrupted vertical lines) treatment for at least three months with individually adapted doses of oxprenolol. Indicated on the abscissa is the average fall in pulse rate. (Adapted from: MARSHALL and BARRITT[43])

23

toms. During recent years, major campaigns have been conducted on a broad front with a view to preventing late complications of hypertensive disease by ensuring that hypertension is diagnosed and treated as early as possible in precisely these young patients whose symptoms are so mild that they themselves would never bother to consult a doctor in order to obtain relief from them. The stronger the current trend towards very early diagnosis as a means of warding off the delayed complications of hypertension, the greater the tendency to favour treatment, not with very potent antihypertensives, but with drugs which, though less highly active, are extremely well tolerated. In this connection, it is obvious that patients who – like almost all these younger hypertensives – are still fully fit for work can scarcely be expected to undergo medication, year in and year out, with drugs which cause more symptoms and more discomfort than the disease itself – quite apart from the fact that, if one did attempt to impose such treatment upon them, they would rarely adhere to it. What such patients generally require instead is a drug which neither provokes orthostatic side effects nor interferes with their normal way of life by producing sedation or other functional disturbances. Moreover, if it is a drug in response to which quite a considerable percentage of the patients actually feel better than before, then there is every likelihood of their persevering with the treatment.

Another advantage of the beta-blockers, though one which has not yet been conclusively demonstrated, lies in the possibility that they may also diminish the coronary complications of hypertensive disease; this would constitute a decisive advance in the management of hypertension. Although antihypertensive agents have previously been available with which it was possible drastically to reduce the incidence of secondary heart failure and of delayed cerebral and renal complications in patients suffering from hypertension[10, 25], the frequency of coronary complications has remained unchanged or has even shown a relative increase[3, 4], if only because patients receiving effective treatment have been living longer thanks to the prevention of cerebrovascular accidents and renal insufficiency. Among the cardiac complications of hypertensive disease, angina pectoris and myocardial infarction account for a very appreciable percentage[10]. With the help of beta-blockers, stress-induced strain on the heart can now be reliably eliminated[76], as TAGGART et al.[74] have most impressively demonstrated in persons subjected to the mental strain of speaking before an audience. In this study it was found that beta-blockers are capable of preventing not only emotionally induced tachycardia but also severe E.C.G. changes in patients with a history of myocardial infarction.

In cases of angina pectoris the therapeutic action of beta-blockers is essentially due to an indirect reduction in myocardial oxygen consumption. In hypertensive patients the same type of effect is brought about by a decrease in the blood pressure and heart rate, i. e. in the two variables upon which myocardial oxygen consumption chiefly depends[9, 41]. But detailed long-term studies on large numbers of patients will first have to be performed before it can be claimed, with greater conviction than is justified by these theoretical considerations, that the addition of beta-blockers to conventional antihypertensive regimes does indeed also exert a favourable influence as regards the risk of coronary complications. The treatment of disturbances in cardiac rhythm and of angina pectoris will be dealt with later by other speakers, so that I need not refer to them here.

Experience of cases receiving long-term therapy for hypertension over periods ranging from months to years has been obtained with beta-blockers prescribed in combination with diuretics, together in some instances with other antihypertensives as well. As a rule, clinically satisfactory reductions in blood pressure which persisted throughout the treatment were obtained in 70–85% of the patients, the tolerability of the beta-blockers remaining consistently good over the whole period in question[2, 6, 15, 16, 22, 32, 57, 70, 88].

Contra-indications and side effects of the beta-blockers

When considering to what extent the beta-blockers constitute an advance in the treatment of hypertension, one must bear in mind not only their possible advantages but also their contra-indications and side effects. They have certain general contra-indications which also apply to their use in hypertensive patients, namely, asthma and related conditions – which great care should be taken to exclude – as well as bradycardia and cardiac insufficiency. These contra-indications, however, are so well known that it would be superfluous to discuss them in detail here.

As I have already mentioned, the beta-blockers are well tolerated, and the incidence of side effects is therefore relatively low. In studies conducted on quite a large scale, in which beta-blockers were employed in combination with other drugs and their contra-indications carefully adhered to, approximately 85% of the patients treated reported no side effects at all[15, 22]. Also of interest in this connection are figures for the percentages of cases in which treatment had to be withdrawn owing to the occurrence of side effects: in 2,300 hypertensives treated with oxprenolol and cyclopenthiazide (Figure 3) the number of withdrawals worked out at 4%[22], as compared with about 8% in 311 patients on combined treatment with propranolol[88] and roughly 13% in 195 cases in which combinations with pindolol were employed[15]. In all these studies, side effects of the type accounting for contra-indications to the use of beta-blockers very seldom necessitated discontinuation of the treatment. In the 2,300 patients receiving oxprenolol, for example, the medication had to be withdrawn only five times because of cardiac insufficiency, twice because of bradycardia (heart rate below 55 beats per minute), and three times because of bronchospasm. Side effects of this kind may be encountered with all the beta-blockers, and bronchospasm has been observed in rare instances even in patients treated with the cardioselective beta-blocker practolol[84, 86]. Relatively common as side effects are also central nervous manifestations ranging from nightmares, sleep disturbances, and dizziness to occasional occurrences of hallucinations[15, 22, 30, 70, 88]. Other side effects include gastro-intestinal upsets, tiredness or fatigability, and a feeling of coldness in the extremities. In rare cases

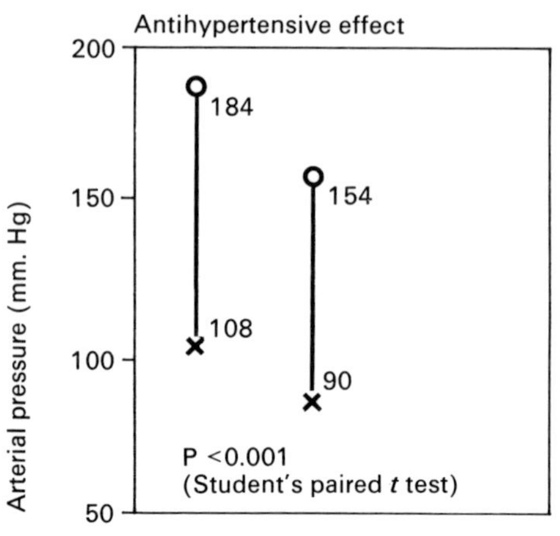

Severe side effects

Signs and/or symptoms	Treatment withdrawn	
Central nervous	61	(2.7%)
Gastro-intestinal	16	(0.7%)
Cardiovascular:		
Heart failure	5	(0.2%)
Bradycardia	2	(0.1%)
Bronchospasm	3	(0.1%)
Skin rash	2	(0.1%)
Total	89	(3.99%)

Fig. 3. Response to four weeks' treatment with oxprenolol plus cyclopenthiazide, and incidence of side effects severe enough to necessitate withdrawal of medication, in a total of 2,300 patients with mild to moderate hypertension (diastolic pressures: 95–120 mm. Hg). (Adapted from: FORREST[22])

there have also been reports of drug rash and of hypertensive reactions occurring in the course of treatment with beta-blockers[17,45,49]. Though these side effects are common to all beta-blockers, there do appear to exist certain differences between these drugs with respect to the side effects they produce: for instance, gastro-intestinal disturbances[70] are alleged to be particularly frequent in response to practolol, and central nervous symptoms[45] or muscle cramp[70] in response to pindolol. A syndrome resembling lupus erythematosus has also been reported, but only following practolol[60].

It should be added, however, that generally speaking all the side effects of beta-blockers can easily be brought under control by withdrawing the medication and taking appropriate countermeasures.

Mechanism of action underlying the antihypertensive effect of beta-blockers

The fourth and last of the questions raised in the introduction to this paper concerns the mechanism by which beta-blockers lower elevated blood pressure. Three main hypotheses have been advanced to account for their antihypertensive effect:

The first is based on the influence exerted by beta-blockade upon the haemodynamics of hypertension. Beta-blockers, including propranolol in particular, have the effect of reducing cardiac output[24,42,59]. In short-term experiments the blood pressure remains unchanged, because the decrease in cardiac output is offset by reflex-induced vasoconstriction. In response to a prolonged diminution in cardiac output, on the other hand, adaptational changes in the cal-

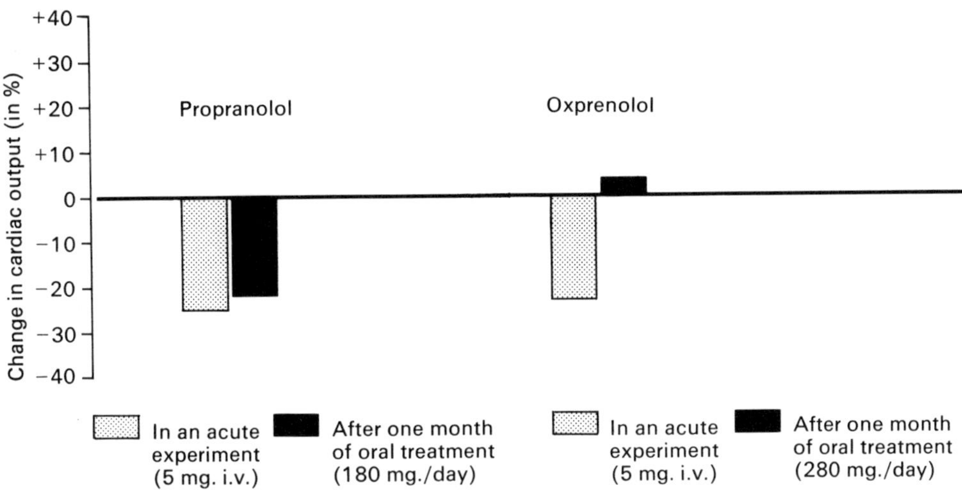

Fig. 4. Effect on cardiac output exerted by propranolol in six patients following a single intravenous injection and in seven following one month of oral treatment, compared with the effect of oxprenolol in six patients following similar acute and chronic medication. (Adapted from: WILSON et al.[85])

ibre of the blood vessels – changes which may perhaps be due to metabolically mediated control mechanisms affecting the blood supply to the organs – appear to preponderate. According to observations reported by TARAZI and DUSTAN[75], peripheral resistance then ultimately diminishes to pre-treatment levels while at the same time, owing to the sustained reduction in cardiac output, a decrease in blood pressure occurs. There is one major argument against this explanation for the antihypertensive effect of the beta-blockers, namely, the fact that certain drugs of this type, such as timolol, alprenolol, or oxprenolol, reduce cardiac output to a much smaller degree than propranolol and sometimes even fail to reduce it at all (this is borne out by the findings presented in Figure 4, from which it can be seen that, although cardiac output showed a marked decrease in response to oxprenolol in an acute experiment, it showed no change as compared with the base-line level after one month of oral treatment with oxprenolol); these beta-blockers that have far less effect on cardiac output nevertheless produce just the same antihypertensive response[24, 34]. Thus, although the course of events upon which the above-mentioned hypothesis is founded may well account to a greater or lesser extent for the overall effect of the beta-blockers, this would hardly seem to be the sole mechanism involved. Be that as it may, one conclusion appears at all events to be justified: since beta-blockers do not lower peripheral resistance, their use in combination with a peripheral vasodilator, e. g. with an hydralazine compound, obviously offers particular advantages. The primary effect exerted by a beta-blocker takes the form of a decrease in cardiac output, which leads in turn to a reflex-induced rise in peripheral vascular resistance. A vasodilator, when administered as monotherapy, acts primarily by dilating the blood vessels; this results in a fall in blood pressure, which triggers off a reflex mechanism causing an increase in cardiac output and heart rate. This increase often leads to subjectively unpleasant palpitation and, in patients whose coronary reserve is limited, may even provoke attacks of angina pectoris. In a regime in which a beta-blocker is added to a vasodilator, on the other hand, these side effects are attenuated thanks to inhibition of the reflex-induced adaptational processes[1, 29, 33, 36, 65]. Suppression of palpitation is so marked, in fact, as to make it impossible to conduct a double-blind trial with the individual components of such a combination.

Another hypothesis which has recently attracted much discussion emphasises the inhibitory effect exerted by beta-blockers on renin secretion. Among the factors known to influence the secretion of renin are not only the total exchangeable sodium, the perfusion pressure in the kidneys, and intrarenal mechanisms originating in the macula densa, but also the sympathetic nervous system. It is possible to step up the output of renin from the kidneys by subjecting the splanchnic nerve to electrical stimulation or by administering injections of isoprenaline[72, 73, 80]. Proof has also been obtained that the sympathetic control mechanism – though admittedly only one of the mechanisms governing renin secretion – can be inhibited by means of beta-blockers[11, 12, 44]. Here, once again, however, it is not known precisely what contribution this inhibitory effect makes to the action by which beta-blockers lower the blood pressure. This, too, is a question which will be examined in detail by other speakers.

At this point I would merely emphasise that the inhibitory effect of beta-blockers on renin secretion may play an important role where these drugs are employed in combination with certain other types of antihypertensive. Both diuretics and vasodilators increase renin secretion, an increase which can be combated by concomitant administration of a beta-blocker. The resultant diminution in renin secretion leads to a decrease in the activity of the renin-angiotensin-aldosterone system as a whole[12]. Consequently, in patients to whom a beta-blocker is given in combination with a vasodilator, the indirectly induced decrease in aldosterone secretion serves to counteract the sodium retention to which vasodilators give rise. Similarly, in the case of combined treatment with a beta-blocker and a diuretic, the beta-blocker also serves to diminish the secondary hyperaldosteronism which occurs as a compensatory reaction in response to loss of sodium provoked by the diuretic. This, too, is one of the mechanisms by which the additional use of a beta-blocker can be expected to ensure a stronger antihypertensive effect, because sodium retention – and the resultant increase in blood volume – would otherwise tend to counteract the fall in blood pressure.

A third hypothesis that has been discussed relates to findings obtained following intracerebroventricular injections in dogs[14, 71], cats[18, 37], and rabbits[19, 62] – findings which suggest that a central mechanism may be involved in the antihypertensive action of beta-blockers. Intracerebroventricular injections of propranolol were observed to produce a prolonged decrease in blood pressure which was accompanied by bradycardia. These changes were frequently preceded by a brief initial rise in pressure which was associated with tachycardia.

Just where and in what manner the beta-adrenergic receptors intervene in central mechanisms responsible for regulating the cardiovascular system is a question to which no clear answers have yet been forthcoming. The whole problem is complicated by the fact that under somewhat modified conditions (e.g. when experiments are performed on anaesthetised[27] as opposed to non-anaesthetised[18] cats) intracerebroventricular administration of isoprenaline, i.e. stimulation of the beta-receptors, is likewise capable of eliciting a fall in blood pressure, although in this instance the fall is accompanied by an increase in the heart rate. In anaesthetised cats, intracerebroventricular doses of beta-blockers failed to counteract either increases in blood pressure produced by electrical stimulation of the medullary reticular formation or reflex-induced increases in blood pressure, whereas such doses did inhibit the tachycardia that had been provoked by the same means[69]. It should be pointed out, on the other hand, that only limited conclusions regarding the possible central activity of drugs can be drawn from the results of experiments in which intracerebroventricular injections are employed, because various other factors – such as the way in which the drug diffuses before it reaches its site of attack, local anaesthetic effects, etc. – greatly add to the difficulties of interpretation.

Finally, it should be mentioned that attempts to account for the antihypertensive effect of the beta-blockers have also been made by reference to other phenomena, e.g.: a resetting of the baroceptors[57]; a decrease in plasma volume, as observed in response to intravenous administration of propranolol[35]; or a guanethidine-like inhibition of post-ganglionic sympathetic transmission, which has been demonstrated in isolated organs[47]. It is not yet clear, however, to what extent such factors are indeed involved in producing the fall in blood pressure which beta-blockers elicit.

References

1 ÄNISHÄNSLIN, W., PESTALOZZI-KERPEL, J., DUBACH, U.C., IMHOF, P.R., TURRI, M.: Antihypertensive therapy with adrenergic beta-receptor blockers and vasodilators. Europ. J. clin. Pharmacol. *4*, 177 (1972)
2 ANGERVALL, G., BYSTEDT, U.: The effect of alprenolol and alprenolol in combination with saluretics in hypertension. Acta med. scand. Suppl. 554: 39 (1974)
3 BARNETT, A.J., SILBERBERG, F.G.: Long-term results of treatment of severe hypertension. Med. J. Aust. *60/ii*, 960 (1973)
4 BEEVERS, D.G., FAIRMAN, M.J., HAMILTON, M., HARPUR, J.E.: The influence of antihypertensive treatment over the incidence of cerebral vascular disease. Postgrad. med. J. *49*, 905 (1973)
5 BENGTSSON, C.: Comparison between alprenolol and propranolol as antihypertensive agents. Acta med. scand. *192*, 415 (1972)
6 BENGTSSON, C.: Long-term effect of alprenolol as antihypertensive agent. Acta med. scand. Suppl. 554: 9 (1974)
7 BHARGAVA, K.P., MISHRA, N., TANGRI, K.K.: An analysis of central adrenoceptors for control of cardiovascular function. Brit. J. Pharmacol. *45*, 596 (1972)

8 BODEM, G., BRAMMELL, H. L., WEIL, J. V., CHIDSEY, C. A.: Comparative evaluation of adrenergic blockade with practolol and propranolol in hypertension. Clin. Res. *20*, 169 (1972); abstract of paper

9 BRAUNWALD, E.: Control of myocardial oxygen consumption. Amer. J. Cardiol. *27*, 416 (1971)

10 BRECKENRIDGE, A., DOLLERY, C. T., PARRY, E. H. O.: Prognosis of treated hypertension. Quart. J. Med., N.S., *39*, 411 (1970)

11 BÜHLER, F. R., LARAGH, J. H., BAER, L., DARRACOTT VAUGHAN, E., Jr., BRUNNER, H. R.: Propranolol inhibition of renin secretion. New Engl. J. Med. *287*, 1209 (1972)

12 BÜHLER, F. R., LARAGH, J. H., VAUGHAN, E. D., Jr., BRUNNER, H. R., GAVRAS, H., BAER, L.: Antihypertensive action of propranolol. Specific antirenin responses in high and normal forms of essential, renal, renovascular and malignant hypertension. Amer. J. Cardiol. *32*, 511 (1973)

13 BULPITT, C. J., DOLLERY, C. T.: Side effects of hypotensive agents evaluated by a self-administered questionnaire. Brit. med. J. *iii*, 485 (1973)

14 CARTER, J. K., MITCHELL, H. W., POYSER, R. H.: Comparison of some haemodynamic changes between central and intravenous administration of $(\pm)$-propranolol in anaesthetized dogs. Proc. Brit. pharmacol. Soc., Southampton, 28th and 29th March, 1974, p. 34

15 COLLINS, I. S., KING, I. W.: Pindolol (Visken, LB46), a new treatment for hypertension: report of a multicentric open study. Curr. ther. Res. *14*, 185 (1972)

16 COMERFORD, M. B., PRINGLE, A.: A long-term study of the antihypertensive effect of alprenolol. Acta med. scand. Suppl. 554: 15 (1974)

17 CROOK, B. R. M., RAFTERY, E. B.: Treatment of "nonaccelerated" essential hypertension with oxprenolol. Circulation *46*, Suppl. II: 142 (1972)

18 DAY, M. D., ROACH, A. G.: β-Adrenergic receptors in the central nervous system of the cat concerned with control of arterial blood pressure and heart rate. Nature new Biol. (Lond.) *242*, 30 (1973)

19 DOLLERY, C. T., LEWIS, P. J., MYERS, M. G., REID, J. L.: Central hypotensive effect of propranolol in the rabbit. Brit. J. Pharmacol. *48*, 343P (1973); abstract of paper (cf. also [62])

20 DORPH, S., BINDER, C.: Evaluation of the hypotensive effect of beta-adrenergic blockade in hypertension. Acta med. scand. *185*, 443 (1969)

21 FELTHAM, P. M., WATSON, O. F., PEEL, J. S., DUNLOP, D. J., TURNER, A. S.: Pindolol in hypertension: a double-blind trial. N. Z. med. J. *76*, 167 (1972)

22 FORREST, W. A.: Oxprenolol and a thiazide diuretic together in the treatment of essential hypertension – a large general practice study. Brit. J. clin. Pract. *27*, 331 (1973)

23 FORREST, W. A.: Treatment of moderately severe essential hypertension in general practice with a combination of cyclopenthiazide with potassium chloride (Navidrex K) and oxprenolol (Trasicor 80 mg): a report on 554 patients. J. int. med. Res. *2*, 7 (1974)

24 FRANCIOSA, J. A., FREIS, E. D., CONWAY, J.: Antihypertensive and hemodynamic properties of the new beta adrenergic blocking agent timolol. Circulation *48*, 118 (1973)

25 FREIS, E. D.: Effectiveness of drug therapy in hypertension: present status. Circulation Res. *28/29*, Suppl. II: 70 (1971)

26 FURBERG, C., MICHAELSON, G.: Effect of Aptin, a β-adrenergic blocking agent, in arterial hypertension. Acta med. scand. *186*, 447 (1969)

27 GAGNON, D. J., MELVILLE, K. I.: Centrally mediated cardiovascular responses to isoprenaline. Int. J. Neuropharmacol. *6*, 245 (1967)

28 GILFRICH, H. J., RAHN, K. H., SCHMAHL, F. W.: A comparison of the antihypertensive action of propranolol and the optical isomers of N-isopropyl-p-nitrophenylethanolamine (INPEA). Pharmacol. clin. *2*, 30 (1969)

29 GOTTLIEB, T.B., KATZ, F.H., CHIDSEY, C.A.: Combined therapy with vasodilator drugs and beta-adrenergic blockade in hypertension. Circulation *45*, 571 (1972)
30 GREENBLATT, D.J., KOCH-WESER, J.: Adverse reactions to propranolol in hospitalized medical patients: a report from the Boston Collaborative Drug Surveillance Program. Amer. Heart J. *86*, 478 (1973)
31 HANSSON, L.: Beta-adrenergic blockade in essential hypertension. Effects of propranolol on hemodynamic parameters and plasma renin activity. Acta med. scand. *194*, Suppl. 550 (1973)
32 HANSSON, L., MALMCRONA, R., OLANDER, R., ROSENHALL, L., WESTERLUND, A., ÅBERG, H., HOOD, B.: Propranolol in hypertension. Report on 158 patients treated up to one year. Klin. Wschr. *50*, 364 (1972)
33 HANSSON, L., OLANDER, R., ÅBERG, H., MALMCRONA, R., WESTERLUND, A.: Treatment of hypertension with propranolol and hydralazine. Acta med. scand. *190*, 531 (1971)
34 JOHNSSON, G., GUZMAN, M. DE, BERGMAN, H., SANNERSTEDT, R.: The haemodynamic effects of alprenolol and propranolol at rest and during exercise in hypertensive patients. Pharmacol. clin. *2*, 34 (1969)
35 JULIUS, S., PASCUAL, A.V., ABBRECHT, P.H., LONDON, R.: Effect of beta-adrenergic blockade on plasma volume in human subjects. Proc. Soc. exp. Biol. (N.Y.) *140*, 982 (1972)
36 KATILA, M., FRICK, M.H.: Combined dihydralazine and propranolol in the treatment of hypertension. Int. J. clin. Pharmacol. Ther. Toxicol. *4*, 111 (1970)
37 KELLIHER, G.J., BUCKLEY, J.P.: Central hypotensive activity of dl- and d-propranolol. J. pharm. Sci. *59*, 1276 (1970)
38 KINCAID-SMITH, P.: Management of severe hypertension. Amer. J. Cardiol. *32*, 575 (1973)
39 KOCH-WESER, J.: Correlation of pathophysiology and pharmacotherapy in primary hypertension. Amer. J. Cardiol. *32*, 499 (1973)
40 LEISHMAN, A.W.D., THIRKETTLE, J.L., ALLEN, B.R., DIXON, R.A.: Controlled trial of oxprenolol and practolol in hypertension. Brit. med. J. *iv*, 342 (1970)
41 LOGUE, R.B., ROBINSON, P.H.: Medical management of angina pectoris. Circulation *46*, 1132 (1972)
42 LYDTIN, H., KUSUS, T., DANIEL, W., SCHIERL, W., ACKENHEIL, M., KEMPTER, H., LOHMÖLLER, G., NIKLAS, M., WALTER, I.: Propranolol therapy in essential hypertension. Amer. Heart J. *83*, 589 (1972)
43 MARSHALL, A.J., BARRITT, D.W.: Oxprenolol in hypertension. Brit. J. clin. Pract. *27*, 337 (1973)
44 MICHELAKIS, A.M., McALLISTER, R.G.: The effect of chronic adrenergic receptor blockade on plasma renin activity in man. J. clin. Endocr. *34*, 386 (1972)
45 MORGAN, T.O., LOUIS, W.J., DAWBORN, J.K., DOYLE, A.E.: The use of prindolol (Visken) in the treatment of hypertension. Med. J. Aust. *59/ii*, 309 (1972)
46 MURPHY, J.E., STANDEN, S.M., FORREST, W.A.: The addition of oxprenolol to hypertensive patients treated with methyldopa – a general practice study. J. int. med. Res. *2*, 1 (1974)
47 MYLECHARANE, E.J., RAPER, C.: Further studies on the adrenergic neuron blocking activity of some β-adrenoceptor antagonists and guanethidine. J. Pharm. Pharmacol. *25*, 213 (1973)
48 NEFF, K., STUMPE, K.O.: Behandlung der arteriellen Hypertonie mit einem Betarezeptorenblocker in Kombination mit einem Saluretikum. Therapiewoche *24*, 429 (1974)
49 NIES, A.S., SHAND, D.G.: Hypertensive response to propranolol in a patient treated with methyldopa – a proposed mechanism. Clin. Pharmacol. Ther. *14*, 823 (1973)
50 O'BRIEN, E.T., MacKINNON, J.: Propranolol and polythiazide in treatment of hypertension. Brit. Heart J. *34*, 1042 (1972)

51 PAPE, J.: The effect of alprenolol in combination with hydralazine in essential hypertension: a double-blind, crossover study and a long-term follow-up study. Acta med. scand. *195*, Suppl. 554: 55 (1974)

52 PATERSON, J.W., DOLLERY, C.T.: Effect of propranolol in mild hypertension. Lancet *ii*, 1148 (1966)

53 PEARSON, R.M., BULPITT, C., GEORGE, C.F., HOLE, D., BRECKENRIDGE, A.: A trial of the combination of guanethidine and oxprenolol in hypertension. Scot. med.J. *19*, 45 (1974)

54 PRICHARD, B.N.C.: Hypotensive action of pronethalol. Brit. med.J. *i*, 1227 (1964)

55 PRICHARD, B.N.C.: Variation in the modification of cardiovascular responses by sympathetic inhibitory drugs. Proc. roy. Soc. Med. *62*, 84 (1969)

56 PRICHARD, B.N.C., GILLAM, P.M.S.: The use of propranolol (Inderal) in the treatment of hypertension. Brit. med.J. *ii*, 725 (1964)

57 PRICHARD, B.N.C., GILLAM, P.M.S.: Treatment of hypertension with propranolol. Brit. med.J. *i*, 7 (1969)

58 PRICHARD, B.N.C., GILLAM, P.M.S., GRAHAM, B.R.: Beta receptor antagonism in hypertension; comparison with the effect of adrenergic neurone inhibition on cardiovascular responses. Int.J. clin. Pharmacol.Ther.Toxicol. *4*, 131 (1970)

59 PRICHARD, B.N.C., SHINEBOURNE, E., FLEMING, J., HAMER, J.: Haemodynamic studies in hypertensive patients treated by oral propranolol. Brit. Heart J. *32*, 236 (1970)

60 RAFTERY, E.B., DENMAN, A.M.: Systemic lupus erythematosus syndrome induced by practolol. Brit. med.J. *ii*, 452 (1973)

61 RAHN, K.H., GILFRICH, H.J.: Untersuchungen zum Mechanismus der antihypertensiven Wirkung von Betareceptorenblockern. Paper presented at IXth Symp. Ges. Nephrol., Basle 1973

62 REID,J.L., LEWIS, P.J., MYERS, M.G., DOLLERY, C.T.: Cardiovascular effects of intracerebroventricular d-, l- and dl-propranolol in the conscious rabbit. J. Pharmacol. exp.Ther. *188*, 394 (1974); cf. also [19]

63 RICHARDSON, D.W., FREUND, J., GEAR, A.S., MAUCK, H.P., Jr., PRESTON, L.W.: Effect of propranolol on elevated blood pressure. Circulation *37*, 534 (1968)

64 SAFAR, M., WEISS, Y., SOBEL, A., LAGRUE, G., MILLIEZ, P.: Action anti-hypertensive d'un bêta-bloqueur, le pindolol. Nouv. Presse méd. *2*, 2685 (1973)

65 SANNERSTEDT, R., STENBERG, J., JOHNSSON, G., WERKÖ, L.: Hemodynamic interference of alprenolol with dihydralazine in normal and hypertensive man. Amer. J. Cardiol. *28*, 316 (1971)

66 SANNERSTEDT, R., STENBERG, J., VEDIN, A., WILHELMSSON, C., WERKÖ, L.: Chronic beta adrenergic blockade in arterial hypertension. Amer. J. Cardiol. *29*, 718 (1972)

67 SCHAFFALITZKY DE MUCKADELL, O.B., GYNTELBERG, F.: The antihypertensive effect of a new beta-blocking agent pindolol compared with chlorthalidone. Europ. J. clin. Pharmacol. *5*, 210 (1973)

68 SCHRÖDER, G., WERKÖ, L.: Nethalide, a beta adrenergic blocking agent. Clin. Pharmacol.Ther. *5*, 159 (1964)

69 SHARE, N.N.: "Alpha" and "beta" adrenergic receptors in the medullary vasomotor center of the cat. Arch. int. Pharmacodyn. *202*, 362 (1973)

70 SIMPSON, F.O.: β-Adrenergic receptor blocking drugs in hypertension. Curr.Ther. *14*, 10: 91 (1973)

71 SRIVASTAVA, R.K., KULSHRESTHA, V.K., SINGH, N., BHARGAVA, K.P.: Central cardiovascular effects of intracerebroventricular propranolol. Europ.J. Pharmacol. *21*, 222 (1973)

72 STARKE, K.: Beziehungen zwischen dem Renin-Angiotensin-System und dem vegetativen Nervensystem. Klin.Wschr. *50*, 1069 (1972)

73 STEIN, J.H., FERRIS, T.F.: The physiology of renin. Arch. intern. Med. *131*, 860 (1973)

74 TAGGART, P., CARRUTHERS, M., SOMERVILLE, W.: Electrocardiogram, plasma catecholamines and lipids, and their modification by oxprenolol when speaking before an audience. Lancet *ii*, 341 (1973)

75 TARAZI, R. C., DUSTAN, H. P.: Beta adrenergic blockade in hypertension. Amer. J. Cardiol. *29*, 633 (1972)

76 TAYLOR, S. H., MEERAN, M. K.: Different effects of adrenergic beta-receptor blockade on heart rate response to mental stress, catecholamines, and exercise. Brit. med. J. *iv*, 257 (1973)

77 TIBBLIN, G., ÅBLAD, B.: Antihypertensive therapy with alprenolol, a β-adrenergic receptor antagonist. Acta med. scand. *186*, 451 (1969)

78 TRAUB, Y., SHAVER, J. A., McDONALD, R. H., Jr., SHAPIRO, A. P.: Effects of practolol on pressor responses to noxious stimuli in hypertensive patients. Clin. Pharmacol. Ther. *14*, 165 (1973)

79 TUCKMAN, J., MESSERLI, F., HODLER, J.: Treatment of hypertension with large doses of the β-adrenergic blocking drug oxprenolol, alone, and in combination with the vasodilator dihydralazine. Clin. Sci. molec. Med. *45*, Suppl. 1: 159s (1973)

80 VANDER, A. J.: Effect of catecholamines and the renal nerves on renin secretion in anesthetized dogs. Amer. J. Physiol. *209*, 659 (1965)

81 WAAL, H. J.: Hypotensive action of propranolol. Clin. Pharmacol. Ther. *7*, 588 (1966)

82 WAAL-MANNING, H. J.: Comparative studies on the hypotensive effects of beta-blockers. N. Z. med. J. *71*, 383 (1970); abstract of paper

83 WAAL-MANNING, H. J.: Lack of effect of d-propranolol on blood pressure and pulse rate in hypertensive patients. Proc. Univ. Otago med. Sch. *48*, 80 (1970)

84 WAAL-MANNING, H. J., SIMPSON, F. O.: Practolol treatment in asthmatics. Lancet *ii*, 1264 (1971); corresp.

85 WILSON, D. F., WATSON, O. F., PEEL, J. S., LANGLEY, R. B., TURNER, A. S.: Some haemodynamic effects of Trasicor (CIBA 39,089-Ba). N. Z. med. J. *68*, 145 (1968)

86 WOOD, R. A., FORRESTER, T. M., JOHNSTON, A. W., PALMER, K. N. V.: Management of hypertension: a trial of practolol (Eraldin). Clin. Trials J. *10*, 53 (1973)

87 ZACEST, R., GILMORE, E., KOCH-WESER, J.: Treatment of essential hypertension with combined vasodilatation and beta-adrenergic blockade. New Engl. J. Med. *286*, 617 (1972)

88 ZACHARIAS, F. J., COWEN, K. J., PRESTT, J., VICKERS, J., WALL, B. G.: Propranolol in hypertension: a study of long-term therapy, 1964–1970. Amer. Heart J. *83*, 755 (1972)

Discussion

A. ZANCHETTI: I wish to comment very briefly on only one aspect of the excellent review which Dr. BRUNNER has just given us of what is currently known about the antihypertensive effect of the beta-blockers. In the course of his paper, he pointed out that there is little or no clear-cut relationship between the degree of the antihypertensive response to a beta-blocker and the size of the dose administered. The flat pattern of the dose-response curve to which he referred is, I think, mainly attributable to the fact that the plasma concentration attained following a given dose tends to vary very considerably from patient to patient.

In studies that we have carried out in Milan during the past year, we administered propranolol to the same patients in doses increasing stepwise from 10 mg. four times daily to 80 mg. four times daily. The dosage increments were made at weekly intervals and the concentrations of propranolol in the plasma were measured using a highly sensitive gas-chromatographic method developed in Milan by CHIDSEY and his group*. When we compared the plasma concentrations with the doses administered**, we found that, in the 20 or more patients studied, there was an enormous scatter. Sometimes the plasma concentrations were very low and sometimes they were very high, particularly in response to the larger doses. On the other hand, when we studied the relationship between the logarithm of the plasma concentrations reached and the antihypertensive responses recorded***, we observed quite a close correlation. Admittedly, the dose-response curve was not very steep, but this is hardly surprising, because propranolol does not of course exert a very potent antihypertensive effect when employed as monotherapy; nevertheless, we did get a highly significant correlation with respect to both the systolic and the diastolic blood pressures: a measurable hypotensive response started to appear at concentrations of about 20 ng. propranolol per millilitre of plasma, and an 18 to 20% decrease in the blood pressure levels occurred at concentrations as high as 400 ng./ml.

H. BRUNNER: How do you account for the fact, Dr. ZANCHETTI, that the curve plotted for the correlation between plasma concentration and response is steeper in the case of relief from emotionally induced tachycardia than it is in the case of the drug's antihypertensive effect?

A. ZANCHETTI: This is a point that I hope to be able to revert to in a later discussion. All I should like to say now is that, in our experience, you already get a marked decrease in the heart rate at low plasma concentrations of propranolol, which suggests that the correlation curve is rather steep in its initial portion. But, at these low concentrations, you do not yet observe much effect on the blood pressure. Subsequently, however, at higher plasma concentrations, the heart rate correlation curve flattens out, whereas the correlation curve for the blood pressure response begins to rise later and thereafter continues to rise.

M. IKEDA: I should like briefly to describe the results of a study which TERASAWA and I carried out in healthy elderly subjects in an attempt to classify various beta-blockers

---

* DI SALLE, E., BAKER, K. M., BAREGGI, F. R., WATKINS, W. D., CHIDSEY, C. A., FRIGERIO, A., MORSELLI, P. L.: A sensitive gas chromatographic method for the determination of propranolol in human plasma. J. Chromatogr. *84,* 347 (1973)
** MORSELLI, P. L., MORGANTI, A., BIANCHETTI, G., DI SALLE, E., LEONETTI, G., ZANCHETTI, A., CHIDSEY, C. A.: Blood levels and pharmacokinetic studies of propranolol during chronic treatment in hypertensive patients. Europ. J. clin. Invest. (printing)
*** LEONETTI, G., MAYER, G., MORGANTI, A., TERZOLI, L., ZANCHETTI, A., MORSELLI, P. L., DI SALLE, E., CHIDSEY, C. A.: Suppression of plasma renin activity and hypotensive action of propranolol in essential hypertensive patients. Europ. J. clin. Invest. (printing)

by reference to their dose-response relationships. The preparations investigated, which were all given intravenously, comprised propranolol, oxprenolol (®Trasicor), Kö 1366, and practolol, and six subjects were studied per beta-blocker. The haemodynamic measurements performed were carried out after the subjects had rested in the supine position for 20–30 minutes in order to ensure stabilisation of their heart rate and blood pressure. Cardiac output was measured by the dilution method, using an earpiece photocell and Evans blue.

Shown in Figure 1, in which the dosages of the four different beta-blockers are expressed on a logarithmic scale, are the dose-response relationships for the decrease in

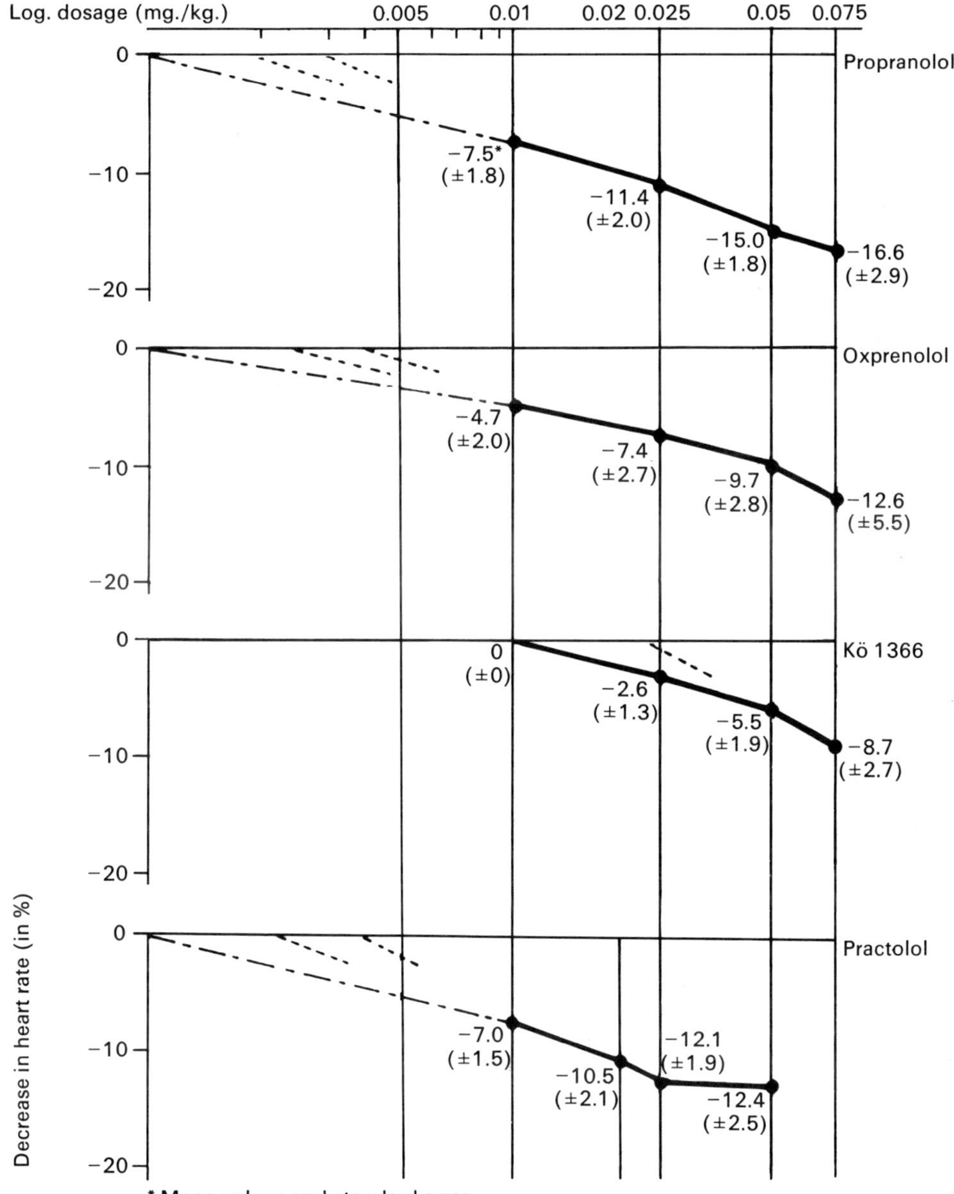

Fig. 1. Dose-response relationship in respect of the decrease in heart rate produced by intravenous doses of four beta-blockers. For further details, see text.

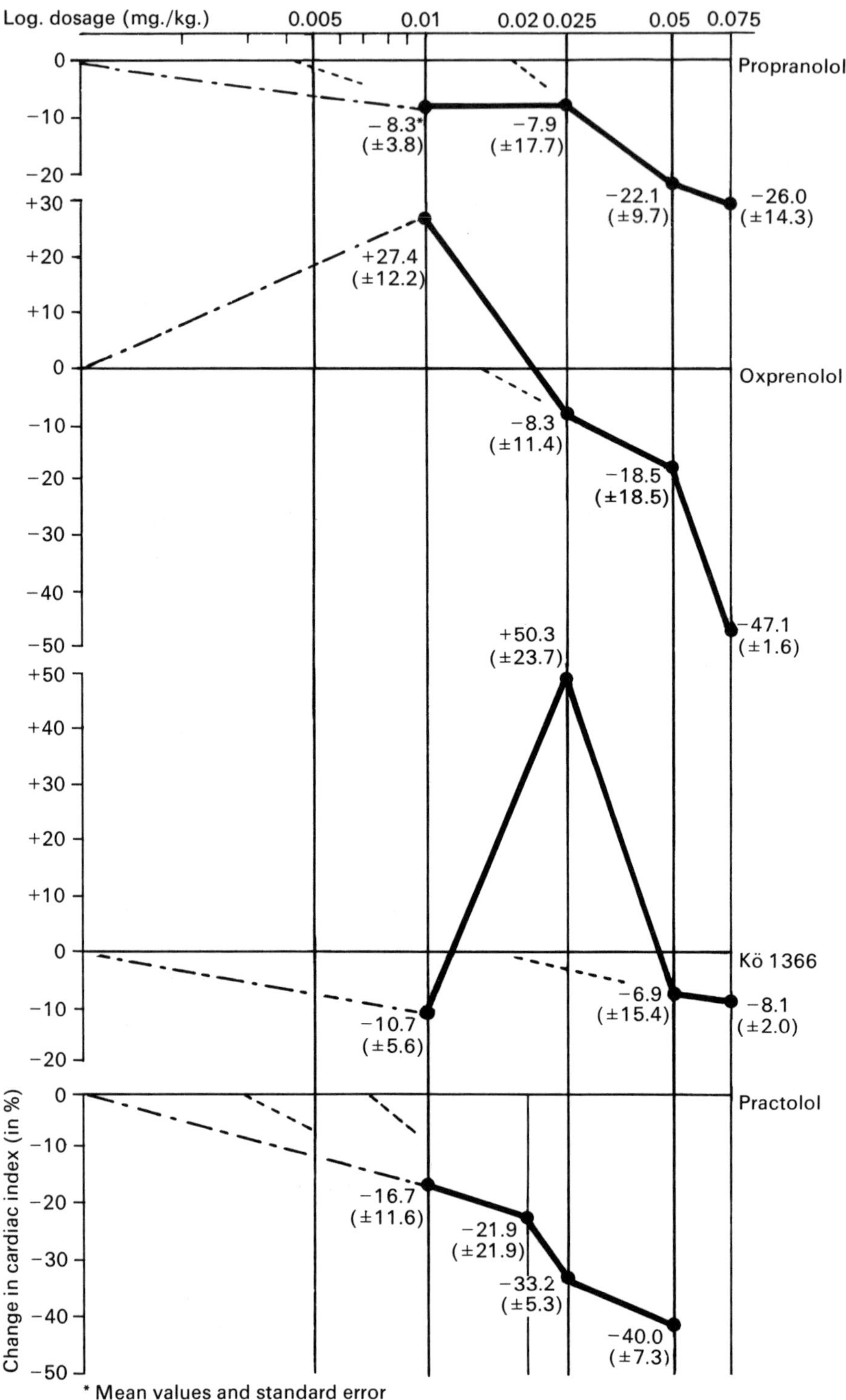

Fig. 2. Dose-response relationship in respect of the changes in cardiac index produced by intravenous doses of four beta-blockers.

heart rate. On a milligramme for milligramme basis, propranolol was found to exert
the most potent effect.
The dose-response relationships in respect of the cardiac index are outlined in Figure 2.
The increase in the cardiac index which oxprenolol and Kö 1366 produced at a low
dosage level is attributable to an increase in stroke index resulting from their intrinsic
sympathomimetic action, the subsequent decrease being due to their negative inotropic
activity in higher doses. This response pattern contrasts with that observed in the
case of propranolol and practolol, which elicited only a decrease in the cardiac index.

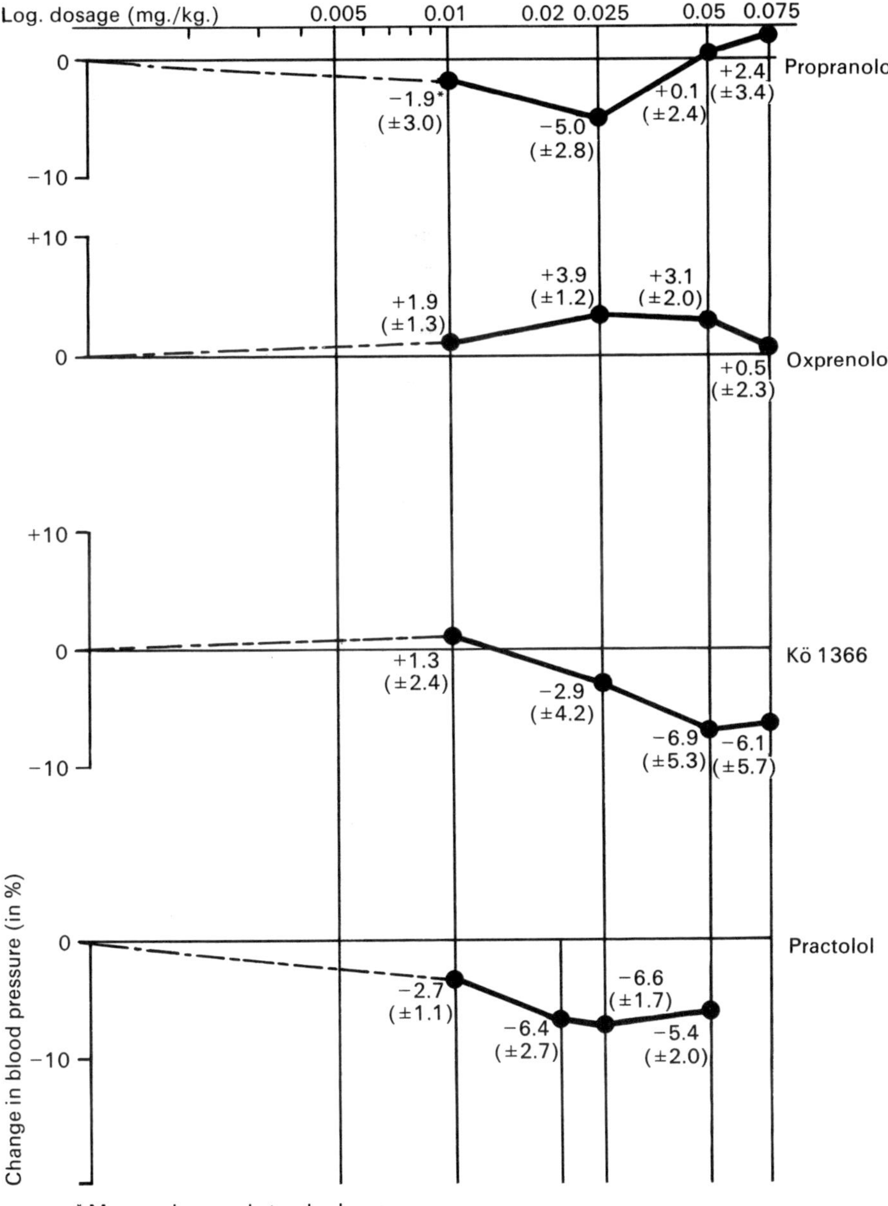

Fig. 3. Dose-response relationship in respect of the changes in mean blood pressure
produced by intravenous doses of four beta-blockers.

Finally, indicated in Figure 3 are the dose-response relationships for the changes in blood pressure induced by the four beta-blockers.

These findings would appear to suggest that beta-blockers exhibiting intrinsic sympathomimetic activity may safely be employed to treat arrhythmia and angina pectoris in elderly patients. They do not provide any indication, however, as to which type of beta-blocker is preferable for the treatment of hypertension. Moreover, to arrive at an assessment of the relative merits of the various beta-blockers when employed for long-term therapy, it will be necessary to make similar evaluations of the haemodynamic changes occurring in response to prolonged oral medication. A detailed report on the study to which I have referred here is due to be published by TERASAWA et al.*.

K. H. WINTERHALTER: In connection with some data that I shall be presenting later on during this symposium, may I ask Dr. BRUNNER whether there is any known relationship between the antihypertensive effect of the beta-blockers under discussion and the changes which they produce in the patient's blood volume and/or in his haematocrit?

H. BRUNNER: When beta-blockers are given alone, the blood volume normally increases owing to the retention of sodium which these drugs produce in most, but not all, cases. As for their influence on the haematocrit, this is a question that I'm afraid I cannot answer.

J. SCHWARTZ: In your paper, Dr. BRUNNER, you indicated the ED_{50} for several beta-blockers. It is clear, of course, that the ED_{50} depends, firstly, on the individual degree of responsiveness to the isoprenaline you took as a reference substance, and, secondly, on the dose of isoprenaline injected. You did not, however, include in your Table 1 the pA_2 values, which are more precise than the ED_{50}. Admittedly, the pA_2 (that is, the negative logarithm of the concentration of an antagonist which makes it necessary to double the dose of an agonist to produce the same effect in the presence of the antagonist as was previously observed in its absence) varies as a function of time, the values recorded at two minutes being different from those recorded at 30 minutes. In other words, the availability of the beta-blocker at the receptor site changes as the minutes pass.

Thus, though it may be questionable whether the dosage schedules of beta-blockers should really be based on the pA_2 values of these drugs, it nevertheless strikes me as paradoxical that no account should be taken of the kinetics of beta-blockers when establishing the interval between doses.

H. BRUNNER: The standard dose of isoprenaline that we used in our experiments was 0.5 mcg./kg. i.v., and it was invariably injected five minutes after intravenous administration of the beta-blocker. In our experience, this is the time by which the effect of the beta-blocker has reached its maximum. We have also determined the pA_2 values of the beta-blockers I mentioned – not, however, in the intact animal, but only *in vitro*, i.e. in isolated guinea-pig atria using isoprenaline as the agonist. If you take these values, you find that the order of the various beta-blockers is the same as that based on their ED_{50} as shown in my Table 1. In this table of mine I merely wished to demonstrate that there is a rough correlation between beta-blocking activity as determined in animal experiments and doses required to produce a clinical antihypertensive effect. I agree of course that any evaluation of beta-blockers must take into account not only their activity as such but also their duration of action. One inevitable conclusion to be drawn from an evaluation of this kind is that practolol would have to be given in even higher doses than it is if its duration of action were not substantially longer than that of the other preparations listed.

---

* TERASAWA, F., SUZUKI, T., LIE HON YING, IKEDA, M.: Dose-response relationship of hemodynamic changes in the aged by intravenous adminitration of propranolol, oxprenolol, Kö 1366 and practolol. Tohoku J. exp. Med. (submitted for publication)

S. H. Taylor: On theoretical grounds it could be argued that the cardioselective beta-blockers should prove more effective in lowering blood pressure at rest and especially in preventing rises in blood pressure occasioned by stress. When reviewing the data available on the antihypertensive activity of the beta-blockers, Dr. Brunner, did you find any practical evidence to support this supposition?

H. Brunner: No. Most or all of the beta-blockers lower cardiac output to a greater or lesser degree, and all of them produce a more or less marked increase in peripheral resistance, at least in "acute" experiments. I don't believe that this increase in peripheral resistance is attributable purely and simply to the relative preponderance of alpha tone which results from blockade of the dilating vascular beta-receptors. It is, I think, more in the nature of a reflex mechanism by which enhanced alpha-receptor activity serves to offset the decrease in cardiac output by increasing vascular tone.

Characterisation of beta-blockers as antihypertensive agents in the light of human-pharmacology studies

by P. R. Imhof*

It can now be taken for granted that all beta-blockers display more or less marked antihypertensive properties, but it seems rather difficult to show whether one or other of the many beta-blocking compounds available – which at present total more than 30 – clearly deserves preference in the treatment of hypertension, since not enough practical experience has yet been acquired with them. It is nevertheless tempting to try and predict the antihypertensive activity of these substances on the basis of their pharmacological "activity profile", which differs quite considerably from one class of beta-blocker to another.

In order to define the activity of a beta-blocker in broad outline, it is necessary first of all to determine its potency and its degree of cardioselectivity and also to establish whether or not it exerts an intrinsic sympathomimetic action (I.S.A.); it is not necessary, however, to consider its membrane-stabilising effects, because these have no bearing on the treatment of hypertension. All these properties, which can be assessed relatively easily and reliably in healthy subjects[12, 29, 30], provide a means of classifying the beta-blockers into one of the four categories indicated in Table 1, which comprise cardioselective and non-cardioselective beta-blockers with and without an intrinsic sympathomimetic action. Seventeen beta-blockers are listed in these four categories; those printed in italics are commercially available preparations, and the

Table 1. Classification of 17 beta-blockers according to presence or absence of cardioselectivity and presence or absence of an intrinsic sympathomimetic action.

Cardioselective beta-blockers		Non-cardioselective beta-blockers	
Intrinsic sympathomimetic action	No intrinsic sympathomimetic action	Intrinsic sympathomimetic action	No intrinsic sympathomimetic action
Practolol	Metoprolol (CGP 2175 = H 93/26)[1, 31]	*Oxprenolol*	*Propranolol*
Para-oxprenolol[49]		*Alprenolol*	*Sotalol*
Acebutolol (M & B 17,803A)[6, 12, 14, 25, 37]	Tenormin (ICI 66,082)[2, 5, 26, 28]	*Pindolol* (also known as *prindolol*)	*Timolol* (MK-950)[20-23, 38, 39]
	Tolamolol (UK 6558)[3, 4, 12, 15, 32]	*Toliprolol*	*Bupranolol*
		Nifenalol	*Butidrine*
			AH 5158[8, 17]

* Research Department (Human Pharmacology), Pharmaceuticals Division, CIBA-GEIGY LIMITED, Basle, Switzerland.

others experimental compounds which deserve interest in connection with the treatment of hypertension. Allen & Hanburys' 5158, a non-cardioselective beta-blocker without intrinsic sympathomimetic activity, is unique insofar as it also has alpha-blocking properties, and we are eagerly awaiting news of the first clinical results obtained with this drug. Merck's timolol (MK-950), Pfizer's tolamolol (UK 6558), ICI's Tenormin (66,082), CIBA-GEIGY's metoprolol (CGP 2175, identical with Hässle's H 93/26), and May & Baker's acebutolol (17,803A) are all beta-blockers displaying definite antihypertensive properties. Para-oxprenolol is listed in Table 1 for reasons which will be discussed later.

What is the significance of cardioselectivity?

Cardioselective beta-blockers inhibit the cardiac beta-receptors (β_1-receptors), but exert little influence on the bronchial and vascular beta-receptors (β_2-receptors) when employed in low doses. The fact that they have little or no effect on the peripheral beta-receptors would theoretically have two advantages: firstly, these cardioselective blockers would be safer to employ in cases where the patient was suffering, or had suffered in the past, from some broncho-spastic condition and, secondly, they would be particularly suitable for the treatment of hypertension because their use would involve no inhibition of the peripheral (vasodilator) beta-receptors. In practice, however, since their cardioselectivity diminishes as the dosage is raised, and since hypertensive patients generally have to be given far larger doses than are required simply to block the β_1-receptors, this cardioselectivity is only relative and offers little if any real advantage.

The hypothesis advanced by CHIDSEY and his group[9] on the basis of haemodynamic studies, according to which the decrease in blood pressure produced by practolol is due, not to a reduction in cardiac output, but to peripheral vasodilatation, is probably not tenable – at least, not in this generalised form. Several authors[18, 24] have in fact observed marked decreases in cardiac output in response to practolol administered in fairly high doses, even though these decreases were not so pronounced as with a beta-blocker such as propranolol; and any diminution in cardiac output inevitably leads to a reflex-induced rise in peripheral resistance, which would appear to be brought about via stimulation of the alpha-receptors. Moreover, in practice, practolol too is liable to precipitate or aggravate bronchospastic conditions[7, 10, 35, 50], and – compared with other, non-cardioselective, beta-blockers – its antihypertensive effect seems to be not more, but if anything, less potent[9, 36].

Assuming, then, that a non-cardioselective beta-blocker is prescribed as treatment for hypertension, what repercussions is its lack of cardioselectivity likely to have? At first sight, the fact that such a drug will raise peripheral resistance by blocking the vascular beta-receptors appears to be a disadvantage, because it will counteract the fall in blood pressure. Experience has shown, however, that, when oxprenolol (®Trasicor)[47] for example – or even propranolol[46] – is administered over a prolonged period, the rise in peripheral resistance is later followed by a gradual decrease, whereas the other haemodynamic variables show very little further change. From this it may be concluded that the increase in peripheral resistance observed in response to treatment with a non-cardioselective beta-blocker is only of a temporary nature and that, where the medication is continued for several months, there is probably no difference between non-cardioselective and cardioselective beta-blockers as far as their influence on peripheral resistance is concerned.

What is the significance of an intrinsic sympathomimetic action?

On theoretical grounds it may be postulated that beta-blockers displaying an intrinsic sympathomimetic action are less likely to induce cardiac insufficiency than beta-blockers devoid of such activity, since the former will diminish the force of contraction of the heart muscle to a lesser extent[13]. But no therapeutic studies offering conclusive statistical proof of the correctness of this supposition have yet been published. Of interest in this connection, however, is the finding that beta-blockers exhibiting no intrinsic sympathomimetic activity, such as propranolol, cause an increase in pulmonary capillary pressure (wedge pressure) and in cardiopulmonary blood volume, whereas those endowed with an intrinsic sympathomimetic action, such as oxprenolol and pindolol, either exert no influence on pulmonary capillary pressure or actually lower it while at the same time producing less of an increase in cardiopulmonary blood volume[40, 41, 44]. Beta-blockers possessing an intrinsic sympathomimetic effect lower the heart rate less markedly[30] and thus entail less risk of provoking excessive bradycardia.

A *pronounced* intrinsic sympathomimetic action is definitely undesirable in the treatment of hypertension. This conclusion is impressively supported by observations we have made with para-oxprenolol: we found that this beta-blocker, which exerts a marked cardiostimulant effect, actually induced an increase in blood pressure and heart rate, despite the fact that it still displayed beta-blocking properties in the isoprenaline test (IMHOF, unpublished data). A *moderate* intrinsic sympathomimetic action, by contrast, such as is exhibited by oxprenolol, alprenolol, and pindolol, does not impair the antihypertensive effect, as has been demonstrated by extensive clinical experience. SIMPSON and WAAL-MANNING[45], for example, when they administered either the non-sympathomimetic beta-blocker propranolol or the sympathomimetic beta-blocker oxprenolol in equivalent doses to 15 hypertensives, found no statistically significant difference in the fall in blood pressure produced by the two drugs; the decrease in the heart rate in these patients, however, was significantly less marked in response to oxprenolol. KINCAID-SMITH[33], too, believes that the degree of intrinsic sympathomimetic activity has no great bearing on the antihypertensive effect of the beta-blockers.

The importance of intrinsic sympathomimetic activity is often considered solely from the standpoint of the effect exerted by beta-blockers on the cardiac beta-receptors. But their intrinsic sympathomimetic activity does, of course, likewise result in stimulation of the peripheral beta-receptors; this applies not only to the non-cardioselective, but also to the cardioselective beta-blockers as soon as the latter are employed in high dosages that are no longer cardioselective. In the case of the bronchial beta-receptors this means that at least some of the undesirable effects of beta-blockade, in the form of inhibition of bronchodilatation, are offset by the drug's intrinsic sympathomimetic action. This is probably also the reason why – in contrast to what might have been expected on purely theoretical grounds – oxprenolol, for example, very seldom indeed gives rise to bronchial complications. In the case of the vascular beta-receptors, on the other hand, an intrinsic sympathomimetic action has the effect of diminishing the increase in peripheral vascular resistance induced by beta-blockade. In patients who are particularly sensitive to sympathomimetic stimulation, such as those suffering from hyperthyroidism, blockade of the peripheral beta-receptors may sometimes become masked by the intrinsic sympathomimetic effect of a beta-blocker. This phenomenon has been observed with pindolol, a non-cardioselective beta-blocker with an intrinsic sympathomimetic action, but was wrongly interpreted as evidence of cardioselectivity[16]. Oxprenolol, too, has for the same reason wrongly been said to display cardioselectivity[36]. In man, beta-blockers with an intrinsic sympathomimetic action may thus certainly *simulate* cardioselectivity; in other words, the differences between a cardioselective beta-blocker with no intrinsic sympathomimetic effect and a non-cardioselective beta-blocker which does exhibit such an effect tend to become blurred the higher the dosages administered. The cardioselectivity of a beta-blocker is dose-dependent and decreases or disappears altogether when larger doses are used, whereas – even at high dosage levels – the intrinsic sympathomimetic effect remains of course undiminished.

In the light of all these considerations, the ideal beta-blocker for the treatment of hypertension would have to be one that exerts a cardioselective effect irrespective of the dosage in which it is prescribed; in addition, it would also have to possess an intrinsic sympathomimetic action to which the peripheral beta-receptors would be more susceptible than the cardiac beta-receptors. No such beta-blocker exists or, for that matter, is ever likely to exist. Experience in the use of beta-blockers as treatment for hypertension, though still relatively limited, suggests that of the beta-blocking compounds at present available on the market no one preparation in particular deserves obvious preference when considered purely and simply from the standpoint of antihypertensive activity. On the other hand, certain differences do appear to exist between the various beta-blockers as regards their safety and tolerability and their liability to produce untoward side effects. Propranolol, for example, is probably more likely to give rise to cardiac decompensation and to bronchospasm than the other preparations. Pindolol seems to have more side effects of a central nature, and practolol displays rather poor gastro-intestinal tolerability in higher doses, besides which it may also provoke skin reactions and systemic lupus erythematosus syndrome[43]. Oxprenolol and alprenolol, by contrast, are remarkably well tolerated and are in fact the two beta-blockers to which I personally accord preference in the treatment of hypertension.

There are two additional points which have not yet been adequately investigated, but which might well have a major bearing on the use of beta-blockers in hypertension, namely, the question of fibrinolysis and of platelet aggregation. Since arteriosclerosis constitutes a particularly serious hazard in hypertensives, long-term treatment with beta-blockers that are capable of activating fibrinolysis and diminishing platelet aggregation would be extremely valuable, because it would eliminate or at least partially counteract two of the factors involved in the pathogenesis of arteriosclerosis. According to findings reported by Ponari et al.[42], the various beta-blockers differ appreciably with regard to their influence on fibrinolysis: these authors claim that oxprenolol activates fibrinolysis, whereas propranolol and pindolol display no such effect. Propranolol, on the other hand, does inhibit platelet aggregation[27]; the action of other beta-blockers on platelet aggregation has not yet been studied.

Although – without indulging in speculation – it is impossible at present to single out from the many new experimental beta-blockers, such as timolol, tolamolol, acebutolol, etc., any one compound in particular as meriting pride of place in the treatment of hypertension, it seems safe to assume that beta-blocker therapy in general will have a great future in this indication, especially if it were to prove possible with the aid of beta-blockers also to reduce the incidence of myocardial infarction which, despite conventional antihypertensive treatment, is still the chief cause of death in hypertensives[11]. The outlook in this respect certainly appears promising, since it has already been demonstrated that in patients suffering from angina pectoris treatment with beta-blockers does in fact help to reduce the number of deaths due to myocardial infarction[34].

References

1 Åblad, B., Carlsson, E., Ek, L.: Pharmacological studies of two new cardio-selective beta-receptor antagonists. Life Sci. *12/I*, 107 (1973)

2 Amery, A., Billiet, L., Fagard, R.: Beta receptors and renin release. New Engl. J. Med. *290*, 284 (1974); corresp.

3 Aronow, W.S., Harding, P.R., Nelson, W.H., Vangrow, J.S., Johnson, L.L., Khursheed, M.: Treatment of arrhythmias with tolamidol. Clin. Pharmacol. Ther. *13*, 856 (1972)

4 Augstein, J., Cox, D.A., Ham, A.L., Leeming, P.R., Snarey, M.: β-Adrenoceptor blocking agents. 1. Cardioselective 1-aryloxy-3-(aryloxyalkylamino)propan-2-ols. J. med. Chem. *16*, 1245 (1973)

5 Barrett, A.M., Carter, J., Fitzgerald, J.D., Hull, R., Le Count, D.: A new type of cardioselective adrenoceptive blocking drug. Brit. J. Pharmacol. *48*, 340P (1973); abstract of paper

6 Basil, B., Jordan, R., Loveless, A.H., Maxwell, D.R.: β-Adrenoceptor blocking properties and cardioselectivity of M & B 17,803A. Brit. J. Pharmacol. *48*, 198 (1973)

7 Bernecker, C., Roetscher, I.: The beta-blocking effect of practolol in asthmatics. Lancet *ii*, 662 (1970)

8 Boakes, A.J., Prichard, B.N.C.: The effect of AH 5158, pindolol, propranolol, D-propranolol on acute exercise tolerance in angina pectoris. Brit. J. Pharmacol. *47*, 673P (1973); abstract of paper

9 BODEM, G., BRAMMELL, H. L., WEIL, J. V., CHIDSEY, C. A.: Comparative evaluation of adrenergic blockade with practolol and propranolol in hypertension. Clin. Res. *20*, 169 (1972); abstract of paper

10 BONN, J. A., TURNER, P., HICKS, D. C.: Beta-adrenergic-receptor blockade with practolol in treatment of anxiety. Lancet *i*, 814 (1972)

11 BRECKENRIDGE, A., DOLLERY, C. T., PARRY, E. H. O.: Prognosis of treated hypertension. Quart. J. Med. *39*, 411 (1970)

12 BRIANT, R. H., DOLLERY, C. T., FENYVESI, T., GEORGE, C. F.: Assessment of selective β-adrenoceptor blockade in man. Brit. J. Pharmacol. *49*, 106 (1973)

13 CHOQUET, Y., CAPONE, R. J., MASON, D. T., AMSTERDAM, E. A., ZELIS, R.: Comparison of the beta adrenergic blocking properties and negative inotropic effects of oxprenolol and propranolol in patients. Amer. J. Cardiol. *29*, 257 (1972); abstract of paper

14 COLEMAN, A. J., LEARY, W. P.: Cardiovascular effects of acebutolol (M & B 17803A) in exercising man; a comparative study with practolol and propranolol. Curr. ther. Res. *14*, 673 (1972)

15 DAVEY, M. J.: The pharmacological properties of tolamolol – a new cardioselective β-blocker. Naunyn-Schmiedeberg's Arch. Pharmacol. *279*, Suppl.: R 13 (1973)

16 DUFOUR, R., SCAZZIGA, B., SCHELLING, J. L.: Acute circulatory effects of a beta adrenergic blocking agent (LB 46) in patients with sympathetic overstimulation. Int. J. clin. Pharmacol. Ther. Toxicol. *4*, 145 (1970)

17 FARMER, J. B., KENNEDY, I., LEVY, G. P., MARSHALL, R. J.: Pharmacology of AH 5158, a drug which blocks both a- and β-adrenoceptors. Brit. J. Pharmacol. *45*, 660 (1972)

18 FINEGAN, R. E., MARLON, A. M., HARRISON, D. C.: Circulatory effects of practolol. Amer. J. Cardiol. *29*, 315 (1972)

19 FORREST, W. A.: Oxprenolol and a thiazide diuretic together in the treatment of essential hypertension – a large general practice study. Brit. J. clin. Pract. *27*, 331 (1973)

20 FRANCIOSA, J. A., CONWAY, J., FREIS, E. D.: Hemodynamic and antihypertensive effects of a new beta adrenergic agent, MK-950. Clin. Pharmacol. Ther. *13*, 138 (1972); abstract of paper

21 FRANCIOSA, J. A., FREIS, E. D., CONWAY, J.: Antihypertensive and hemodynamic properties of the new beta adrenergic blocking agent timolol. Circulation *48*, 118 (1973)

22 FREIS, E. D., FRANCIOSA, J. A., CONWAY, J.: Hemodynamic and antihypertensive effects of timolol in hypertensive patients. Minerva cardioangiol. *22*, 56 (1974); abstract of paper

23 FROHLICH, E. D.: A comparison of timolol (MK-950) and propranolol in essential hypertension. Minerva cardioangiol. *22*, 58 (1974); abstract of paper

24 FROHLICH, E. D., BHATIA, S.: Hemodynamic effects of propranolol, practolol and sotalol in hypertensive patients. Clin. Pharmacol. Ther. *13*, 138 (1971); abstract of paper

25 GEORGE, C. F., BRIANT, R. H., FENYVESI, T., DOLLERY, C. T.: Pharmacology of M & B 17803 in man and dog. Europ. J. clin. Invest. *1*, 372 (1971)

26 GRAHAM, B. R., LITTLEJOHNS, D. W., PRICHARD, B. N. C., SCALES, B., SOUTHORN, P.: Preliminary observations on the human pharmacology of I.C.I. 66082 in normal volunteers. Brit. J. Pharmacol. *49*, 154P (1973); abstract of paper

27 HAMPTON, J. R., HARRISON, M. J. G., HONOUR, A. J., MITCHELL, J. R. A.: Platelet behaviour and drugs used in cardiovascular disease. Cardiovasc. Res. *1*, 101 (1967)

28 HANSSON, L., ÅBERG, H., JAMESON, S., KARLBERG, B., MALMCRONA, R.: Initial clinical experience with I.C.I. 66.082, a new β-adrenergic blocking agent, in hypertension. Acta med. scand. *194*, 549 (1973)

29 HARRISON, J., TURNER, P.: Comparison of propranolol and I.C.I. 50,172 on iso-prenaline-induced increase in skin temperature in man. Brit. J. Pharmacol. *36*, 177P (1969); abstract of paper

30 IMHOF, P.: Die Herzfrequenz als Messgrösse bei Phase-I-Studien. Arzneimittel-Forsch. (Drug Res.) *23*, 1640 (1973)

31 JOHANSSON, B.: Effects of propranolol and a new "cardioselective" β-blocker, H 93/26, on responses of isolated rat atria to isoprenaline and noradrenaline. Europ. J. Pharmacol. *24*, 194 (1973)

32 KAPPENBERGER, L., FELLMANN, H., NAGER, F.: Die Therapie der Angina pectoris mit β-Rezeptoren-Blockern und einem Isosorbiddinitrat-Depotpräparat. Schweiz. med. Wschr. *103*, 1789 (1973)

33 KINCAID-SMITH, P.: Beta-blocking drugs in the treatment of hypertension. IXth Int. Congr. Angiol., Florence 1974

34 LAMBERT, D. M. D.: Beta-blockers and life expectancy in ischaemic heart-disease. Lancet *i*, 793 (1972); corresp.

35 LANSER, K. G., SIEMSSEN, S., SILL, V.: Practolol (ICI-50172), ein kardioselektiver Beta-Blocker. Z. Kardiol. *62*, 80 (1973)

36 LEISHMAN, A. W. D., THIRKETTLE, J. L., ALLEN, B. R., DIXON, R. A.: Controlled trial of oxprenolol and practolol in hypertension. Brit. med. J. *iv*, 342 (1970)

37 LEWIS, B. S., BAKST, A., MITHA, A. S., PURDON, K., GOTSMAN, M. S.: Haemo-dynamic effects of a new beta-blocking agent "Sectral" (M & B 17803 A). Brit. Heart J. *35*, 743 (1973)

38 LOHMOELLER, G., FROHLICH, E. D.: Antihypertensive effects of a new beta-adrenergic antagonist (MK 950). Comparison with propranolol in man. Europ. J. clin. Invest. *3*, 251 (1973); abstract of paper

39 LOUIS, W. J.: The evaluation of MK-950 as an anti-hypertensive agent. Minerva cardioangiol. *22*, 91 (1974); abstract of paper

40 MAJID, P. A., SAXTON, C., STOKER, J. B., TAYLOR, S. H.: Comparison of the haemo-dynamic effects of acute intravenous and oral therapy with propranolol and ox-prenolol in hypertensive patients. Cardiovasc. Res. *6* (1970), VIth World Congr. Cardiol., London 1970, Abstr., p. 208

41 PINTO, B. G.: Radiocardiographic study on cardio-pulmonary hemodynamics, changes induced by adrenergic beta-blockade on normal subjects, loc. cit. [40], p. 248

42 PONARI, O., CIVARDI, E., POTÌ, R.: Action of some β-blockers on plasma fibrinolysis in vitro and in vivo in man. Arzneimittel-Forsch. (Drug Res.) *22*, 629 (1972)

43 RAFTERY, E. B., DENMAN, A. M.: Systemic lupus erythematosus syndrome induced by practolol. Brit. med. J. *ii*, 452 (1973)

44 RIVIER, J.-L., NISSIOTIS, E., JAEGER, M.: Comparaison des effets hémodynamiques immédiats de trois médicaments béta-bloqueurs. Thérapie *25*, 245 (1970)

45 SIMPSON, F. O., WAAL-MANNING, H. J.: Hypertension and β-adrenergic blockade. In: Symp. Beta adrenergic blocking agents, Sydney, Australia, 1970, p. 59

46 TARAZI, R. C., DUSTAN, H. P.: Beta adrenergic blockade in hypertension. Practical and theoretical implications of long-term hemodynamic variations. Amer. J. Cardiol. *29*, 633 (1972)

47 TURNER, A. S., PEEL, J. S.: Clinical and haemodynamic studies with β-blockade, loc. cit. [45], p. 51

48 TURNER, P.: Alprenolol and propranolol in hyperthyroid tachycardia. Brit. J. Pharmacol. *40*, 146P (1970); abstract of paper

49 VAUGHAN WILLIAMS, E. M., BAGWELL, E. E., SINGH, B. N.: Cardiospecificity of β-receptor blockade. Cardiovasc. Res. *7*, 226 (1973)

50 WAAL-MANNING, H. J., SIMPSON, F. O.: Practolol treatment in asthmatics. Lancet *ii*, 1264 (1971); corresp.

Discussion

K.D. Bock: In connection with what has been said by Dr. Brunner and Dr. Imhof, I should like to refer very briefly to a clinical trial I conducted together with Dr. M. Anlauf and Dr. P. Merguet in our unit. We compared three beta-blockers in 77 hospitalised patients suffering from mild or moderate essential or renal hypertension. The three drugs in question were: oxprenolol (®Trasicor) and propranolol – i.e. two non-cardioselective beta-blockers, one with and the other without intrinsic sympathomimetic activity – and metoprolol (CGP 2175, identical with H 93/26), which is a cardioselective blocker showing no intrinsic sympathomimetic activity. From Table 1 it can be seen that the groups of patients treated were very similar with regard to age, to the proportion of cases with labile and established hypertension, and to the blood pressure levels recorded during the control period lasting 8–21 days prior to treatment. Indicated in Table 2 are the dosage ranges employed and the percentages of cases in which the systolic and/or diastolic pressures decreased to normal levels in the course of the medication, which averaged three weeks in duration. Shown in Figure 1 are the

Table 1. Age, proportion of cases with labile and stable hypertension, and mean control blood pressure levels in three groups of hypertensive patients (N = number of cases) treated with three beta-blockers.

	N	Age in years ($\bar{x}$ and range)	Labile/stable	Mean control blood pressure levels (mm. Hg)
Propranolol	32	44 (17–71)	10/22	177.5/105.6
Oxprenolol	25	39 (17–59)	8/17	173.6/106.1
Metoprolol	20	43 (22–62)	6/14	171.2/109.2

Table 2. Average daily dose and percentage of cases in which systolic and/or diastolic blood pressure returned to normal levels in response to treatment with three beta-blockers.

	Dose in mg./day ($\bar{x}$ and range)	Systolic and/or diastolic blood pressure restored to normal values
Propranolol	201 (40–600)	18/32 = 56%
Oxprenolol	189 (80–360)	13/25 = 52%
Metoprolol	214 (80–400)	9/20 = 45%

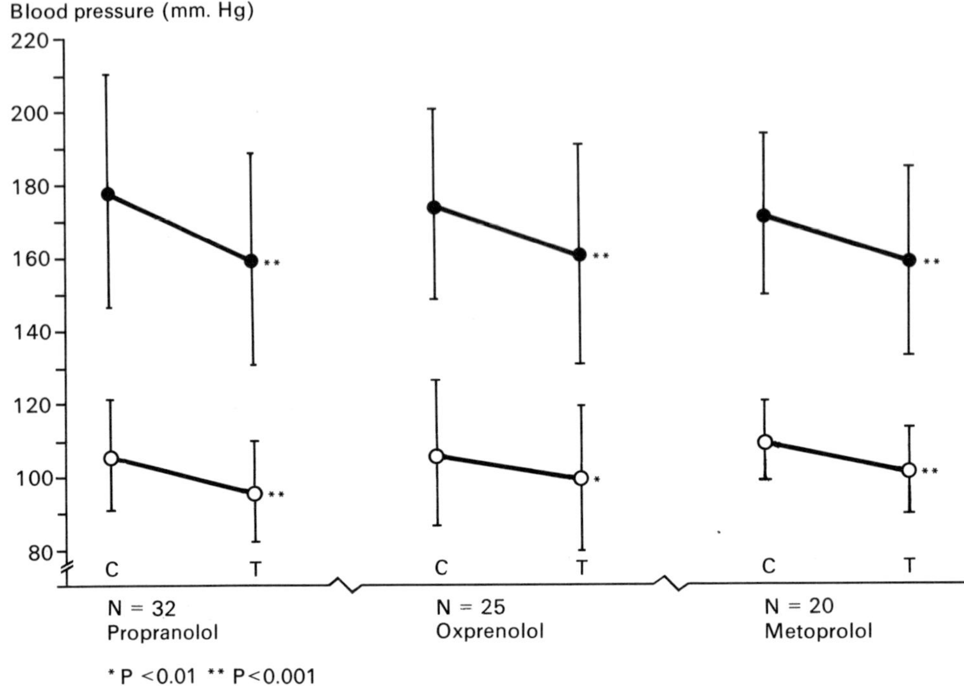

Fig. 1. Systolic ($\bullet$) and diastolic ($\circ$) blood pressure levels ($\bar{x} \pm s$) before (C) and after (T) treatment with three beta-blockers.

mean blood pressure levels measured in the recumbent position at the end of the control and treatment periods, respectively. Though the mean blood pressures were significantly lowered by the treatment, the differences in the percentages of patients responding in each of the three groups were not statistically significant. These results appear to indicate that neither the presence nor the absence of cardioselectivity or of intrinsic sympathomimetic activity has an important bearing on the antihypertensive effect of beta-blockers. For the moment, at least, it would therefore seem that the choice of a suitable beta-blocker for antihypertensive therapy will continue to depend on other factors, including especially the extent to which the drug can safely be used on a long-term basis.

B. N. C. PRICHARD: Dr. IMHOF mentioned that he would be interested to hear something more about Allen & Hanburys' preparation AH 5158. This drug is effective in angina pectoris when given intravenously*; administered in larger doses it also produces a postural fall in blood pressure – as indeed might be expected. The postural decrease in blood pressure is no doubt due to its blocking effect on the alpha-receptors. Though it is also effective in cases of hypertension, once the dosage reaches a certain height you begin to get a postural decrease in blood pressure; this postural hypotension, which is quite unlike the response elicited by a pure beta-blocking action, is once again probably attributable to the drug's alpha-blocking properties.

P. R. IMHOF: This is quite interesting, Dr. PRICHARD. The alpha-blocking activity of AH 5158 is rather weak, i.e. equivalent to about one-fifth to one-seventh of that of

---

* BOAKES, A. J., PRICHARD, B. N. C.: The effect of AH 5158, pindolol, propranolol, D-propranolol on acute exercise tolerance in angina pectoris. Brit. J. Pharmacol. *47*, 673P (1973); abstract of paper

phentolamine. When administering a beta-blocker together with phentolamine, we have not found that this combination of beta and alpha blockade gives rise to any postural hypotension – as I am sure Dr. Taylor will agree.

S. H. Taylor: Yes, I do agree. At this point I should like to mention another drug, namely tolamolol, which also displays alpha-blocking activity in addition to its effects on the beta-receptors. This drug, too, produces responses in hypertensive patients similar to those mentioned by Dr. Prichard with AH 5158.
May I ask Dr. Imhof a question. The non-cardioselective beta-blockers, such as propranolol and oxprenolol, also inhibit lipolysis and renin secretion. Don't you think that by virtue of these properties such drugs may perhaps offer advantages over the cardioselective beta-blockers in the long-term treatment of hypertension?

P. R. Imhof: I think that observations reported by Amery et al.* may be relevant in this connection. These authors found that practolol did not lower plasma renin activity.

A. Zanchetti: I do not believe that the rise occurring in peripheral resistance in response to a beta-blocker – even a non-cardioselective beta-blocker – can be interpreted as being due to an inhibition of peripheral vasodilatation, because so far we have no evidence that a normal tonic vasodilatation exists which is maintained through activation of vascular beta-receptors. Consequently, it does not puzzle me that you get an increase in peripheral resistance both with practolol as well as with propranolol and oxprenolol. I think the change in peripheral resistance can be quite satisfactorily accounted for in terms of autoregulation: in other words, when a decrease in cardiac output occurs, this is counterbalanced for some time by a rise in peripheral resistance. As for the question of plasma renin activity, I don't think that at the moment we can say that practolol does not block renin activity, because there is in fact quite a good deal of evidence to suggest that, on the contrary, it can and does block renin, although maybe not so strongly as propranolol. Incidentally, I am not personally very much in favour of referring here to one effect as "central" and the other as "peripheral". The term "cardioselective" may, admittedly, not be a perfect one, but I suggest that it is at all events better than "central", which I should prefer to reserve for effects exerted on the brain.

W. Schweizer: We shall be discussing the question of renin later on in our proceedings. May I therefore ask if there are any more questions or observations directly relevant to Dr. Imhof's paper.

J. Schwartz: *In vitro*, practolol exhibits a certain degree of cardioselectivity, inasmuch as its effect on the beta-receptors of the heart is several times more pronounced than its effect on the peripheral beta-receptors. *In vivo*, the difference is only slight, and any cardioselectivity the drug might display is virtually abolished if a tablet strength of 100 mg. is used. Is there any evidence that the other cardioselective substances you listed, Dr. Imhof, really are cardioselective *in vivo*?

P. R. Imhof: I don't think that at present there is any beta-blocker available which displays greater cardioselectivity than practolol. But this cardioselectivity, as I have tried to indicate in my paper, is a very relative concept and one that holds good only for low doses; sometimes – particularly, for example, when employing high doses as treatment for hypertension – it is actually misleading to speak of cardioselectivity.

H. Brunner: Just one final question for Dr. Prichard and Dr. Taylor. Is the alpha-blocking action of drugs such as AH 5158 and tolamolol merely something additional which is akin to a side effect, or does it actually enable the drug to produce an antihypertensive response that is on the whole more pronounced than that obtainable with the other beta-blockers?

---

* Amery, A., Billiet, L., Fagard, R.: Beta receptors and renin release. New Engl. J. Med. *290*, 284 (1974); corresp.

B. N. C. Prichard: I don't believe we know at the moment, for the simple reason that there isn't enough evidence. Personally, I should be rather inclined to regard this additional property as being probably just a side effect, but further investigations will have to be carried out before your question can be answered properly, Dr. Brunner.

S. H. Taylor: We found that in hypertensive patients the increase in blood pressure occurring during exercise was less marked with tolamolol than with practolol, propranolol, and oxprenolol. This was perhaps to be expected, because the factor of vasoconstriction obviously plays a bigger role in circulatory adjustments at high levels of exercise.

Selection of hypertensive patients for treatment with beta-blockers alone or in combination

by R. Sannerstedt*

When antihypertensive agents for the active treatment of arterial hypertension first appeared on the scene in the early 1950s, these drugs were prescribed with one essential objective in mind – to bring down the patient's elevated blood pressure at all costs, even if the achievement of this goal involved inconvenient dosage schedules and the risk of troublesome side effects.

During the 20 years that have meanwhile elapsed, this situation has changed dramatically for the better. The introduction of new and more effective antihypertensive agents with well-defined pharmacological profiles and more specific points of attack has enabled us to broaden the indications for active antihypertensive therapy, so that they now also include patients not immediately at risk from the disease. At the same time, a great deal more has been learned about the pathophysiology of hypertension, and knowledge of its biochemical background has also improved.

This progress has created a solid platform on the basis of which it should today be possible to practise *effective antihypertensive treatment founded upon clear-cut pathophysiological principles* and tailored to meet the needs of the individual patient[9, 24, 26, 37].

It therefore follows that the use of beta-blockers, i. e. of agents which block the beta-adrenergic receptors, as drugs for the active treatment of arterial hypertension can be viewed from different angles, depending upon such factors as:

- the various patterns of *haemodynamic alterations* encountered in hypertensive cardiovascular disease and their susceptibility to the influence of beta-blockers;
- the *endocrinological disorders* associated with chronically elevated blood pressure and the extent to which these can be corrected by beta-blockers;
- the *different types of beta-blocker* available for clinical use and their optimal employment in the various categories of hypertensive patient;
- the *interactions* occurring between beta-blockers and other antihypertensive agents and the exploitation of these interactions in the routine treatment of hypertension.

From the haemodynamic aspect, it is primarily upon the pumping action of the heart that the beta-blockers exert their effect, this effect resulting in a lowering of the heart rate, a decrease in cardiac output, and a reduction in myocardial oxygen consumption[21]. In addition, there is evidence indicating

---

* Medicinska kliniken I, Sahlgrenska sjukhuset, University of Gothenburg, Sweden.

Table 1. Stages of arterial hypertension as proposed by the World Health Organisation.

Stage 1	High blood pressure without evidence of organic changes in the cardiovascular system
Stage 2	High blood pressure with cardiovascular hypertrophy but without other evidence of organ damage
Stage 3	High blood pressure with evidence of organ damage attributable to the hypertensive disease

that during long-term treatment with beta-blockers the peripheral circulation is also affected as the result of a gradual diminution in systemic vascular resistance.

Although it is true that a majority of hypertensive patients will derive benefit from a beta-blocker, irrespective of the stage their hypertension has reached, it should be borne in mind that the haemodynamic repercussions of beta-blockade also vary depending on the stage of the disease. The various stages of hypertensive cardiovascular disease, as defined for example by the World Health Organisation[47] (Table 1), have in fact been shown to be characterised by differing haemodynamic patterns[27, 36]. Thus, for instance, when patients assignable to *W.H.O. Stage 1* (corresponding to what might be called *"latent hypertension"*) are considered as a group, a typical haemodynamic feature of their condition is found to be a hyperkinetic circulation at rest: heart rate and cardiac output are elevated, whereas the calculated systemic vascular resistance does not differ numerically from normal (Figure 1)[27, 36].

The logical approach to treatment in this group of patients would be to aim at normalising the circulation by lowering heart rate and cardiac output. Since recent studies have demonstrated that in these patients with latent, borderline hypertension – many of whom are young subjects – sympathetic nervous over-drive goes hand in hand with reduced vagal tonus[22], as well as with increased peripheral renin activity[8, 10, 25, 46], the successful development of well-tolerated beta-blockers has now provided us with a virtually specific form of therapy for this type of case. As illustrated in Figure 2, it has been established that in patients with borderline hypertension intravenous treatment with beta-blockers is capable of restoring cardiac output to normal[38]. Good results have also been reported in such cases in response to oral medication, relatively low dosages of beta-blockers being usually sufficient to achieve a satisfactory effect[11, 14, 15, 28, 45].

Arterial hypertension classifiable as *W.H.O. Stage 2* (corresponding to what may be described as *"established hypertension"*) is characterised clinically by signs of cardiac hypertrophy, i.e. X-ray and electrocardiographic evidence of left-ventricular hypertrophy, as well as by Grade II retinal vascular changes on the Keith-Wagener-Barker scale. Typical of the haemodynamic picture is a normokinetic circulation at rest, coupled with increased systemic vascular resistance (Figure 3)[36].

In cases of this kind, therefore, the main therapeutic attack should be directed against the increased systemic vascular resistance, the aim being to reduce the

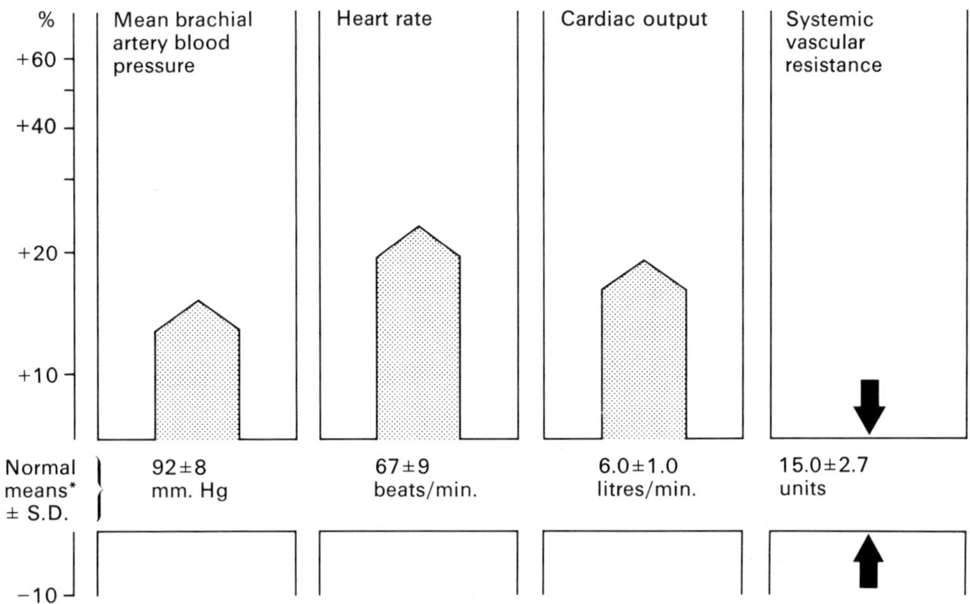

*Adjusted to age and physical characteristics of the hypertensive patient group

Fig. 1. W.H.O. Stage 1 arterial hypertension. Haemodynamic pattern at rest in untreated men.

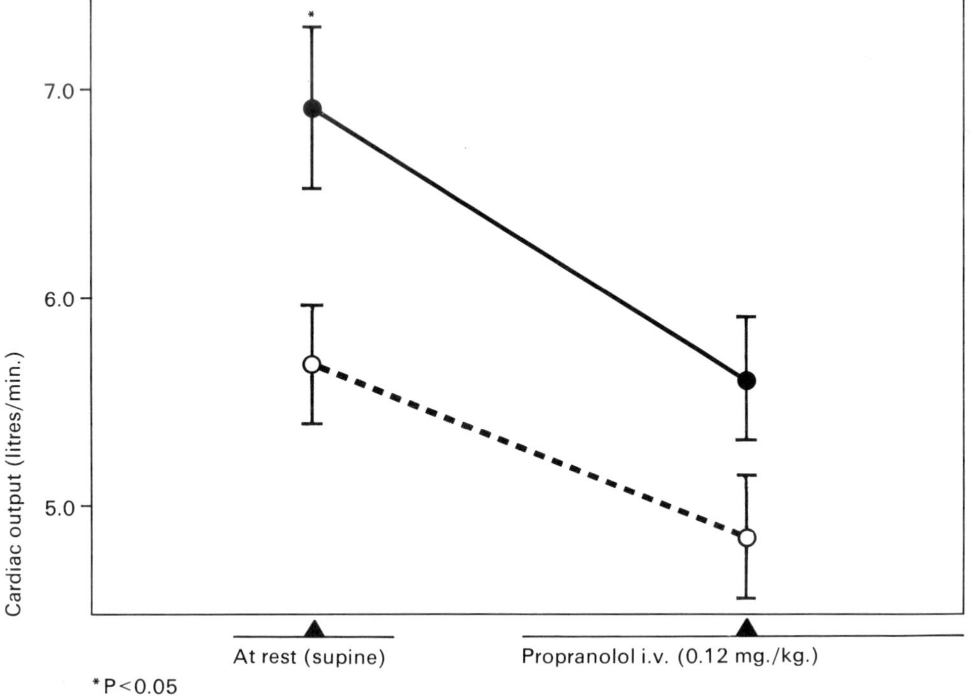

*P<0.05

Fig. 2. W.H.O. Stage 1 arterial hypertension. Cardiac output before and after acute beta-blockade in 17 hypertensive patients (●——●) and in 18 healthy controls (O‑‑‑O). Mean values ± S.E.M.

latter to normal levels both at rest and when the patient is exposed to mental or physical stress. From the pathophysiological standpoint, potent peripheral vasodilators such as hydralazines would accordingly be the drugs of choice here. Via the baroceptors, however, such agents will lead to a compensatory increase in heart rate and cardiac output – a reflex by which the body endeavours to counteract the blood-pressure lowering effect of the drug and one which will also give rise to such subjectively unpleasant side effects as palpitation.

As revealed in both animal and human experimental studies, this reflex can be effectively inhibited by beta-blockers without interfering with the antihypertensive action of the peripheral vasodilator (Figure 4)[3, 6, 12, 39, 40]. What is more, clinical studies have shown that, where a beta-blocker is given in combination with a peripheral vasodilator, it is possible to achieve not merely an additive but sometimes even a synergistic antihypertensive effect (Figure 5)[12, 48]. In addition, the frequency of such subjectively disagreeable side effects as palpitation decreases[18, 23, 41, 48], and the beta-blocker also counteracts the increase in renin secretion provoked by the peripheral vasodilator[32, 34].

At this point, reference should be made to the possibility of employing – as an alternative to the peripheral vasodilators – agents which block the alpha-adrenergic receptors[2, 29]. At present, however, the therapeutic value of those

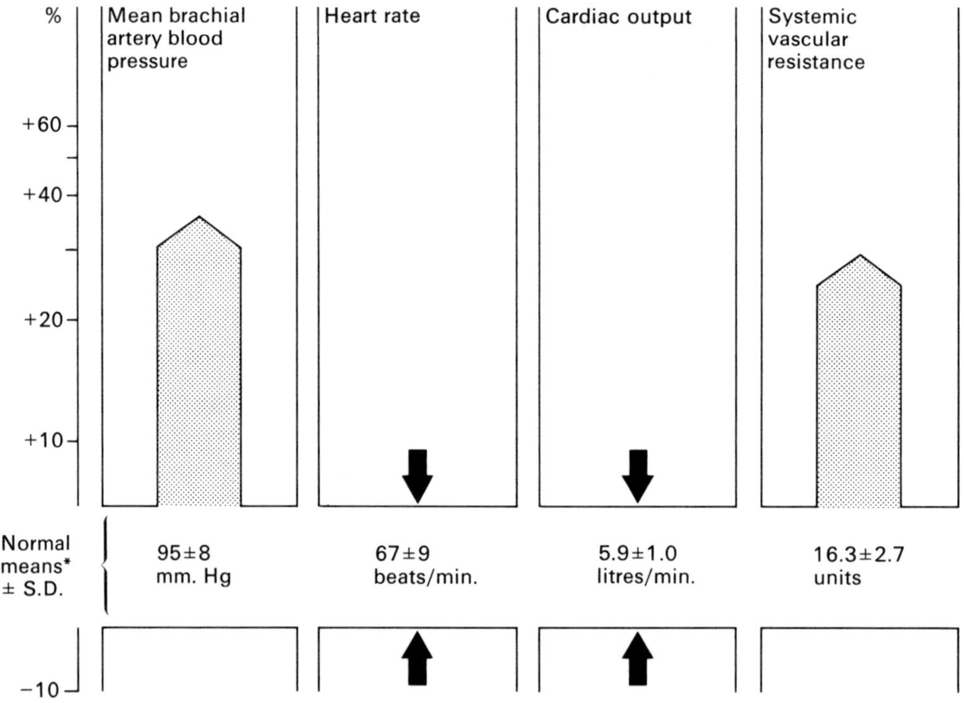

*Adjusted to age and physical characteristics of the hypertensive patient group

Fig.3. W.H.O. Stage 2 arterial hypertension. Haemodynamic pattern at rest in untreated men.

alpha-blockers that are available for clinical use as treatment for hypertension in combination with beta-blockers has yet to be confirmed. Meanwhile, the administration of a peripheral vasodilator such as hydralazine together with a beta-blocker has at all events been proven, both theoretically and clinically, to provide a safe and effective form of combined therapy for the treatment of established hypertension[1, 13, 18, 23, 31, 33, 40, 48].

Such medication can be initiated simultaneously with both drugs, the doses of which should, if necessary, be increased step by step to a maximum dosage

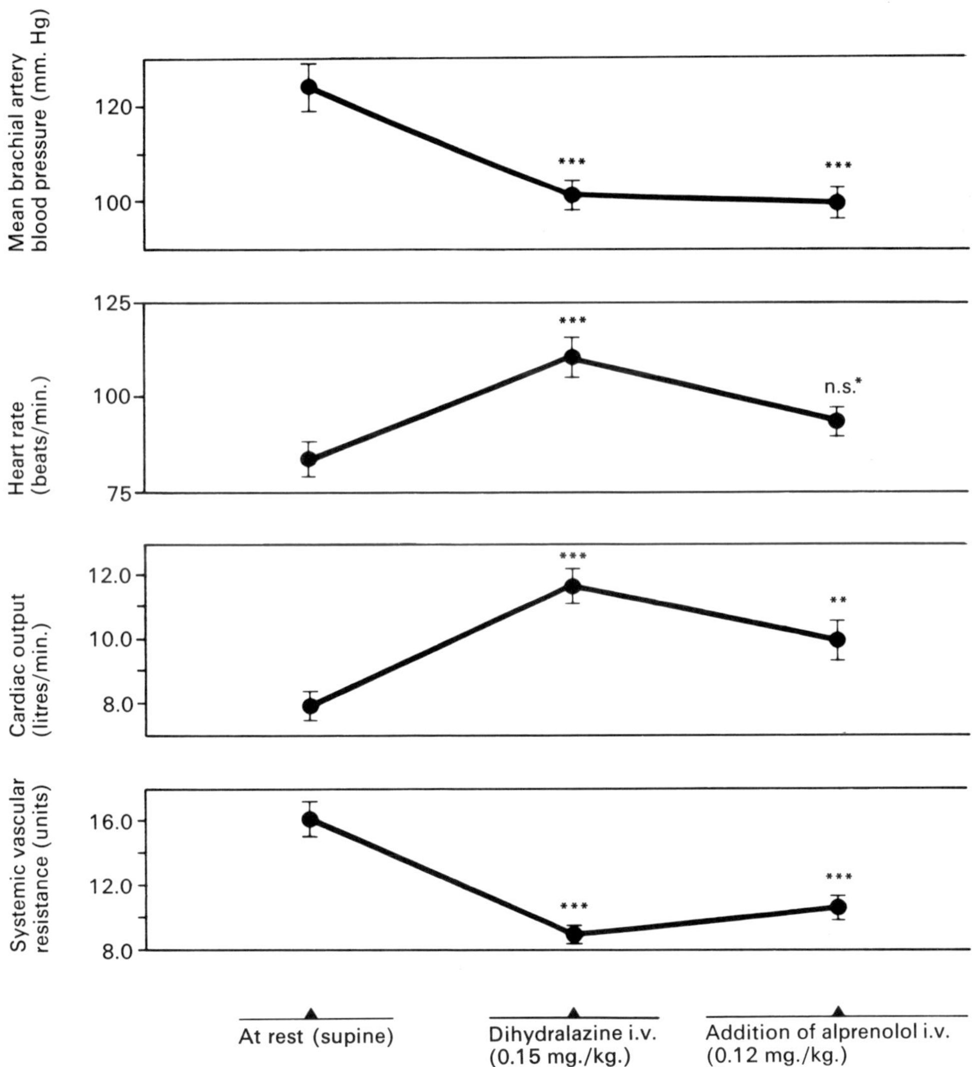

*n.s. = difference from initial mean not significant
**P<0.05   ***P<0.001

Fig. 4. W.H.O. Stages 1–2 arterial hypertension. Effect of acute beta-blockade on haemodynamic changes produced by a peripheral vasodilator. Mean values ± S.E.M. (N = 10).

amounting generally to 200 mg. hydralazine daily plus, say, 300–400 mg. oxprenolol (®Trasicor) daily. The two drugs should be given together in divided doses administered, if possible, not more frequently than three times a day. If it is considered preferable to commence with one drug only, it is with the beta-blocker that the treatment should be started. Beginning, for example, with 80 mg. oxprenolol two or three times daily, one should successively raise the dosage until an optimal antihypertensive effect has been obtained. Once effective beta-blockade has been achieved, one can then, if need be, add hydralazine to the regimen, starting directly with daily doses of 75–100 mg.

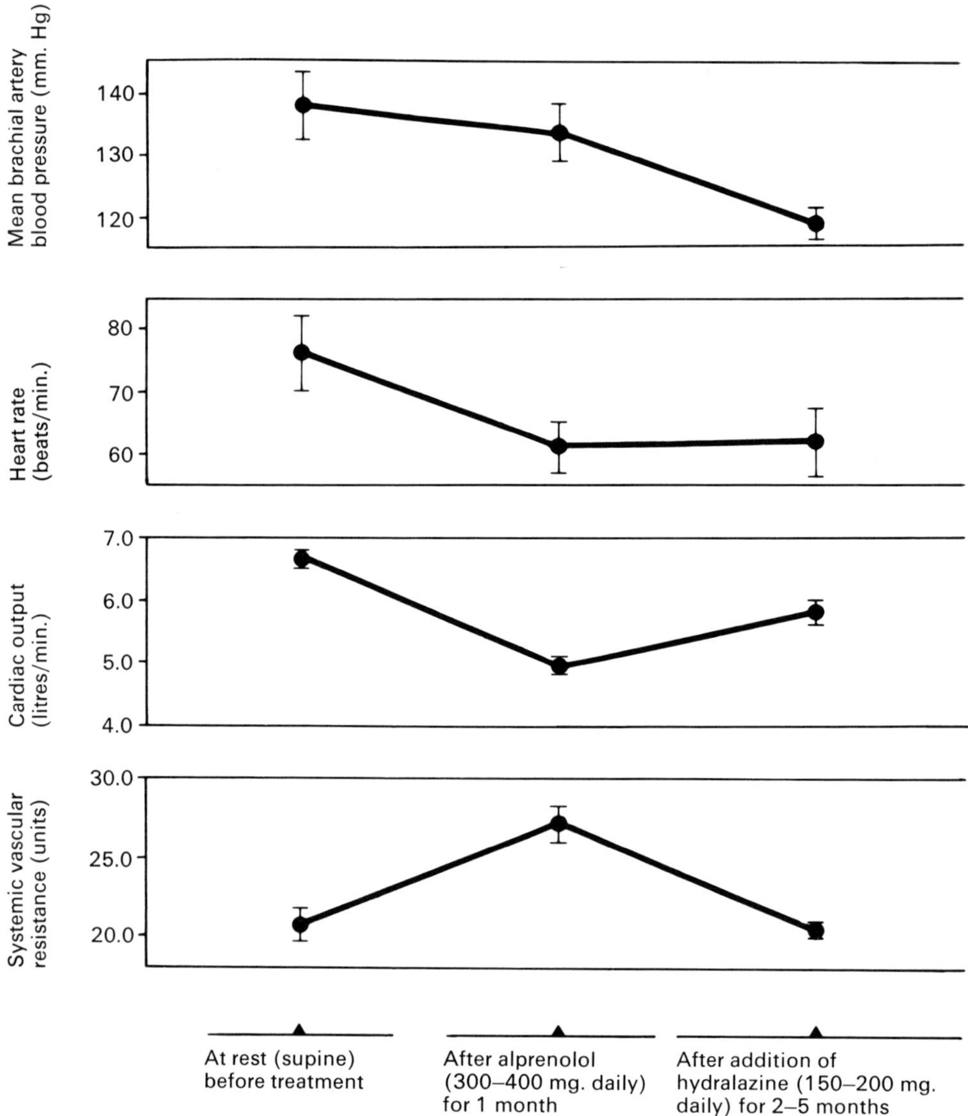

Fig. 5. W.H.O. Stage 2 arterial hypertension. Haemodynamic effects of sustained treatment with a beta-blocker, given first alone and then in combination with a peripheral vasodilator. Mean values ± S.E.M. (N = 3).

Beta-blockers have, incidentally, also been extensively used as monotherapy in established hypertension[17, 19, 35, 43].

It should be noted that parenterally administered beta-blockers may or may not produce an immediate antihypertensive effect of moderate degree and that, particularly where a non-cardioselective beta-blocker with no intrinsic sympathomimetic action is given, the already elevated systemic vascular re-

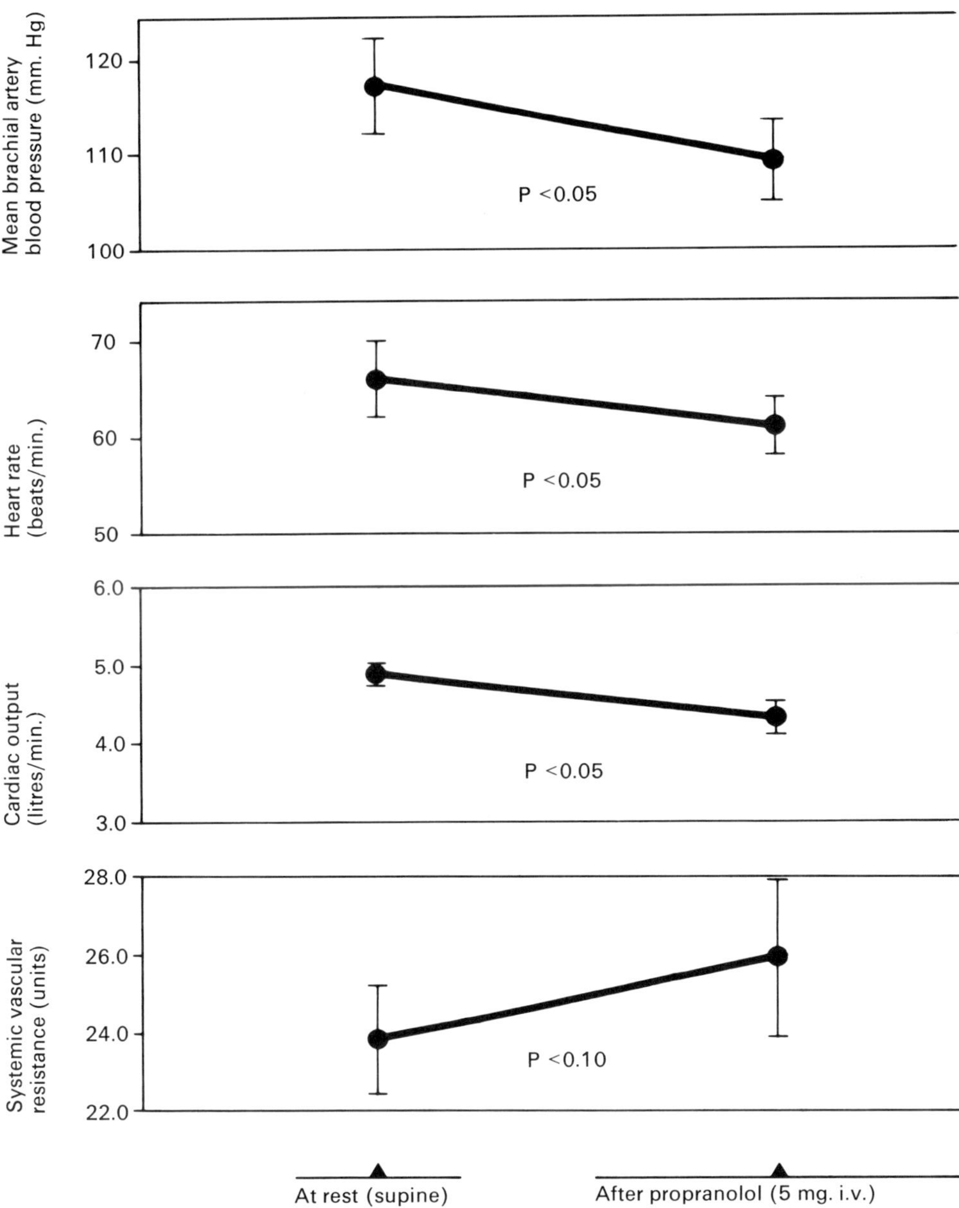

Fig. 6. W.H.O. Stage 2 arterial hypertension. Haemodynamic effects of acute beta-blockade. Mean values ± S.E.M. (N = 5). The P values indicated are in respect of paired differences.

sistance may, if anything, tend to increase even further owing to the relative preponderance of alpha-receptor activity in the peripheral vascular beds (Figure 6)[16, 20, 30].

The antihypertensive effect observed in response to short-term treatment with beta-blockers may be associated with an unchanged or increased systemic

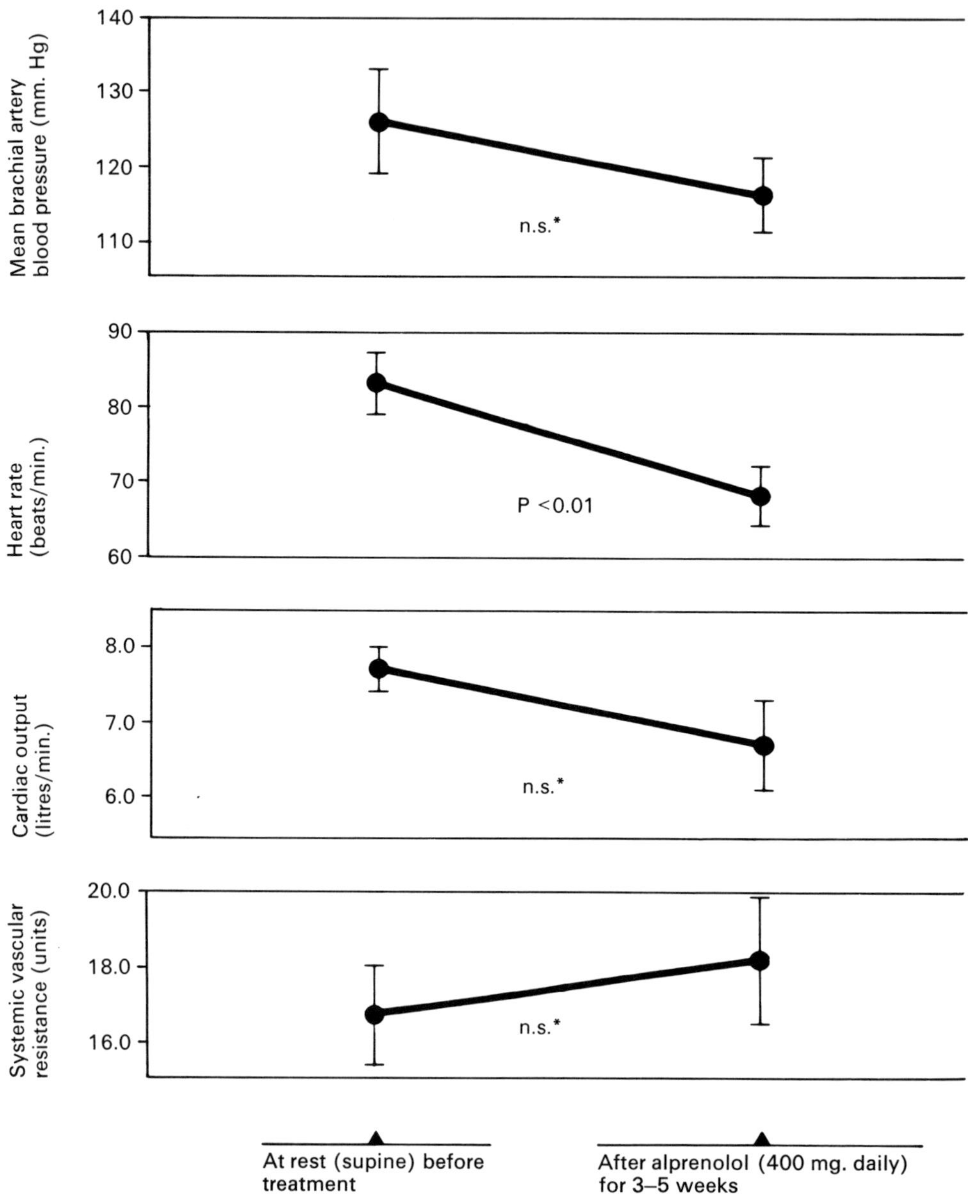

*n.s. = difference not significant (P >0.10)

Fig. 7. W.H.O. Stages 1–2 arterial hypertension. Haemodynamic effects of oral treatment with a beta-blocker. Mean values ± S.E.M. (N = 6). The P values indicated are in respect of paired differences.

vascular resistance, the moderate decrease in blood pressure being due solely to a diminution in cardiac output (Figure 7)[16, 30, 40]. Recent findings, on the other hand, indicate that in the course of long-term oral therapy the systemic vascular resistance may diminish again while the reduction in blood pressure is maintained (Figure 8)[41]. Although, when beta-blockers are employed as monotherapy in established hypertension, high doses may sometimes be re-

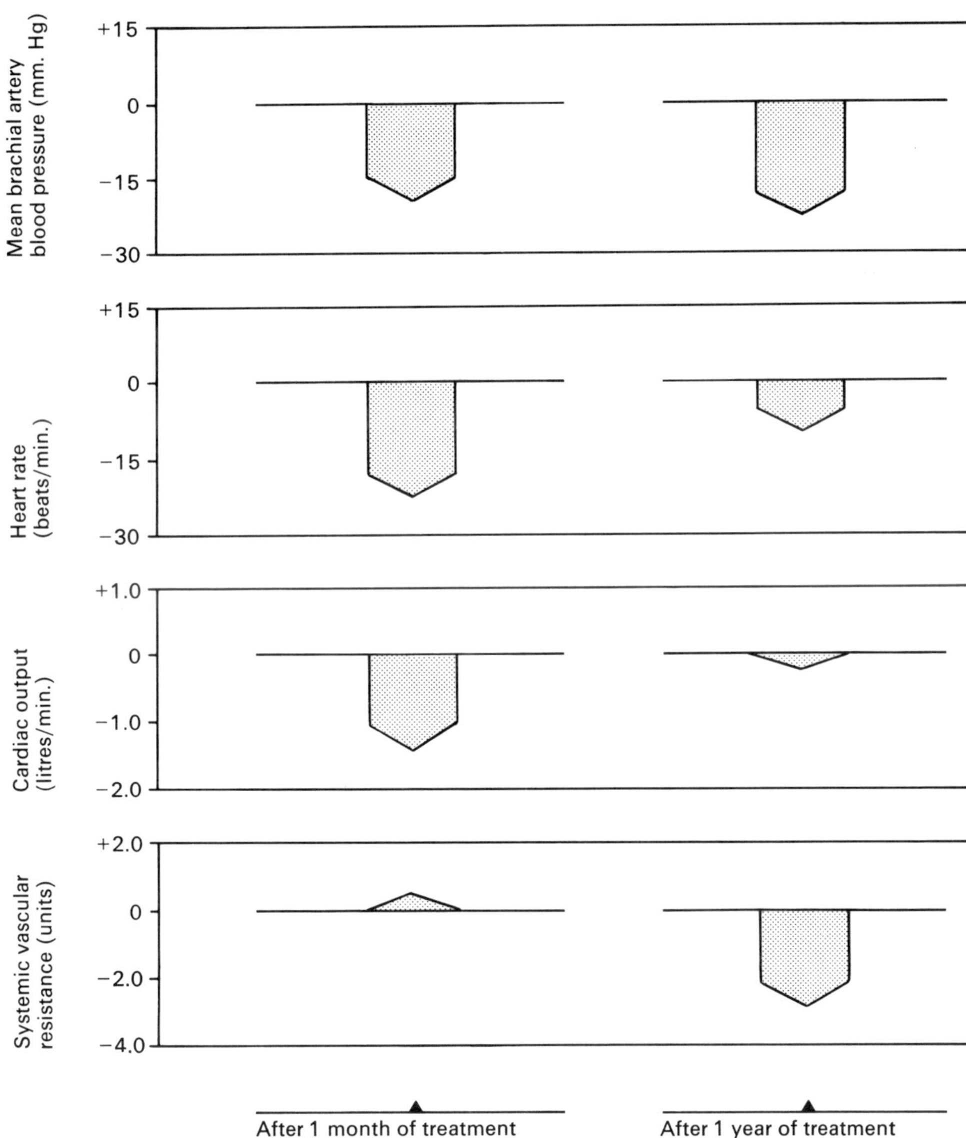

Fig. 8. Comparison between the haemodynamic effects of beta-blockade after treatment for one month and for one year in a hypertensive man receiving a cardioselective beta-blocker (H 93/26) in a dose of 240 mg. daily for two months, followed by 150 mg. daily for ten months. The four haemodynamic variables in question were recorded at rest in the supine position.

quired to elicit an optimal antihypertensive response, in our experience it is only in certain carefully selected cases that it proves rewarding to resort to huge doses – as originally advocated by investigators in Great Britain – in order to obtain a greater effect[35]. In the majority of patients, in fact, a daily dosage of, say, 300–400 mg. oxprenolol will suffice to produce a well-nigh optimal antihypertensive effect.

In cases where arterial hypertension has reached *W.H.O. Stage 3* (corresponding to what may be referred to as *"advanced hypertension"*), signs of cardiovascular hypertrophy are also associated with evidence of organic damage to the brain, heart, or kidneys attributable to the patient's high blood pressure. Considered as a group, patients with this advanced form of hypertension have a hypokinetic circulation at rest, with a decreased cardiac output and a grossly elevated systemic vascular resistance (Figure 9)[36]. In addition, the arteriovenous oxygen difference and plasma volume tend to increase as an indication of impending heart failure.

Here, treatment ought ideally to be aimed at reversing all these circulatory changes. In many cases, however, this is no longer possible. The main objective must therefore be to lower the blood pressure so as to reduce the burden on the heart and to protect vital organs from further damage. For this purpose, diuretic agents may be regarded as the drugs of first choice; they can

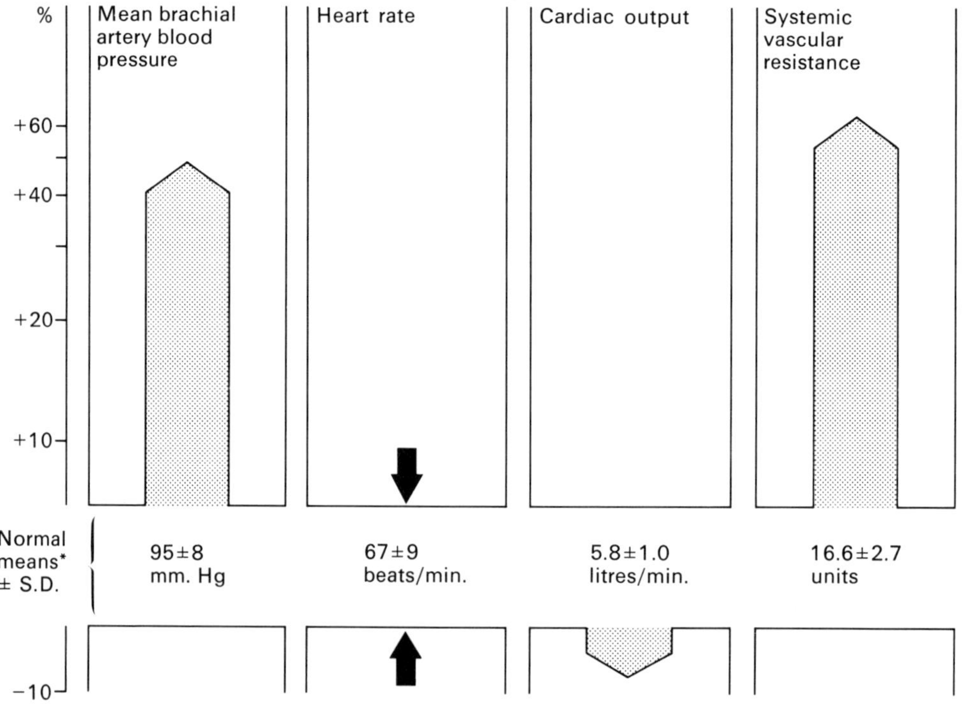

Fig.9. W.H.O. Stage 3 arterial hypertension. Haemodynamic pattern at rest in untreated men.

be expected to decrease the blood volume and to lower the blood pressure, thereby lightening the load on the heart. The addition of a beta-blocker as second drug will serve to enhance the antihypertensive effect and to counter-act any increase in renin secretion provoked by the diuretic, besides exerting a beneficial influence on such symptoms of angina pectoris as may also be present.

From the purely haemodynamic aspect, a diuretic plus a beta-blocker may admittedly be an unfavourable combination, since it will probably lead to a further decrease in the already diminished cardiac output. Nevertheless, if effectively applied, it may well afford vital organs valuable protection against further pressure-induced damage. The low cardiac output as such will probably not prove deleterious, because – as in other conditions associated with a re-duced cardiac output, such as mitral valvular disease[44] – it tends to be offset by autoregulatory mechanisms and redistribution of the blood flow.

Attention has recently been drawn to the important influence exerted by a high arterial impedance in contributing towards decompensation of a mildly abnormal heart[5]. In order to decrease systemic vascular resistance and to reduce the afterload, a peripheral vasodilator such as hydralazine may there-fore be added to the twofold combination of a diuretic and a beta-blocker, particularly since the beta-blocker will effectively counteract the tendency of the peripheral vasodilator to increase cardiac work.

Among the various beta-blockers available, those displaying moderate intrinsic sympathomimetic activity and a less pronounced cardiodepressant effect might be preferable in this type of case, especially if myocardial function is impaired. Clinical studies have shown that such beta-blockers are usually well tolerated, also by elderly patients with reduced cardiac function[7].

Peripheral renin activity in relation to urinary excretion of aldosterone and sodium has recently been suggested as a sensitive parameter for use in selecting those pa-tients with essential hypertension who would make suitable candidates for treatment with beta-blockers[26]. Patients displaying high levels of renin activity were found to respond well to beta-blockers[4]. Those with low renin activity, on the other hand, showed quite a satisfactory decrease in blood pressure under treatment with chlortalidone or an aldosterone antagonist (cf. Table 2)[42].

However, even if adequately controlled prospective studies were to confirm that modest variations in renin activity do indeed provide strong clues to the degree of therapeutic efficacy which a given type of drug is likely to exhibit in a given case, the practical value of such a screening procedure would still remain questionable. Bearing in mind the very large number of subjects re-quiring consideration as candidates for active antihypertensive therapy, rou-tine determinations of renin and aldosterone as a screening procedure prior to treatment would substantially add to the already high costs incurred in providing every hypertensive patient with the adequate work-up, therapy, and follow-up which he should in fact be given. What is more, in most countries, renin and aldosterone determinations will probably continue for many years to involve elaborate procedures, the use of which will remain restricted to a few specialised laboratories with limited capacities. Not even the leading clinics in

the United States currently have the wherewithal to provide such screening for all their hypertensive patients. For the time being, therefore, the practical implications of this approach from the clinical standpoint can perhaps be summarised as follows: if diuretics alone fail to elicit a satisfactory blood pressure response, an attempt can be made at treatment with a beta-blocker, which may well have a good therapeutic effect provided the patient was initially a case of "high-renin essential hypertension"; conversely, a patient who has derived no benefit from a beta-blocker alone may show an excellent response to a diuretic or aldosterone antagonist provided he belongs to the minority group of cases labelled as "low-renin essential hypertension".
At present it would thus seem preferable, and probably as effective, to base the choice of therapy on clinical findings and simple laboratory tests which

Table 2. Schematic review indicating the type of response which diuretics and beta-blockers are likely to elicit, depending on whether the patient is suffering from low, normal, or high renin essential hypertension.

Type of hypertension	Response to treatment with	
	diuretics	beta-blockers

Low-renin essential hypertension
   with volume expansion
   (accounting for approx. 30% of all cases)
	+	−

Normal-renin essential hypertension (approx. 55% of all cases)	±	±

High-renin essential hypertension
   with vasoconstriction
   (approx. 15% of all cases)
	−	+

Table 3. Suggested guidelines for the use of beta-blockers in arterial hypertension.

	Latent hypertension corresponding to W.H.O. Stage 1	Established hypertension corresponding to W.H.O. Stage 2	Advanced hypertension corresponding to W.H.O. Stage 3
Aim of treatment	To normalise the circulation by reducing heart rate and cardiac output	To normalise the circulation by reducing systemic vascular resistance	To prevent cardiac insufficiency, alleviate ischaemic coronary heart disease, and protect vital areas against further pressure-induced damage
Basic therapy	Beta-blockers	Peripheral vasodilators plus beta-blockers	Diuretics plus beta-blockers
Supplementary therapy		Diuretics	Peripheral vasodilators

afford a general idea of the type and degree of the haemodynamic changes underlying the hypertensive state, using for this purpose such relatively simple pointers as the heart rate, the presence or absence of cardiovascular hypertrophy, and indications of organic damage due to hypertension, and relying upon one's overall clinical judgment. On the basis of these findings and of one's general impression of the patient's condition, the treatment selected should be that best calculated to meet his individual needs.

Provided it proves possible to establish close cooperation with the patient so as to ensure life-long, supervised therapy, and provided adequate attention is paid to other non-medicinal therapeutic measures, the restoration of a normal circulation in cases of latent or established hypertension is a goal which need not be regarded as unrealistic (Table 3). If this goal can be achieved, there is good reason to suppose that progression of the disease into the stage of advanced hypertension, with its high morbidity and mortality risks, can be effectively prevented.

In conclusion, it may be stated that beta-blockers, prescribed either alone or in combination with other antihypertensive drugs – preferably peripheral vasodilators and diuretic agents – can be successfully employed for the treatment of the majority of patients in all stages of hypertensive cardiovascular disease.

References

1 ÄNISHÄNSLIN, W., PESTALOZZI-KERPEL, J., DUBACH, U. C., IMHOF, P. R., TURRI, M.: Antihypertensive therapy with adrenergic beta-receptor blockers and vasodilators. Europ. J. clin. Pharmacol. *4*, 177 (1972)

2 BEILIN, L.J., JUEL-JENSEN, B. E.: Alpha and beta adrenergic blockade in hypertension. Lancet *i*, 979 (1972)

3 BRUNNER, H., HEDWALL, P. R., MEIER, M.: Influence of adrenergic beta-receptor blockade on the acute cardiovascular effects of hydralazine. Brit. J. Pharmacol. *30*, 123 (1967)

4 BÜHLER, F. R., LARAGH, J. H., VAUGHAN, E. D., Jr., BRUNNER, H. R., GAVRAS, H., BAER, L.: Antihypertensive action of propranolol. Specific antirenin responses in high and normal renin forms of essential, renal, renovascular and malignant hypertension. Amer. J. Cardiol. *32*, 511 (1973)

5 COHN, J.N.: Vasodilator therapy for heart failure. The influence of impedance on left ventricular performance. Circulation *48*, 5 (1973)

6 DARBY, T.D., BAKER, C.H.: Further evidence for combination therapy in hypertension. Fed. Proc. *30*, 675 (1971); abstract of paper

7 EISALO, A., HEINO, A., MUNTER, J.: The effect of alprenolol in elderly patients with raised blood pressure. Acta med. scand. *195*, Suppl. 554: 23 (1974)

8 ESLER, M.D., NESTEL, P.J.: Essential hypertension with symptoms of hyperkinetic circulation. Med. J. Aust. *60/ii*, 253 (1973)

9 FROHLICH, E.D.: Clinical significance of hemodynamic findings in hypertension. Chest *64*, 94 (1973)

10 FROHLICH, E.D., KOZUL, V.J., TARAZI, R.C., DUSTAN, H.P.: Physiological comparison of labile and essential hypertension. Circulat. Res. *27*, Suppl. I: 55 (1970)

11 FROHLICH, E.D., TARAZI, R.C., DUSTAN, H.P.: Beta-adrenergic blocking therapy in hypertension: selection of patients. Int. J. clin. Pharmacol. Ther. Toxicol. *4*, 151 (1970)

12 GILMORE, E., WEIL, J., CHIDSEY, C.: Treatment of essential hypertension with a new vasodilator in combination with beta-adrenergic blockade. New Engl. J. Med. *282*, 521 (1970)

13 GOTTLIEB, T. B., KATZ, F. H., CHIDSEY, C. A.: Combined therapy with vasodilator drugs and beta-adrenergic blockade in hypertension. A comparative study of minoxidil and hydralazine. Circulation *45*, 571 (1972)

14 GYSLING, E., REGOLI, D.: Oxprenolol: long-term effects in arterial hypertension. Clin. Pharmacol. Ther. *14*, 995 (1973)

15 HAMET, P., KUCHEL, O., CUCHE, J. L., BOUCHER, R., GENEST, J.: Effect of propranolol on cyclic AMP excretion and plasma renin activity in labile essential hypertension. Canad. med. Ass. J. *109*, 1099 (1973)

16 HANSSON, L.: Beta-adrenergic blockade in essential hypertension. Effects of propranolol on hemodynamic parameters and plasma renin activity. Acta med. scand. *194*, Suppl. 550 (1973)

17 HANSSON, L., MALMCRONA, R., OLANDER, R., ROSENHALL, L., WESTERLUND, A., ÅBERG, H., HOOD, B.: Propranolol in hypertension. Report on 158 patients treated up to one year. Klin. Wschr. *50*, 364 (1972)

18 HANSSON, L., OLANDER, R., ÅBERG, H., MALMCRONA, R., WESTERLUND, A.: Treatment of hypertension with propranolol and hydralazine. Acta med. scand. *190*, 531 (1971)

19 JOHNSSON, G. (Editor): The effect of the β-adrenergic blocker alprenolol in hypertension. Acta med. scand. *195*, Suppl. 554 (1974)

20 JOHNSSON, G., GUZMAN, M. DE, BERGMAN, H., SANNERSTEDT, R.: The haemodynamic effects of alprenolol and propranolol at rest and during exercise in hypertensive patients. Pharmacol. clin. *2*, 34 (1969)

21 JORGENSEN, C. R., WANG, K., WANG, Y., GOBEL, F. L., NELSON, R. R., TAYLOR, H.: Effects of propranolol on myocardial oxygen consumption and its hemodynamic correlates during upright exercise. Circulation *48*, 1173 (1973)

22 JULIUS, S., PASCUAL, A. V., LONDON, R.: Role of parasympathetic inhibition in the hyperkinetic type of borderline hypertension. Circulation *44*, 413 (1971)

23 KATILA, M., FRICK, M. H.: Combined dihydralazine and propranolol in the treatment of hypertension. Int. J. clin. Pharmacol. Ther. Toxicol. *4*, 111 (1970)

24 KOCH-WESER, J.: Correlation of pathophysiology and pharmacotherapy in primary hypertension. Amer. J. Cardiol. *32*, 499 (1973)

25 KUCHEL, O., CUCHE, J. L., HAMET, P., BOUCHER, R., BARBEAU, A., GENEST, J.: The relationship between adrenergic nervous system and renin in labile hyperkinetic hypertension. In Genest, J., Koiw, E. (Editors): Hypertension – 1972, p. 118 (Springer, Berlin/Heidelberg/New York 1972)

26 LARAGH, J. H.: Vasoconstriction – volume analysis for understanding and treating hypertension: the use of renin and aldosterone profiles. Amer. J. Med. *55*, 261 (1973)

27 LUND-JOHANSEN, P.: Hemodynamics in early essential hypertension. Acta med. scand. *183*, Suppl. 482 (1968)

28 LUND-JOHANSEN, P.: Hemodynamic changes at rest and during exercise in long-term β-blocker therapy of essential hypertension. Acta med. scand. *195*, 117 (1974)

29 MAJID, P. A., SHARMA, B., MEERAN, M. K., TAYLOR, S. H.: Combined alpha and beta adrenergic blockade in treatment of hypertension. Brit. Heart J. *35*, 561 (1973); abstract of paper

30 MAJID, P. A., SHARMA, B., SAXTON, C., STOKER, J. B., TAYLOR, S. H.: Haemodynamic effects of oxprenolol in hypertensive patients. Postgrad. med. J. *46*, Suppl. (Nov.): 67 (1970)

31 MUIESAN, G., MOTOLESE, M., COLOMBI, A.: Hypotensive effect of oxprenolol in mild hypertension: a co-operative controlled study. Clin. Sci. molec. Med. *45*, Suppl. 1: 163s (1973)

32 O'Malley, K., Velasco, M., McNay, J. L.: Adrenergic mechanism of minoxidil-induced increase in plasma renin activity. Clin. Res. *21*, 953 (1973); abstract of paper

33 Pape, J.: The effect of alprenolol in combination with hydralazine in essential hypertension: a double-blind, crossover study and a long-term follow-up study. Acta med. scand. *195*, Suppl. 554: 55 (1974)

34 Pettinger, W. A., Campbell, W. B., Keeton, K.: Adrenergic component of renin release induced by vasodilating antihypertensive drugs in the rat. Circulat. Res. *33*, 82 (1973)

35 Prichard, B. N. C., Gillam, P. M. S.: Treatment of hypertension with propranolol. Brit. med. J. *i*, 7 (1969)

36 Sannerstedt, R.: Hemodynamic response to exercise in patients with arterial hypertension. Acta med. scand. *180*, Suppl. 458 (1966)

37 Sannerstedt, R.: Differences in haemodynamic pattern in various types of hypertension. Triangle (En.) *9*, 293 (1970)

38 Sannerstedt, R., Julius, S., Conway, J.: Hemodynamic responses to tilt and beta-adrenergic blockade in young patients with borderline hypertension. Circulation *42*, 1057 (1970)

39 Sannerstedt, R., Stenberg, J., Johnsson, G., Werkö, L.: Hemodynamic interference of alprenolol with dihydralazine in normal and hypertensive man. Amer. J. Cardiol. *28*, 316 (1971)

40 Sannerstedt, R., Stenberg, J., Vedin, A., Wilhelmsson, C., Werkö, L.: Chronic beta adrenergic blockade in arterial hypertension. Amer. J. Cardiol. *29*, 718 (1972)

41 Tarazi, R. C., Dustan, H. P.: Beta adrenergic blockade in hypertension. Practical and theoretical implications of long-term hemodynamic variations. Amer. J. Cardiol. *29*, 633 (1972)

42 Vaughan, E. D., Jr., Laragh, J. H., Gavras, I., Bühler, F. R., Gavras, H., Brunner, H. R., Baer, L.: Volume factor in low and normal renin essential hypertension. Treatment with either spironolactone or chlorthalidone. Amer. J. Cardiol. *32*, 523 (1973)

43 Waal-Manning, H. J.: Comparative studies on the hypotensive effects of beta-adrenergic receptor blockers. In Simpson, F. O. (Editor): Beta-adrenergic receptor blocking drugs, Proc. Symp., Auckland 1970, p. 64 (Australasian Drug Inform. Serv. 1970)

44 Werkö, L.: Mitral valvular disease. Hemodynamic studies of the consequences for the circulation (Almqvist & Wiksell, Uppsala 1964)

45 Werning, C.: Blutdrucksenkung und Reninsuppression durch Beta-Rezeptorenblockade mit Visken bei Patienten mit Grenzwerthypertonie. Med. Klin. *68*, 1559 (1973)

46 Werning, C., Fischer, N., Kaip, E., Stiel, D., Trübestein, G. K., Vetter, H.: Erhöhte Reninstimulation nach Orthostase bei labiler oder Grenzwerthypertonie. Dtsch. med. Wschr. *97*, 1038 (1972)

47 World Health Organisation: Arterial hypertension and ischaemic heart disease. Preventive aspects. Wld Hlth Org. techn. Rep. Ser. No. 231 (1962)

48 Zacest, R., Gilmore, E., Koch-Weser, J.: Treatment of essential hypertension with combined vasodilatation and beta-adrenergic blockade. New Engl. J. Med. *286*, 617 (1972)

Discussion

W. SOMERVILLE: I feel that a high percentage of Stage 3 patients will still prove resistant to therapy irrespective of the dose or type of beta-blocker used and even if diuretics and peripheral vasodilators such as hydralazine are also added to the regimen.

R. SANNERSTEDT: You may be right, Dr. SOMERVILLE. The problem is complicated by the fact that it is very hard to define Stage 3 hypertension, especially in view of the difficulty of assessing in a given case how much of the syndrome is due to hypertension *per se* and how much to other causes, such as atherosclerosis for example. What I have tried to offer in my paper is not a final solution, but a sort of working hypothesis indicating how antihypertensive therapy can be tailored to meet the patient's needs in the light of certain haemodynamic principles. I am perfectly aware of the fact that one may encounter difficulties in individual cases and also that there are other types of drug which can be resorted to besides diuretics, beta-blockers, and the others that I mentioned. But I did not enlarge on this problem, because I was chiefly concerned with trying to define the role that beta-blockers in particular can play in the management of hypertension.

R. C. TARAZI: I believe it is very important to attempt, as Dr. SANNERSTEDT has done, to introduce an element of rationality into the selection of drugs for the treatment of hypertension. With regard to the haemodynamic principles he has just mentioned, however, I am not too happy about what I suspect to be the implication that beta-blockers act as effective antihypertensives only in states characterised by a high cardiac output. I personally do not know of any single haemodynamic function on the basis of which one could make a reliable guess as to whether a beta-blocker will or will not prove effective as an antihypertensive agent. One conclusion that seemed to emerge from the papers presented by Dr. BRUNNER and Dr. IMHOF, and a conclusion with which I think most of us would agree, is that the antihypertensive effect of beta-blockers is not a direct or necessary result of the reduction they cause in cardiac output, but depends rather on an adaptation of peripheral resistance to whatever other haemodynamic changes have occurred. Consequently, though I find Dr. SANNERSTEDT's classification theoretically attractive – in the sense that in Stage 1 hypertension, for example, you are dealing with a high cardiac output, which you proceed to reduce by giving a beta-blocker – I don't know whether the basis on which the choice of drug is made is one with which everyone would fully agree. Nevertheless, I still consider it extremely important to try to develop some sort of rational basis for the selection of treatment.

R. SANNERSTEDT: Dr. TARAZI and his colleagues in Cleveland have convincingly demonstrated that in the course of time hypertensive patients receiving a beta-blocker do in fact eventually, in response to sustained therapy, show a decrease – or a reversal of the increase – in peripheral resistance. But can we afford to be content with this? Since it may take one year of treatment, or even longer, before this decrease occurs, wouldn't it be more logical from the purely pathophysiological standpoint to try to counteract the rise in peripheral resistance from the very beginning?

R. C. TARAZI: The relatively long delay in recording the eventual decrease in peripheral resistance was due to the method we employed. It is conceivable that – with some beta-blockers or under some conditions, such as association with diuretics or vasodilators – a drop in peripheral resistance may occur within as little as one week or one month.

S. H. TAYLOR: I should like to make the point that "systemic vascular resistance" is in fact an entity which we cannot measure. Simply dividing the mean blood pressure by the mean cardiac output and assuming this to be a measurement is meaningless. In reality, the relationship between pressure and flow in the circulation is a far more complex function than is implied by dividing one measurement by another. It should take into account the *arterial* blood volume, which of course we cannot measure, and *phasic* pressure and *phasic* flow. Although from the two measurements of phasic pressure and flow one can calculate instantaneous aortic impedance, this is something that no author has yet talked about. I would therefore suggest that it is unnecessary and totally misleading to talk about, or base any argument upon, a measurement called "systemic or total vascular resistance", which in fact is no measurement at all.

R. SANNERSTEDT: Yes, Dr. TAYLOR, we are not of course measuring systemic vascular resistance *directly*, but merely calculating it. Despite your well-founded objection, I still feel that for general purposes and for presenting, as I have tried to do, some therapeutic guidelines for doctors not particularly concerned with haemodynamic subtleties, this is a working term which can still justifiably be used. Or could you perhaps propose a better alternative?

S. H. TAYLOR: In this particular instance, I don't see any point in speculating about something we cannot measure. This is a totally misleading concept, because the vascular resistance differs in every regional vascular territory in the body; it varies with time and it varies in an instantaneous manner with all stages of the cardiac cycle. Furthermore, when the blood pressure and cardiac output change – not to mention the changes that may be occurring in arterial blood volume – then I consider that calculations of mean aortic resistance have no counterpart in the actual changes in resistance that may be taking place in the regional vascular beds.

Renin-aldosterone interaction, beta-blockade, and differing antihypertensive efficacy of beta-blockers in high, normal, or low renin essential hypertension

by F. R. Bühler*

In the ten years that have elapsed since their introduction into clinical use[42, 45] the beta-blockers have come to be an essential component of antihypertensive therapy. That they exert an excellent antihypertensive effect which does not diminish in the course of time, and that they are incomparably better tolerated than other antihypertensive agents, has been documented in various publications[27, 57, 63]. But, although a vast number of clinical and experimental studies have been published on the beta-blockers, the mechanism by which they exert their antihypertensive effect is still a matter for discussion today.

Owing to the apparently paradoxical effect of the beta-blockers on the peripheral blood vessels[26], it was for years maintained by some[4, 25, 35] that the antihypertensive action of these drugs was connected with the reduction they produce in cardiac output. This hypothesis, however, was not accepted by other authors[5, 52], and the likelihood of its proving correct diminished when it was found that beta-blockers other than propranolol – such as alprenolol and timolol – have similar effects on blood pressure but exert hardly any influence on cardiac output[24, 29]. A fairly recent paper on haemodynamic variations, moreover, has shown that the antihypertensive efficacy of propranolol, apart from being linked to a decrease in peripheral vascular resistance, has no connection with any other cardiovascular variable[52]. Hence, the question that now has to be elucidated is why beta-blockade produces a reduction in peripheral vascular resistance, and thus in blood pressure, in only some patients.

Systematic analysis of the renin-angiotensin-aldosterone system in relation to urinary sodium excretion has made it possible to subdivide essential hypertension into high, normal, and low renin forms[31, 32]. Furthermore, in both clinical[39, 59] and experimental[2, 3, 58] studies it has been demonstrated that renin secretion is largely – though not exclusively – regulated by the autonomic nervous system and can be suppressed by beta-blockers.

It was therefore a logical step to treat patients of differing renin status with beta-blockers under similar conditions. Trials conducted along these lines revealed that propranolol exerted an excellent antihypertensive effect in high-renin patients and a varying effect – which was nevertheless good in about half the cases – in normal-renin patients[11, 13]. The reduction in diastolic blood pressure clearly correlated with the decrease in plasma renin activity[11]. In addition, beta-blockade also diminished aldosterone secretion, though to a lesser degree than would have been expected in view of the extent to which renin activity was suppressed.

---

 * Departement für Innere Medizin der Universität, Kantonsspital, Basle, Switzerland.

In striking contrast to these findings, propranolol failed to reduce the diastolic blood pressure in low-renin patients. Various studies have shown that this type of hypertension tends to be volume-dependent[28, 30, 61] and responds better to long-acting diuretics or spironolactone[1, 16, 19, 48, 55]. More sophisticated hormone analyses have demonstrated that in this low-renin type of hypertension the extent to which aldosterone secretion can be suppressed by salt loading is reduced[18, 34]. Moreover, low-renin essential hypertension differs from the other forms in that the aldosterone levels are relatively high in comparison with the renin levels[12].

The ability or inability of the beta-blockers to exert an antihypertensive effect in these three subtypes of essential hypertension, as well as characteristic endogenous differences in the renin-aldosterone interaction associated with the three subtypes, also suggest that the role played by the beta-adrenergic nervous system in the pathogenesis of the hypertension may differ from one subtype to another. With the aim of re-examining these relationships, the present study was designed to compare the antihypertensive and metabolic effects of propranolol with the results of a component analysis of the renin-aldosterone system[12] in patients with essential hypertension of the high, normal, and low renin type.

Methodology

1. Patients and test procedures
The two studies reported on in the present paper relate to altogether 125 patients with essential hypertension and 27 normal subjects, and they were all conducted in the Hypertension Center or Metabolism Unit of the Columbia Presbyterian Medical Center. Antihypertensive therapy, including the administration of diuretics, was discontinued at least four weeks prior to the studies. In all the patients, plasma renin activity was plotted against 24-hour urinary sodium excretion, and the resultant index was compared with a standard curve established for 52 normal subjects in a similar investigation[9].

In the first study[13] 74 patients were tested either in the Metabolism Unit (N = 19) or on an ambulatory basis (N = 55). Following determination of their renin-sodium index, these patients were given propranolol in a daily dosage of 80–540 mg., administered in 3–4 fractional doses. The aim was to achieve a dosage level of 2–5 mg./kg. body weight, i.e. the level usually required to produce beta-blockade[62]. In the case of the out-patients, propranolol medication was continued for periods ranging from six weeks to two years; at the end of this time the renin and aldosterone measurements were repeated. In nine patients with essential hypertension of the normal-renin type, oral propranolol medication in the above-mentioned dosage was resumed for a further 5–7 days following a control period of 5–7 days during which the patients received a diet of known and unvarying sodium, potassium, and calorie content. At the end of this treatment period the renin and aldosterone measurements were repeated. Blood pressure was determined by sphygmomanometry or with the aid of an automatic recorder (Physiometrics SR 1). The blood pressure values reported are mean values from at least three individual measurements performed on the day on which plasma renin activity was determined.

In the second study[12] the renin-aldosterone-sodium system was first examined in 27 normal subjects aged between 20 and 54 years. Sixteen of them had been receiving an unvarying diet of low, normal, or high sodium content for 4–5 days, or, in five cases, for up to 12 days. The other 11 subjects were studied initially without any prior dietary restrictions and then after their salt intake had been slightly reduced. In this second study, 51 hypertensives aged between nine and 59 years were admitted as in-patients to the Metabolism Unit and subdivided into the three renin groups. In both the normal subjects and the hypertensives plasma renin activity, plasma aldosterone concentration, and plasma electrolytes were determined in the same blood sample, which was taken before lunch with the patient in the sitting position and, as a rule, on the fourth or fifth day of a regulated diet. Finally, the aldosterone and sodium levels were also measured in the 24-hour urine, which was collected at the same time.

2. Laboratory methods

Plasma renin activity and urinary aldosterone levels were both determined with the aid of a radioimmunoassay method, the details of which have been published[46, 47]. For measurement of the plasma aldosterone concentration use was made of a new and rapid radioimmunoassay procedure[14, 15]. In order to calculate significance levels in the case of paired and unpaired data, recourse was had to Student's *t* test. The results are indicated in the form of mean values ±S.E.M.

Results

Beta-blockade induced by propranolol

1. Antihypertensive effect

The changes in blood pressure recorded in individual patients are shown in the upper portion of Figure 1. In 14 of the 19 high-renin cases the diastolic pressure fell to 95 mm. Hg or below. In the normal-renin patients the response was good, but not uniform, the diastolic pressure falling to 95 mm. Hg or below in 25 of the 38 cases. In none of the low-renin patients was this target level of 95 mm. Hg attained.

In the patients with high-renin hypertension the diastolic pressure fell on the average by 32.19 ± 2.01 mm. Hg, i.e. from 122.98 ± 2.66 to 90.79 ± 2.33 mm. Hg ($P < 0.001$). In the normal-renin patients the mean reduction amounted to 18.82 ± 1.17 mm. Hg, i.e. the diastolic pressure fell from 112.63 ± 1.54 to 93.82 ± 1.74 mm. Hg ($P < 0.001$). In the low-renin group, by contrast, the mean fall in pressure of 4.71 ± 1.25 mm. Hg, i.e. from 119.71 ± 2.52 to 115.00 ± 2.91 mm. Hg, was not statistically significant.

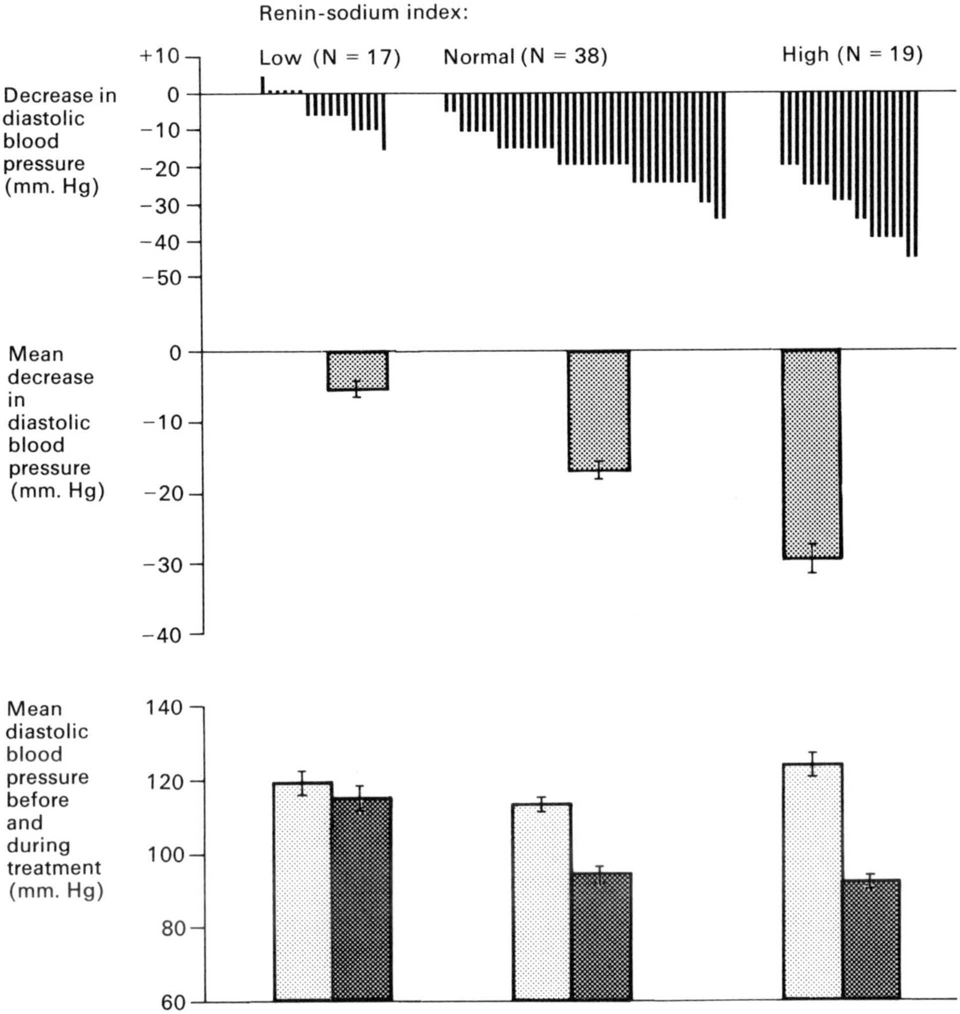

Fig. 1. Changes in diastolic blood pressure produced by propranolol in 74 patients with essential hypertension of the low, normal, and high renin type. *Above:* effect in the individual patients. *Middle:* mean fall in pressure. *Below:* mean diastolic pressures before (lighter columns) and during (darker columns) propranolol treatment[13].

2. *Effect on plasma renin activity*

In virtually all cases, treatment with propranolol led to a reduction in plasma renin activity, the mean decrease being 63%; in no instance was an increase observed in response to propranolol. The mean decrease in plasma renin activity was 9.85 ± 2.16 ng./ml./hr in the high-renin group (N = 13), 1.75 ± 0.17 ng./ml./hr in the normal-renin patients (N = 22), and 0.48 ± 0.17 ng./ml./hr in the low-renin cases (N = 12). The correlation between the change in diastolic blood pressure and the decrease in plasma renin activity is indicated in Figure 2.

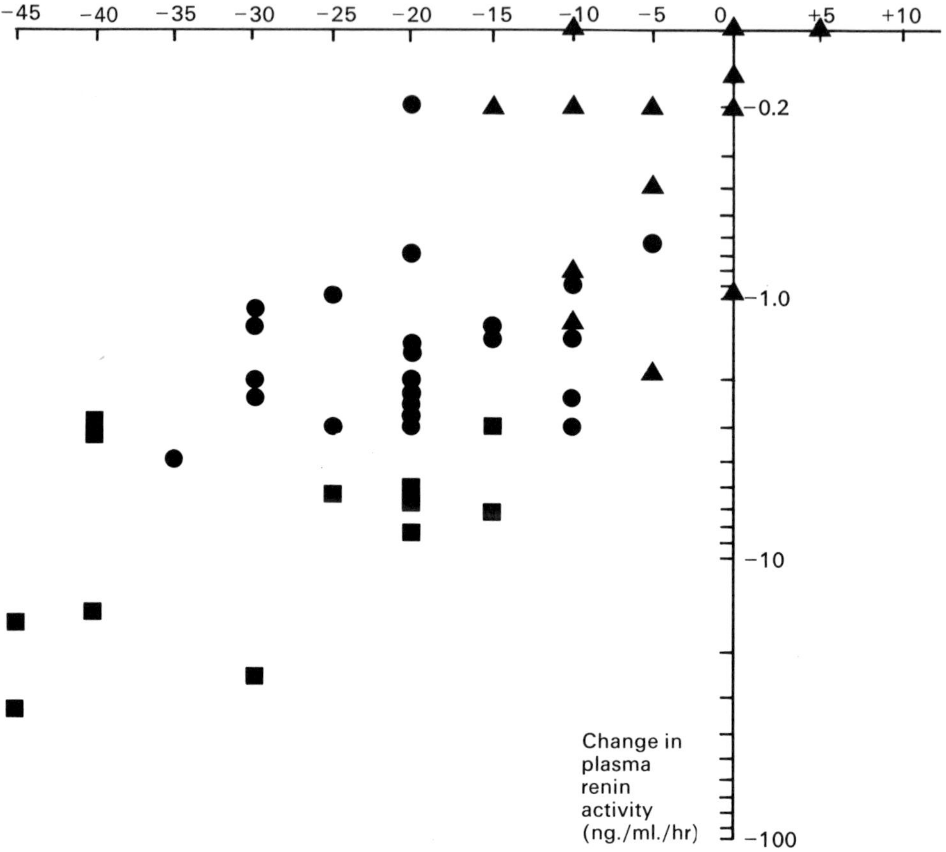

Fig. 2. Changes in diastolic blood pressure (abscissa) and decrease in renin activity (ordinate) induced by propranolol in 47 patients with hypertension of the low (▲), normal (●), or high (■) renin type[11].

3. Effect on aldosterone levels

In nine patients with essential hypertension of the normal-renin type, the effect exerted by propranolol on blood pressure, plasma renin activity, aldosterone excretion rate, sodium balance, and plasma potassium levels was investigated under metabolic balance conditions (Figure 3). In cases where the sodium and potassium intake was kept at an unvarying, normal level propranolol did not cause any change in the sodium balance, as revealed by the fact that sodium excretion in the 24-hour urine remained constant. At the same time, however, propranolol reduced the mean diastolic blood pressure in these patients from 110.00 ± 3.82 to 90.00 ± 3.23 mm. Hg; moreover, plasma renin activity fell by 66%, i.e. from 3.82 ± 0.62 to 1.28 ± 0.57 ng./ml./hr $(P < 0.001)$, and the aldosterone excretion rate decreased by 28%, i.e. from 12.86 ± 2.13 to

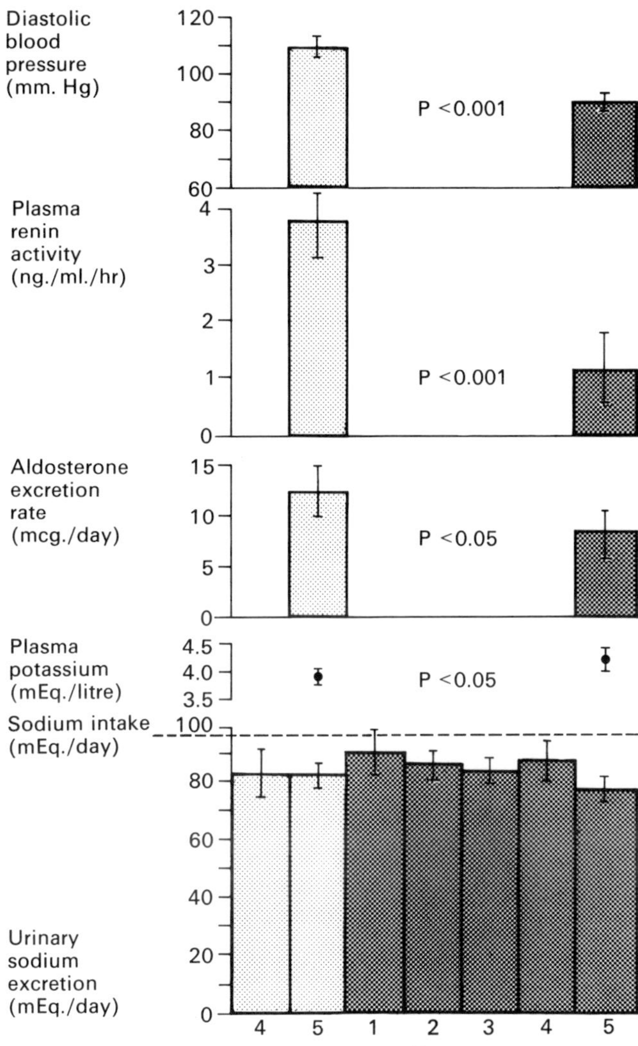

Fig. 3. Nine patients with normal-renin hypertension. Under metabolic balance conditions urinary sodium excretion *(below)* remained constant during propranolol treatment (darker columns) as compared with the pretreatment period (lighter columns), although the diastolic blood pressure *(above)* decreased. The aldosterone excretion rate diminished along with the plasma renin activity, but to a lesser extent *(middle)*. This quantitative difference seems to be accounted for by the propranolol-induced rise in plasma potassium which was also observed[13].

9.13 ± 2.02 mcg./day (P < 0.05). This reduction in the aldosterone excretion rate was accompanied by a rise in plasma potassium from 3.82 ± 0.11 to 4.26 ± 0.21 mEq./litre (P < 0.05).

The propranolol-induced effect of potassium on renin and aldosterone was studied in a further 12 patients. The plasma potassium levels rose in seven cases (+ 0.59 ± 0.15 mEq./litre), fell in four (−0.38 ± 0.18 mEq./litre), and remained unchanged in one. As Figure 4 shows, in the cases in which plasma potassium rose plasma aldosterone, which was measured at the same time, was suppressed to a lesser extent than plasma renin activity, with the result that the ratio indicated on the ordinate becomes less than 1. This means that, expressed in percent, plasma aldosterone decreased less than plasma renin activity.

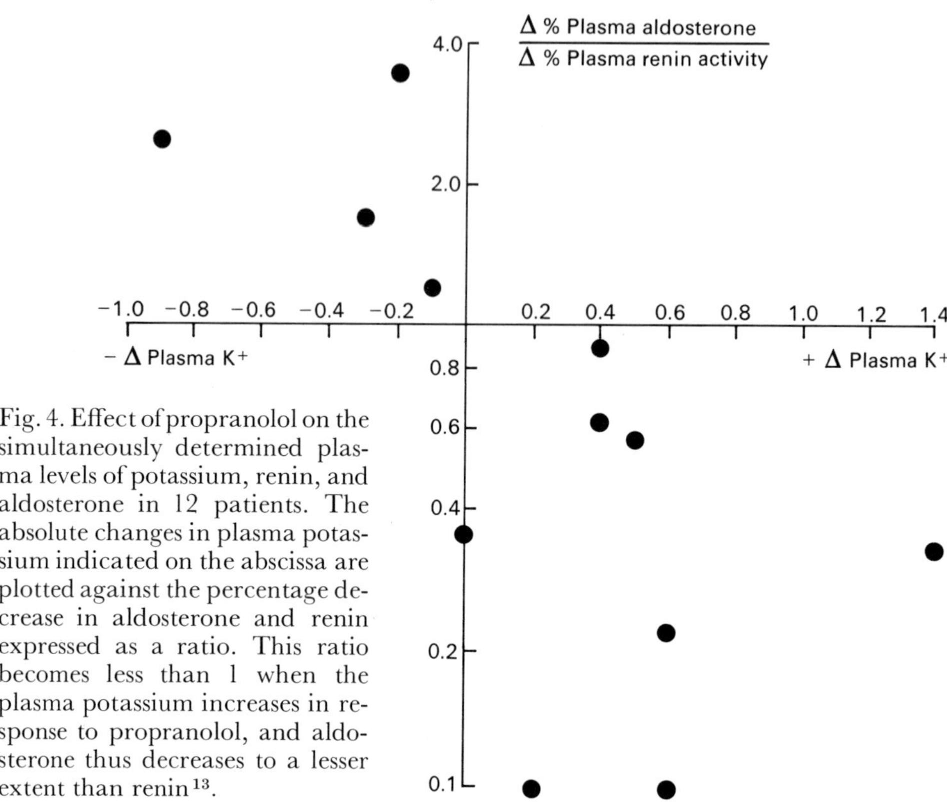

Fig. 4. Effect of propranolol on the simultaneously determined plasma levels of potassium, renin, and aldosterone in 12 patients. The absolute changes in plasma potassium indicated on the abscissa are plotted against the percentage decrease in aldosterone and renin expressed as a ratio. This ratio becomes less than 1 when the plasma potassium increases in response to propranolol, and aldosterone thus decreases to a lesser extent than renin[13].

Interaction of plasma renin and plasma aldosterone

1. Plasma aldosterone and sodium balance
In Figure 5 the plasma aldosterone concentrations found in 27 normal subjects are plotted against the same subjects' urinary sodium excretion rates. The curve reveals that the aldosterone concentration increased sharply as the urinary sodium decreased. Studies performed under metabolic balance conditions (N = 113), as well as studies conducted in subjects on an unrestricted diet (N = 20), yielded results within the same normal range. Plasma aldosterone varied from 25 to 170 ng./100 ml. in cases where the sodium excretion rate was below 25 mEq./day, from 6 to 24 ng./100 ml. in cases where the sodium excretion rate was between 75 and 125 mEq./day, and from 1 to 13 ng./100 ml. in cases where sodium excretion exceeded 200 mEq./day.
Figure 6 provides a summary picture of the aldosterone-sodium relationship in 51 patients with essential hypertension. The 16 patients with essential hypertension of the low-renin type, all displaying a normal aldosterone excretion rate and normal plasma potassium levels, showed in most cases a normal plasma aldosterone concentration. In the presence of a low sodium excretion rate, on the other hand, plasma aldosterone was relatively reduced in half the cases. In the presence of a fairly high sodium excretion rate, by contrast, the

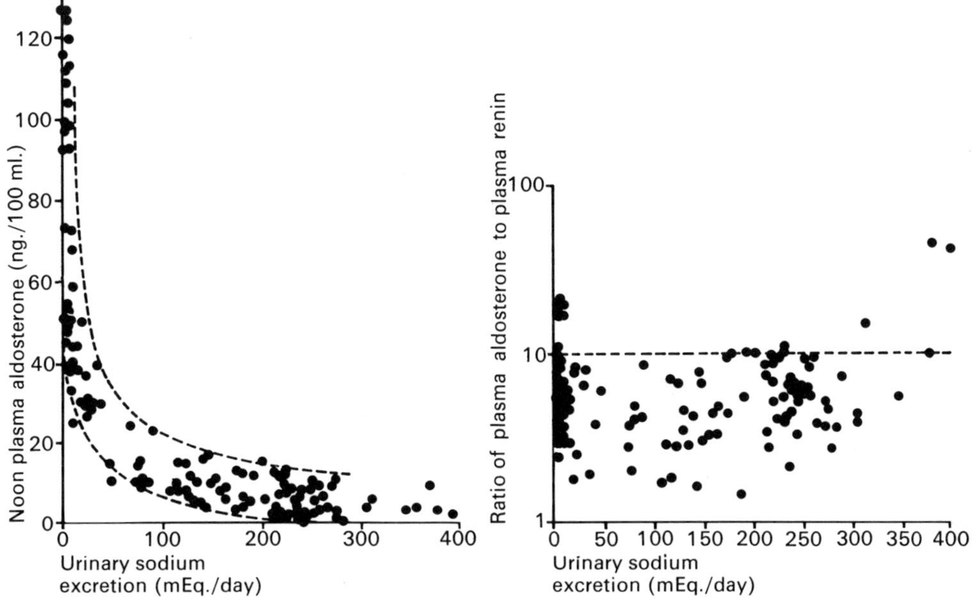

Fig. 5. Studies conducted in 27 normal subjects. *Left:* the plasma aldosterone concentration, measured in the sitting position before lunch, plotted against the 24-hour urinary sodium excretion rate. Aldosterone rises along with the withdrawal of sodium and decreases in response to sodium loading. The limits of the normal range of this function are indicated by the interrupted lines. *Right:* ratios of the simultaneously measured plasma aldosterone and plasma renin plotted against the corresponding urinary sodium excretion rate[12].

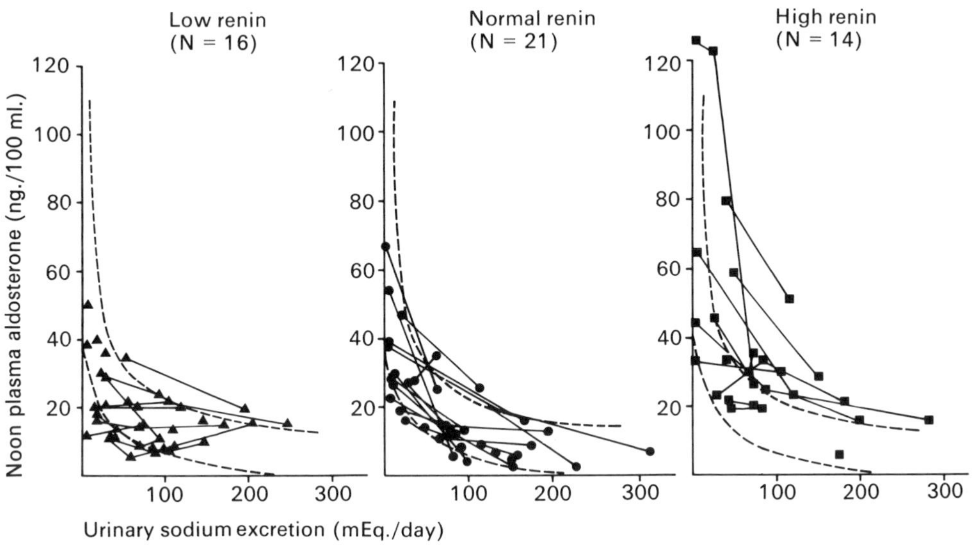

Fig. 6. Plasma aldosterone plotted against daily urinary sodium excretion in low-renin, normal-renin, and high-renin patients. The normal range (interrupted lines) was established in 27 normal subjects (cf. Figure 5)[12].

aldosterone concentration was either in the upper part of the normal range or slightly elevated, thus reflecting a decrease in aldosterone suppressibility.

In the 21 patients with hypertension of the normal-renin type the aldosterone-sodium relationship was found to be normal. In ten out of the 14 high-renin patients the plasma aldosterone concentration was increased in line with the increase in renin. In these two groups, therefore, aldosterone behaved in the same way as renin.

2. *Relationship of renin and aldosterone to the sodium balance*

In order to make a detailed analysis of the renin-aldosterone interaction, the plasma renin activity and plasma aldosterone concentration, measured simultaneously and expressed as an aldosterone-renin ratio, were plotted against the 24-hour urinary sodium excretion rate. In the normal subjects a good correlation was found between aldosterone and renin, inasmuch as the aldosterone-renin ratio was 10 or less over a wide range of sodium excretion rates. Only at the two extremes, i. e. at sodium excretion rates of less than 10 or more than 300 mEq./day, was this ratio of 10 exceeded.

In sharp contrast to these findings, the aldosterone-renin ratio proved to be abnormal in 14 out of 16 patients with low-renin hypertension. Since plasma aldosterone was within the normal range, this abnormality was evidently to be accounted for by the reduced renin activity. As shown by the lines connecting the values obtained in 2–3 balance studies performed in the same patient, various types of reaction are observed in low-renin hypertensives. In nine patients the aldosterone-renin ratio increased in response to sodium loading, whereas in five it decreased. Two patients with the relatively highest renin values displayed a normal aldosterone-renin ratio. These findings suggest that

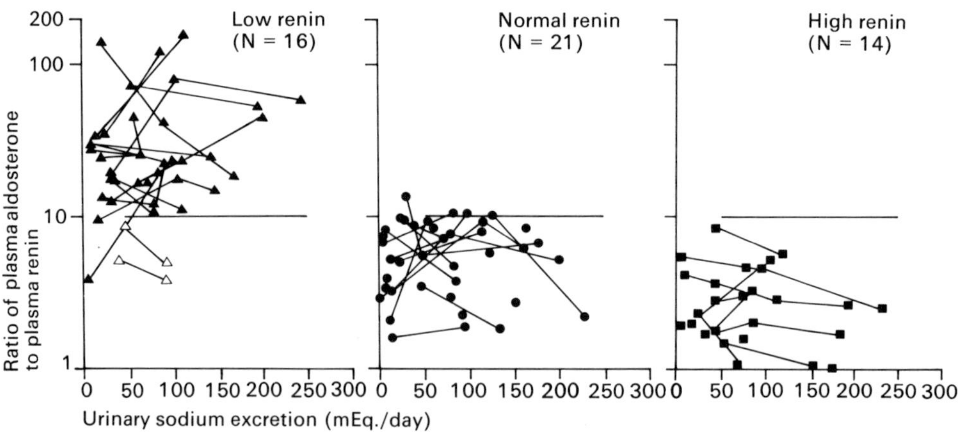

Fig. 7. Ratio of plasma aldosterone to plasma renin (both measured simultaneously) plotted against urinary sodium excretion in low-renin, normal-renin, and high-renin patients. In 14 of the 16 low-renin patients the upper limit, established on the basis of studies performed in normal subjects (cf. Figure 5), is exceeded in the presence of slight sodium loading[12].

essential hypertension of the low-renin type can be further subdivided into various subgroups (Figure 7).

Patients of the normal-renin type showed a normal aldosterone-renin ratio irrespective of whether their sodium excretion rates were normal or high. Similarly, a normal aldosterone-renin ratio was observed in patients of the high-renin type; in some of these cases, however, aldosterone was relatively reduced in comparison with renin. At an aldosterone-renin ratio of less than 2 the plasma potassium levels amounted to 3.3 ± 0.4 mEq./litre (N = 10), as compared with 3.8 ± 0.4 mEq./litre (N = 14) in the patients with a normal aldosterone-renin ratio (P < 0.01). This reduction in the aldosterone-renin ratio in some of the high-renin patients can therefore be explained as being due to the decrease in the plasma potassium levels.

Discussion

Beta-blockade with propranolol produced an excellent and sustained reduction in diastolic blood pressure in most of the patients with a high renin-sodium index. In the patients with hypertension of the normal-renin type, the effect of the drug varied, but was likewise satisfactory and sustained in half of them. In patients of the low-renin type, on the other hand, propranolol failed to exert a significant pressure-lowering effect. In somewhat more than 50% of all the patients with essential hypertension, monotherapy with propranolol reduced the diastolic pressure to 95 mm. Hg or less. Propranolol achieves a reduction in abnormal plasma renin activity and in blood pressure not only in patients with essential hypertension, but also in those with malignant, renovascular, or renal-parenchymal hypertension[10, 13]. It produces a mean decrease of 63% in plasma renin activity. This effect sets in promptly and is well maintained over a period of years. It also persists to a relative degree when salt intake is physiologically reduced. In addition, our studies have shown that there is a direct relationship between reduction in diastolic blood pressure and decrease in plasma renin activity. Along with the suppression of plasma renin activity, the simultaneously measured plasma aldosterone concentration likewise fell, albeit to a less pronounced extent. This somewhat less marked suppression of aldosterone under beta-blockade was accompanied by an increase in plasma potassium levels, which is known to stimulate aldosterone while at the same time suppressing renin secretion[7, 56]. Consequently, propranolol appears to lower the blood pressure, firstly, by reducing vasoconstriction induced by the renin-angiotensin system, and, secondly, also by reducing the effect of aldosterone on volume. This interpretation offers an explanation for the finding reported by other authors that peripheral vascular resistance is reduced – or fails to increase – in response to beta-blockade[52]. A correlation between suppression of plasma renin activity and fall in blood pressure has recently also been found by other authors in studies with alprenolol[17]. In addition, our aldosterone findings may perhaps account for the observation that as a rule neither body weight nor blood volume[11, 53] increases during monotherapy with propranolol. The presence of a mechanism by which renin-induced vasoconstriction is antagonised is also suggested by the fact that, as has recently been demonstrated[8], a competitive angiotensin-receptor antagonist reduces the blood pressure in high-renin but not in low-renin patients.

.This theory that beta-blockade exerts an anti-renin and anti-aldosterone effect does not, however, exclude the possibility that it may also combat hypertension by acting at other or additional sites of attack. Propranolol, for example, is readily taken up by the central nervous system[36] and lowers the blood pressure following intracerebroventricular administration[49] and also following injection into the carotid or vertebral artery[50]. On the other hand, it is also known that stimulation of the vasomotor centre increases renin secretion, an effect which can likewise be blocked by propranolol[41]. The exact mechanism by which propranolol suppresses renin secretion is still not clear; there is evidence in support of both extrarenal[43] and intrarenal[51, 60] factors. The suppression of renin secretion seems at all events to be a function of L-propranolol, i.e. of the beta-blocking substance, and not of the inactive D-isomer[38, 54].

However the suppression of renin secretion is brought about, it appears that not only this effect but also the simultaneous reduction in aldosterone secretion plays an important role in the antihypertensive action of propranolol. The fundamental question thus arises as to whether these albeit only subtle abnormalities in the renin-aldosterone system, which are corrected in propranolol-sensitive forms of hypertension, do not perhaps reflect increased activity on the part of the beta-adrenergic nervous system.

On the other hand, the failure of propranolol to exert an antihypertensive effect in hypertension of the low-renin type should be stressed. These forms of hypertension appear to be governed less by renin-induced vasoconstriction[8] than by a volume factor[28, 30, 61]. Although low-renin hypertension does not constitute a uniform type of the disease, and although known[6, 37] or still unknown[33] mineralocorticoids might be involved as well, it is also possible that the beta-adrenergic nervous system likewise plays a role in the pathogenesis of these forms of hypertension. In the low-renin patients, but not in the high-renin or normal-renin ones, a dissociation between renin and aldosterone was found – a dissociation characterised in particular by the fact that, although aldosterone was normal in terms of absolute values, it was relatively too high. A reduced aldosterone suppressibility of this kind in response to salt loading has also been described by other authors[18, 34]. A similar aldosterone-renin dissociation is observed following beta-blockade with propranolol, the dissociation being apparently modulated by the rise in plasma potassium provoked by the beta-blocker. Precisely this stimulation of aldosterone secretion by potassium in response to beta-blockade may indicate the presence of a subtle disturbance of potassium metabolism in essential hypertension of the low-renin type. The existence of such a disturbance has been suspected on various occasions, but has not so far actually been demonstrated, the reason possibly being that in the chronic state the slight aldosteronism associated with low-renin hypertension counterbalances the hyperpotassaemic effect of beta-adrenergic insufficiency.

Studies in dogs, moreover, have shown that in response to propranolol renal vascular resistance and the filtration fraction increase, whereas sodium excretion decreases[40]. This propranolol-induced increase in the filtration fraction suggests the presence of post-glomerular vasoconstriction. This would result in a decreased hydrostatic and an increased oncotic pressure in the peritubular capillaries, together with an increased reabsorption of sodium in the proximal tubules[22]. The presence of post-glomerular vasoconstriction has also been postulated in cases of low-renin essential hypertension[44]. Hence, beta-blockade could give rise to an increased retention of sodium and fluid both via physical mechanisms operative in the proximal tubules and via relative aldosteronism in the distal tubules. This could mean that patients of the low-renin type who fail to respond to propranolol have a form of hypertension in which the beta-adrenergic nervous system already displays relatively reduced activity for endogenous reasons. It is a fact that a reduction in orthostatic catecholamine excretion in the urine has also been demonstrated in these patients[20].

Differences of this kind in autonomic nervous regulatory mechanisms in the three renin types of essential hypertension might perhaps also account for the contradictory results obtained in various studies on noradrenaline excretion rates or catecholamine metabolism in essential hypertension[21]. In addition, direct correlations have been found between noradrenaline excretion and renin secretion following, for example, physiological stimulation of the autonomic nervous system[23].

The antihypertensive effectiveness or ineffectiveness of beta-blockers in patients with essential hypertension, as well as the metabolic and hormonal changes induced by the beta-blockers, indicate that the renin-aldosterone system, regulated – or perhaps it would be better to say "apparently dysregulated" – by the adrenergic nervous system, plays a pivotal role in the pathogenesis of essential hypertension.

Summary

Suppression of the renin-aldosterone system by beta-blockers and the antihypertensive efficacy of these drugs in essential hypertension associated with high and normal renin levels are compared and contrasted with the characteristically abnormal renin-aldosterone ratio in low-renin hypertension and with the inability of beta-blockers to exert an antihypertensive effect in patients of this type.

In a first study, involving 74 patients, propranolol produced a uniformly satisfactory reduction in diastolic blood pressure in 19 patients with hypertension of the high-renin type, the mean extent of the reduction being 32 mm. Hg; in 38 patients with hypertension of the normal-renin type, the decrease in blood pressure in response to propranolol varied, but amounted on the average to 19 mm. Hg. By contrast, propranolol was practically ineffective in all 17 cases of hypertension of the low-renin type. In addition, a direct correlation was found between the reduction in diastolic pressure and the extent to which renin secretion was suppressed. Along with renin, aldosterone, too, was reduced, a finding which perhaps accounts for the lack of change in the sodium balance in response to propranolol. The suppression of aldosterone, which was somewhat less marked than that of renin, was accompanied by a slight increase in plasma potassium. Consequently, propranolol appears to reduce the blood pressure by influencing both the vasoconstrictor effect of renin and the effect of aldosterone on volume.

In a second study, conducted in 51 patients, no essential deviations from the normal pattern of renin-aldosterone interaction as established in 27 normal subjects was found in the 14 patients of the high-renin type or in the 21 of the normal-renin type. In 14 of the 16 low-renin patients, by contrast, a dissociation was found between aldosterone and renin, which was due in particular to the fact that aldosterone was suppressed to a lesser extent by salt loading. This phenomenon of a somewhat too high aldosterone in relation to renin may be imitated by beta-blockade.

Taken together, these analyses point to a relatively higher activity of the adrenergic nervous system in hypertensions of the high-renin and, in part, of the normal-renin type. Low-renin hypertension, on the other hand, might be marked by a comparatively reduced beta-adrenergic tone, which could possibly explain the relative aldosteronism found in these cases and also their failure to respond to beta-blockade. In addition to the aldosterone-induced effect on the distal tubules, an endogenous beta-adrenergic deficiency could also contribute, via physical mechanisms operative in the proximal tubules, to the expansion of volume which plays a major role in the pathogenesis of low-renin hypertension.

Acknowledgment

The results presented in this paper were obtained by the author while he was working in Dr. J. H. Laragh's group at the Hypertension Center, Columbia University, New York.

References

1 Adlin, E.V., Marks, A.D., Channick, B.J.: Spironolactone and hydrochlorothiazide in essential hypertension. Blood pressure response and plasma renin activity. Arch. intern. Med. *130*, 855 (1972)
2 Alexandre, J.-M., Ménard, J., Chevillard, C., Schmitt, H.: Increased plasma renin activity induced in rats by physostigmine and effects of alpha- and beta-receptor blocking drugs thereon. Europ. J. Pharmacol. *12*, 127 (1970)
3 Assaykeen, T.A., Clayton, P.L., Goldfien, A., Ganong, W.F.: Effect of alpha- and beta-adrenergic blocking agents on the renin response to hypoglycemia and epinephrine in dogs. Endocrinology *87*, 1318 (1970)
4 Bello, C.T., Sevy, R.W., Harakal, C., Hillyer, P.N.: Relationship between clinical severity of disease and hemodynamic patterns in essential hypertension. Amer. J. med. Sci. *253*, 194 (1967)

5 BIRKENHÄGER, W.H., KRAUSS, X.H., SCHALEKAMP, M.A.D.H., KOLSTERS, G., KROON, B.J.M.: Antihypertensive effects of propranolol. Observations on predictability. Folia med. neerl. *14*, 67 (1971)

6 BROWN, J.J., FERISS, J.B., FRASER, R., LEVER, A.F., LOVE, D.R., ROBERTSON, J.I.S., WILSON, A.: Apparently isolated excess deoxycorticosterone in hypertension. A variant to the mineralocorticoid-excess syndrome. Lancet *ii*, 243 (1972)

7 BRUNNER, H.R., BAER, L., SEALEY, J.E., LEDINGHAM, J.G.G., LARAGH, J.H.: The influence of potassium administration and potassium deprivation on plasma renin in normal and hypertensive subjects. J. clin. Invest. *49*, 2128 (1970)

8 BRUNNER, H.R., GAVRAS, H., LARAGH, J.H., KEENAN, R.: Angiotensin-II blockade in man by sar^1-ala^8-angiotensin II for understanding and treatment of high blood pressure. Lancet *ii*, 1045 (1973)

9 BRUNNER, H.R., LARAGH, J.H., BAER, L., NEWTON, M.A., GOODWIN, F.T., KRAKOFF, L.R., BARD, R.H., BÜHLER, F.R.: Essential hypertension: renin and aldosterone, heart attack and stroke. New Engl. J. Med. *286*, 441 (1972)

10 BÜHLER, F.R., BAER, L., JACOB, G., VAUGHAN, E.D., Jr., PATTNER, A.M., SOTELLO, J.E., BRUNNER, H.R., LARAGH, J.H.: Inhibition of renin secretion by propranolol: its diagnostic and therapeutic value in essential, renovascular, malignant and end-stage hypertension. In: Abstr. Vth Int. Congr. Nephrol., Mexico City 1972, p.123, Abstr.681

11 BÜHLER, F.R., LARAGH, J.H., BAER, L., VAUGHAN, E.D., Jr., BRUNNER, H.R.: Propranolol inhibition of renin secretion. A specific approach to diagnosis and treatment of renin-dependent hypertensive diseases. New Engl. J. Med. *287*, 1209 (1972)

12 BÜHLER, F.R., LARAGH, J.H., SEALEY, J.E., BRUNNER, H.R.: Plasma aldosterone-renin interrelationships in various forms of essential hypertension. Studies using a rapid assay of plasma aldosterone. Amer. J. Cardiol. *32*, 554 (1973)

13 BÜHLER, F.R., LARAGH, J.H., VAUGHAN, E.D., Jr., BRUNNER, H.R., GAVRAS, H., BAER, L.: Antihypertensive action of propranolol. Specific antirenin responses in high and normal renin forms of essential, renal, renovascular and malignant hypertension. Amer. J. Cardiol. *32*, 511 (1973)

14 BÜHLER, F.R., SEALEY, J.E., LARAGH, J.H.: Radioimmunoassay of plasma aldosterone. In Laragh, J.H. (Editor): Hypertension manual, p.655 (Yorke Med. Books/Dun-Donnelley, New York 1973)

15 BÜHLER, F.R., SEALEY, J.E., LARAGH, J.H., MANNING, E.L.: Plasma aldosterone radioimmunoassay: rapid preassay purification using celite pipette columns. Biochem. Pharmacol. (printing)

16 CAREY, R.M., DOUGLAS, J.G., SCHWEIKERT, J.R., LIDDLE, G.W.: The syndrome of essential hypertension and suppressed plasma renin activity. Normalization of blood pressure with spironolactone. Arch. intern. Med. *130*, 849 (1972)

17 CASTENFORS, J., JOHNSSON, H., ORÖ, L.: Effect of alprenolol on blood pressure and plasma renin activity in hypertensive patients. Acta med. scand. *193*, 189 (1973)

18 COLLINS, R.D., WEINBERGER, M.H., DOWDY, A.J., NOKES, G.W., GONZALES, G.M., LUETSCHER, J.A.: Abnormally sustained aldosterone secretion during salt loading in patients with various forms of benign hypertension; relation to plasma renin activity. J. clin. Invest. *49*, 1415 (1970)

19 CRANE, M.G., HARRIS, J.J.: Effect of spironolactone in hypertensive patients. Amer. J. med. Sci. *260*, 311 (1970)

20 CRANE, M.G., HARRIS, J.J., JOHNS, V.J., Jr.: Hyporeninemic hypertension. Amer. J. Med. *52*, 457 (1972)

21 DeQUATTRO, V., MIURA, Y.: Neurogenic factors in human hypertension: mechanism or myth? Amer. J. Med. *55*, 362 (1973)

22 EARLEY, L.E., DAUGHARTY, T.M.: Sodium metabolism. New Engl. J. Med. *281*, 72 (1969)

23 ESLER, M. D., NESTEL, P. J.: Renin and sympathetic nervous system responsiveness to adrenergic stimuli in essential hypertension. Amer. J. Cardiol. *32*, 643 (1973)

24 FRANCIOSA, J. A., FREIS, E. D., CONWAY, J.: Antihypertensive and hemodynamic properties of the new beta adrenergic blocking agent timolol. Circulation *48*, 118 (1973)

25 FROHLICH, E. D., TARAZI, R. C., DUSTAN, H. P.: Beta-adrenergic blocking therapy in hypertension: selection of patients. Int. J. clin. Pharmacol. Ther. Toxicol. *4*, 151 (1970)

26 FROHLICH, E. D., TARAZI, R. C., DUSTAN, H. P., PAGE, I. H.: The paradox of beta-adrenergic blockade in hypertension. Circulation *37*, 417 (1968)

27 HANSSON, L., MALMCRONA, R., OLANDER, R., ROSENHALL, A., WESTERLUND, A., ÅBERG, H., HOOD, B.: Propranolol in hypertension. Report on 158 patients treated up to one year. Klin. Wschr. *50*, 364 (1972)

28 HELMER, O. M., JUDSON, W. E.: Metabolic studies on hypertensive patients with suppressed plasma renin activity not due to hyperaldosteronism. Circulation *38*, 965 (1968)

29 JOHNSSON, G., GUZMAN, M. DE, BERGMAN, H., SANNERSTEDT, R.: The haemo-dynamic effects of alprenolol and propranolol at rest and during exercise in hypertensive patients. Pharmacol. clin. *2*, 34 (1969)

30 JOSE, A., CROUT, J. R., KAPLAN, N. M.: Suppressed plasma renin activity in essential hypertension, roles of plasma volume, blood pressure, and sympathetic nervous system. Ann. intern. Med. *72*, 9 (1970)

31 LARAGH, J. H.: Vasoconstriction-volume analysis for understanding and treating hypertension: the use of renin and aldosterone profiles. Amer. J. Med. *55*, 261 (1973)

32 LARAGH, J. H., BAER, L., BRUNNER, H. R., BÜHLER, F. R., SEALEY, J. E., VAUGHAN, E. D., Jr.: Renin, angiotensin and aldosterone system in pathogenesis and management of hypertensive vascular disease. Amer. J. Med. *52*, 633 (1972)

33 LIDDLE, G. W.: Mineralocorticoid hypertension. Verh. Dtsch. Ges. inn. Med. (printing)

34 LUETSCHER, J. A., WEINBERGER, M. H., DOWDY, A. J., BALIKIAN, H., BRODIE, A., WILLOUGHBY, S.: Effects of sodium loading, sodium depletion and posture on plasma aldosterone concentration and renin activity in hypertensive patients. J. clin. Endocr. *29*, 1310 (1969)

35 LYDTIN, H., KUSUS, T., DANIEL, W., SCHIERL, W., ACKENHEIL, M., KEMPTER, H., LOHMÖLLER, G., NIKLAS, M., WALTER, I.: Propranolol therapy in essential hypertension. Amer. Heart J. *83*, 589 (1972)

36 MASUOKA, D., HANSSON, E.: Autoradiographic distribution studies of adrenergic blocking agents. II. ^{14}C-propranolol, a β-receptor-type blocker. Acta pharmacol. (Kbh.) *25*, 447 (1967)

37 MELBY, J. C., DALE, S. L., WILSON, T. E.: 18-Hydroxy-deoxycorticosterone in human hypertension. Circulat. Res. *28/29*, Suppl. II: 143 (1971)

38 MEURER, K.-A.: Die Bedeutung des sympathico-adrenalen Systems für die Renin-freisetzung. Klin. Wschr. *49*, 1001 (1971)

39 MICHELAKIS, A. M., McALLISTER, R. G.: The effect of chronic adrenergic receptor blockade on plasma renin activity in man. J. clin. Endocr. *34*, 386 (1972)

40 NIES, A. S., McNEIL, J. S., SCHRIER, R. W.: Mechanism of increased sodium reabsorption during propranolol administration. Circulation *44*, 596 (1971)

41 PASSO, S. S., ASSAYKEEN, T. A., GOLDFIEN, A., GANONG, W. F.: Effect of a- and β-adrenergic blocking agents on the increase in renin secretion produced by stimulation of the medulla oblongata in dogs. Neuroendocrinology *7*, 97 (1971)

42 PRICHARD, B. N. C., GILLAM, P. M. S.: The use of propranolol (Inderal) in the treatment of hypertension. Brit. med. J. *ii*, 725 (1964)

43 REID, I. A., SCHRIER, R. W., EARLEY, L. E.: An effect of extrarenal beta adrenergic stimulation on the release of renin. J. clin. Invest. *51*, 1861 (1972)

44 Schalekamp, M.A.D.H., Schalekamp-Kuyken, M.P.A., Birkenhäger, W.H.: Abnormal renal haemodynamics and renin suppression in hypertensive patients. Clin. Sci. *38*, 101 (1970)

45 Schröder, G., Werkö, L.: Nethalide, a beta adrenergic blocking agent. Clin. Pharmacol. Ther. *5*, 159 (1964)

46 Sealey, J.E., Bühler, F.R., Laragh, J.H., Manning, E.L., Brunner, H.R.: Aldosterone excretion. Physiological variations in man measured by radioimmunoassay or double-isotope dilution. Circulat. Res. *31*, 367 (1972)

47 Sealey, J.E., Gerten-Banes, J., Laragh, J.H.: The renin system: variations in man measured by radioimmunoassay or bioassay. Kidney int. *1*, 240 (1972)

48 Spark, R.F., Melby, J.C.: Hypertension and low plasma renin activity: presumptive evidence for mineralocorticoid excess. Ann. intern. Med. *75*, 831 (1971)

49 Srivastava, R.K., Kulshrestha, V.K., Singh, N., Bhargava, K.P.: Central cardiovascular effects of intracerebroventricular propranolol. Europ. J. Pharmacol. *21*, 222 (1973)

50 Stern, S., Hoffman, M., Braun, K.: Cardiovascular responses to carotid and vertebral artery infusions of propranolol. Cardiovasc. Res. *5*, 425 (1971)

51 Tanigawa, H., Allison, D.J., Assaykeen, T.A.: A comparison of the effects of various catecholamines on plasma renin activity alone and in the presence of adrenergic blocking agents. In Genest, J., Koiw, E. (Editors): Hypertension – 1972, p. 37 (Springer, Berlin/Heidelberg/New York 1972)

52 Tarazi, R.C., Dustan, H.P.: Beta adrenergic blockade in hypertension. Practical and theoretical implications of long-term hemodynamic variations. Amer. J. Cardiol. *29*, 633 (1972)

53 Tarazi, R.C., Frohlich, E.D., Dustan, H.P.: Plasma volume changes with long-term beta-adrenergic blockade. Amer. Heart J. *82*, 770 (1971)

54 Vandongen, R., Peart, W.S., Boyd, G.W.: Adrenergic stimulation of renin secretion in the isolated perfused rat kidney. Circulat. Res. *32*, 290 (1973)

55 Vaughan, E.D., Jr., Bühler, F.R., Gavras, I., Brunner, H.R., Baer, L., Laragh, J.H.: Spirolactone compared to diuretic therapy in analysis and treatment of normal and low renin hypertensions. Circulation *45/46*, Suppl. II: 83 (1972); abstract of paper

56 Veyrat, R., Brunner, H.R., Grandchamp, A., Müller, A.F.: Inhibition of renin by potassium in man. Acta endocr. (Kbh.) *55*, Suppl. 119: 86 (1967); abstract of paper

57 Waal-Manning, H.J.: Comparative studies on the hypotensive effects of beta-blockers. N.Z. med. J. *71*, 383 (1970); abstract of paper

58 Winer, N., Chokshi, D.S., Walkenhorst, W.G.: Effects of cyclic AMP, sympathomimetic amines, and adrenergic receptor antagonists on renin secretion. Circulat. Res. *29*, 239 (1971)

59 Winer, N., Chokshi, D.S., Yoon, M.S., Freedman, A.D.: Adrenergic receptor mediation of renin secretion. J. clin. Endocr. *29*, 1168 (1969)

60 Winer, N., Walkenhorst, W.G., Helman, R., Lamy, D.: Effects of adrenergic antagonists in states of increased renin secretion. In Assaykeen, T.A. (Editor): Control of renin secretion, Proc. Workshop, Santa Ynez, California, 1971, p. 65 (Plenum Press, New York/London 1972)

61 Woods, J.W., Liddle, G.W., Stant, E.G., Jr., Michelakis, A.M., Brill, A.B.: Effect of an adrenal inhibitor in hypertensive patients with suppressed renin. Arch. intern. Med. *123*, 366 (1969)

62 Zacest, R., Koch-Weser, J.: Relation of propranolol plasma level to β-blockade during oral therapy. Pharmacology (Basle) *7*, 178 (1972)

63 Zacharias, F.J., Cowen, K.J., Prestt, J., Vickers, J., Wall, B.G.: Propranolol in hypertension: a study of long-term therapy, 1964–1970. Amer. Heart J. *83*, 755 (1972)

Discussion

K. H. WINTERHALTER: Did you find, Dr. BÜHLER, that the influence exerted by beta-blockers on plasma potassium differed in the three classes of hypertension you studied?

F. R. BÜHLER: Unfortunately, we haven't enough data in the high and low renin categories to be able to answer this question. That's why in this context I referred only to the normal-renin patients.

F. GROSS: I have two major questions. Firstly, what precisely is implied by your use of the terms "low", "normal", and "high" renin? Do you group your patients on the basis of one casual renin determination or do you – in the case of your low-renin patients, for example – also consider the response of plasma renin to either a diuretic or exercise or to any other possible procedures for stimulating plasma renin activity? In other words, are the plasma renin levels in your low-renin groups identical with those generally considered to be indicative of a suppressed renin response? Now for the second question. In essential hypertension, plasma renin activity is as a rule either normal or slightly reduced. When you have essential hypertensives with high renin levels, what, I wonder, are the reasons for this increased plasma renin activity? Are they already losing sodium through the kidney, or can you perhaps suggest some other explanation for the high renin activity? It seems to me important to find out what induces the kidney to release higher amounts of renin in these special cases of essential hypertension.

F. R. BÜHLER: May I answer the second question first. These patients with essential hypertension still, of course, have normal renal function. This means that there is no overt difference between low, normal, and high renin hypertension. What is it, then, that causes the renin to shoot up to above-normal levels in a proportion of the total essential-hypertensive population amounting to at least 16%? The answer, I'm afraid, is that we just don't know. The fact that beta-blockers lower the blood pressure, restore the elevated renin levels to normal, and may also normalise aldosterone could perhaps help to provide a clue to the answer. What we may possibly be dealing with here is the metabolic basis of a dysregulation which might conceivably originate in the adrenergic nervous system.

Regarding your first question, Dr. GROSS, I can state that we have about the same percentage of low-renin patients as are encountered by other investigators. To detect these patients, however, we prefer to use a more physiological test; we do not stimulate renin secretion by giving a diuretic, because BRUNNER et al.* have shown that, if you administer furosemide in order to stimulate renin, you cannot discriminate so well between low and normal renin patients. What we therefore do is to make first of all a casual renin determination on a casual sodium intake and then put the patients on a mildly reduced sodium diet which moves them over to the left side of the curve and enables us to identify low-renin hypertension more easily. For most of our patients we thus have two renin-sodium indices. Finally, with regard to the practicability of renin measurements, I think that more reliable kits for measuring renin activity will probably become more widely available in the fairly near future.

J. BROD: I should like to know, Dr. BÜHLER, something more about the clinical stage of the hypertension from which your low-renin, normal-renin, and high-renin patients were suffering. The reference you made, in the spoken version of your paper, to the age of the patient as one of the factors possibly determining whether he will belong to the high-renin or low-renin category serves, of course, as a reminder that we are dealing in these two groups with different kinds of patients. It is mainly in patients aged between 20 and 30 or 40 years that we encounter initial hypertension of the W.H.O.

---

* BRUNNER, H. R., SEALEY, J. E., LARAGH, J. H.: Renin subgroups in essential hypertension: further analysis of their pathophysiological and epidemiological characteristics. Circulat. Res. *32*, Suppl. I: 99 (1973)

Stage 1 type. After the age of 50 years this initial form of hypertension is hardly ever met with; instead, these older patients almost all suffer from hypertension of Stage 2 or 3. In this connection, I was particularly struck by your comment to the effect that the patients with essential hypertension upon whom you were reporting all still had normal renal function. I doubt whether, even in the early stage of hypertension, renal function is always still normal. Would you care to comment on these two points, Dr. BÜHLER?

F. R. BÜHLER: When I say that the patients had normal renal function, what I mean is that their blood urea nitrogen and endogenous creatinine clearances were still within normal limits. As regards the question of the patients' ages, in the first study to which I referred there was no difference in age between the three subgroups, nor any statistically significant differences in their pre-treatment blood pressure levels. In the second study, in which the renin-aldosterone interaction was investigated, there was once again no significant difference between the mean ages of the patients in the three subgroups. In a third study conducted in Basle*, however, the high-renin essential hypertensives were indeed significantly younger – a fact which accords with your suggestion that one would expect to find more cases of Stage 1 hypertension in this high-renin group.

M. IKEDA: I'd like to add a comment on the significance of renin as a risk factor in hypertension. In view of the close causal connection existing between chronically elevated blood pressure and cardiovascular complications, one of the main aims of treatment for hypertension is, of course, to lower the pressure in an attempt to prevent morbidity and mortality due to these complications. Sometimes, however, hypertension and cardiovascular disease are of quite differing aetiology. In this context it should, I think, be pointed out that, although renin raises the blood pressure, we do not know exactly to what extent it also contributes towards the maintenance of elevated blood pressure levels in essential hypertension, nor do we know much about the role of renin as a risk factor in connection with the development of cardiovascular complications. I do not think that high plasma renin activity promotes the development of acute lesions of the small arteries, which take the form of fibrinoid necrosis and which, according to our own experimental findings**, constitute the type of lesion fundamentally responsible for cerebral haemorrhages in cases of hypertension. But renin might, on the other hand, be one of the risk factors contributing towards cardiovascular complications by accelerating the development of atheromatous lesions, although no reliable data are available to confirm this.

Since beta-blockers are known to reduce the level of plasma renin activity, their possible value in helping to prevent cardiovascular complications would become clearer if only we could determine the precise significance of renin as a risk factor in this connection. In the light of what Dr. BÜHLER has reported in his paper, I feel that it would, for example, be particularly valuable to undertake further studies in the high plasma renin group of essential hypertensives. Here, I would also like to emphasise the importance attaching to the time at which plasma renin activity is measured following administration of a beta-blocker. To illustrate this point, Table 1 shows the fluctuations in plasma renin activity in five normal subjects recorded at various times after administration of 40 mg. oxprenolol (®Trasicor). A statistically significant decrease occurred after one hour, the plasma renin levels regaining their pre-treatment values after 4–6 hours. The mean decrease in plasma renin activity roughly paralleled the accompanying reduction in heart rate and cardiac output.

---

* BÜHLER, F. R., PATEL, U., MARBET, G.: Ambulante und stationäre Bestimmung des Renin-Natrium-Index zur Unterteilung und gezielten Behandlung der essentiellen Hypertonie. Schweiz. med. Wschr. *104*, 1802 (1974)
** IKEDA, M., FUJII, J., SEKI, A.: Pathogenesis of cerebral hemorrhages in experimental hypertension in rabbits with particular reference to acute vascular lesions (fibrinoid necrosis) of small arteries and arterioles. Jap. Circulat. J. (En.) *37*, 1293 (1973)

Table 1. Plasma renin activity measured in five healthy subjects, aged 19 to 24 years, before and after a single oral dose of 40 mg. oxprenolol.

Subject	Plasma renin activity (ng./ml./hr)				
	Before	After 1 hour	After 2 hours	After 4 hours	After 6 hours
A	0.47	0.20	0.33	0.47	0.73
B	0.87	0.33	0.40	1.07	0.93
C	0.80	0.33	0.40	0.40	0.47
D	1.10	0.53	0.60	0.80	0.80
E	0.60	0.53	0.73	0.93	1.20
Mean±S.D.	0.77 ± 0.244	0.38 ± 0.144*	0.49 ± 0.167	0.73 ± 0.290	0.83 ± 0.268

*P <0.05 as compared with pre-treatment values

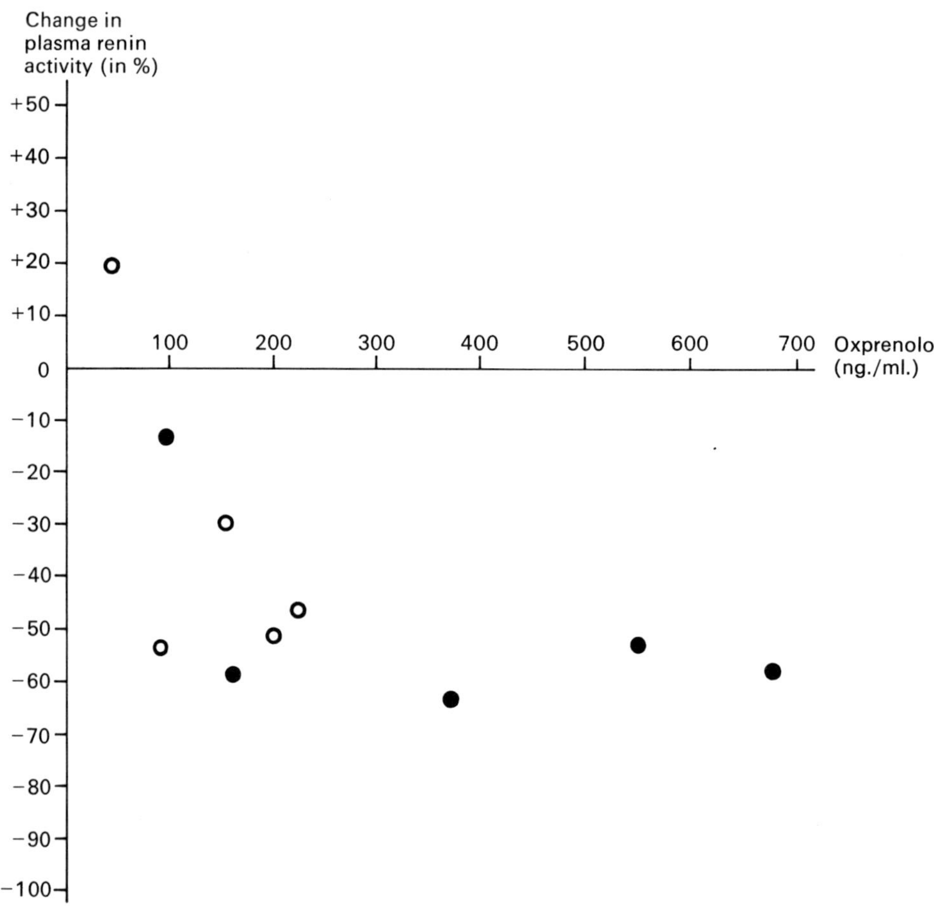

Fig. 1. Changes in plasma renin activity recorded in five healthy subjects, aged 19 to 24 years, in relation to the blood concentrations of oxprenolol measured one (●) and two (O) hours after a single oral dose of 40 mg.

Indicated in Figure 1 is the relationship between the blood level of oxprenolol and the decrease in plasma renin activity. A blood concentration of oxprenolol exceeding 100 ng./ml. is regarded as sufficient to produce a large measure of beta-blockade. The closed circles show the plasma renin levels after one hour, and the open circles the levels after two hours. In none of the five subjects was there any obvious correlation between the percentage decrease in plasma renin activity and the blood concentration of oxprenolol.

A paper by KURAMOTO et al.* referring to this study will be appearing later this year in the journal Cardiology.

J. MÉNARD: May I ask Dr. BÜHLER whether his three groups of patients differed in respect of their heart rates – either at rest or in the upright position?

F. R. BÜHLER: In answer to your question, there was no difference in the control heart rate, nor in the heart rate after beta-blockade.

---

* KURAMOTO, K., KURIHARA, H., MURATA, K., KIMATA, S., MATSUSHITA, S., IKEDA, M., MURAKAMI, M.: Hemodynamic effects and variations in plasma renin activity after single oral administration of oxprenolol. Cardiology (Basle) *59*, 41 (1974)

The effect of beta-blockers and exercise on haemodynamics and plasma renin concentration in healthy subjects and hypertensives

by G. Hitzenberger, W. Waldhäusl, J. Bonelli, A. Korn, and D. Magometschnigg*

Introduction

It has been known for some time that the sympathetic nervous system plays a special role in the regulation of the renin-angiotensin system and that it has a major bearing on the control of blood pressure levels[1, 4]. In this connection, the beta-receptors, along with cyclic adenosine monophosphate, fulfil an important function as transmitters[8-10, 12].

Dysregulation of the renin-angiotensin system has frequently been reported to occur in some forms of essential hypertension[2, 5-7]. Hitherto, however, the behaviour of this system has been studied almost exclusively under resting conditions, the literature containing in fact hardly any reports on investigations conducted during physical exercise.

We therefore studied the behaviour of the renin-angiotensin system, both at rest and during physical exercise, in healthy volunteers and in hypertensives. The tests on the healthy subjects, which were designed to provide a basis for comparison, have now been more or less completed, but it has not so far been possible to perform more than a few random tests in hypertensives. Consequently, the results obtained in these patients can only be regarded as provisional, and they in no way permit valid conclusions to be drawn at this stage.

Material and methods

A. Healthy subjects

Six male volunteers aged from 19 to 27 years were hospitalised for five days, during which they received a standardised diet containing 120 mEq. salt daily. On the first two days of this period they trained on the bicycle ergometer so that their submaximal work capacity could be assessed (heart rate over 170 beats/min., arterial pH not less than 7.25). On the control day (the third day in hospital) their urine was collected in two-hourly portions under resting conditions in order to determine the urinary catecholamine concentrations[3]. Urine was collected again in the same way on the following day, on which the volunteers exercised on the ergometer (fourth day in hospital).

On this fourth day cardiac output, total peripheral resistance, and pulmonary artery pressure were determined at rest and in the maximal steady state during

---

* Abteilung für klinische Pharmakologie, I. Medizinische Universitätsklinik, Vienna, Austria.

exercise (from the third to the sixth minute of exercise on the ergometer). The determinations, carried out in the lying position, were performed using a Swan-Ganz thermodilution catheter and recording blood pressure by direct intra-arterial measurement. The time course of the plasma renin concentration in peripheral venous blood was also measured, employing the method described by WALDHÄUSL and LEWANDOWSKI[11].

This experimental procedure was repeated under the same conditions after the volunteers had been treated on an ambulant basis with 40 mg. propranolol four times daily for 14 days.

B. Hypertensives
Identical studies – i.e. one control study and one study undertaken after 14 days' treatment with 40 mg. propranolol four times daily – have so far been carried out in three hypertensives (two men with labile hypertension and one woman with fixed essential hypertension), using the same experimental procedure.

Results

A. Healthy subjects

a) Changes in haemodynamic and biochemical variables at rest
As shown in Figures 1 and 2, in response to propranolol cardiac output fell to a significant extent – i.e. from 2.65 ± 0.43 to 2.50 ± 0.45 litres $(P < 0.05)$ – owing to the reduction in heart rate from 71 ± 11.37 to 63 ± 11.36 beats/min. $(P < 0.05)$. Stroke volume tended to increase (from 37.16 ± 3.31 to 39.56 ± 3.72 ml.), but this difference was not significant. Mean arterial pressure, however, showed a significant decrease from 100 ± 6.96 to 84 ± 5.6 mm. Hg, not only as a result of the reduction in cardiac output, but also because of a statistically significant diminution in peripheral vascular resistance from $3,081 \pm 363$ to $2,683 \pm 365$ dynes sec. cm.$^{-5}$/m.^{2} $(P < 0.005)$. In Figure 3 these changes are shown as percentage deviations from the control values.

Although the resting plasma renin concentration tended to decrease following treatment with propranolol, the difference (0.69 ± 0.46 Goldblatt units $\times 10^{-4}$/ml. after treatment, as compared with 0.79 ± 0.53 Goldblatt units $\times 10^{-4}$/ml. before) was not significant (Figure 5). One reason for this absence of a significant decrease may have been that, in view of the relatively large scatter of the values in this low range, the number of tests performed was too small. Urinary catecholamine excretion likewise remained unchanged after treatment with propranolol (Figure 5).

b) Changes in haemodynamic and biochemical variables upon physical exercise
As Figure 1 shows, the increase in cardiac output following exercise on the bicycle ergometer was less marked after treatment with propranolol than before, owing to the pronounced fall in heart rate from 186 ± 8.5 to 149 ± 29

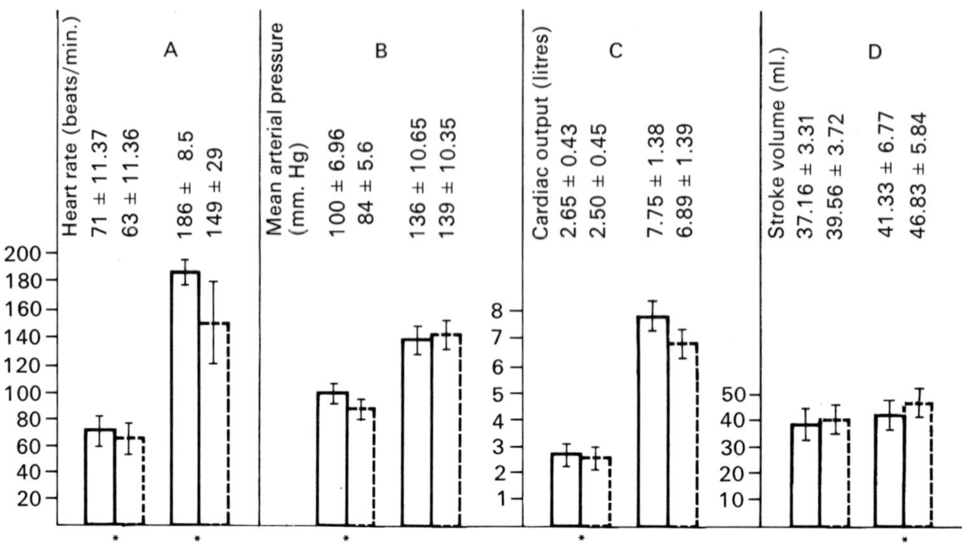

Fig. 1. Mean values ± S.E. for heart rate (A), mean arterial pressure (B), cardiac output (C), and stroke volume (D) in six healthy volunteers before (———) and after (– – –) treatment with propranolol at rest (first pair of columns in each case) and during exercise (second pair of columns).

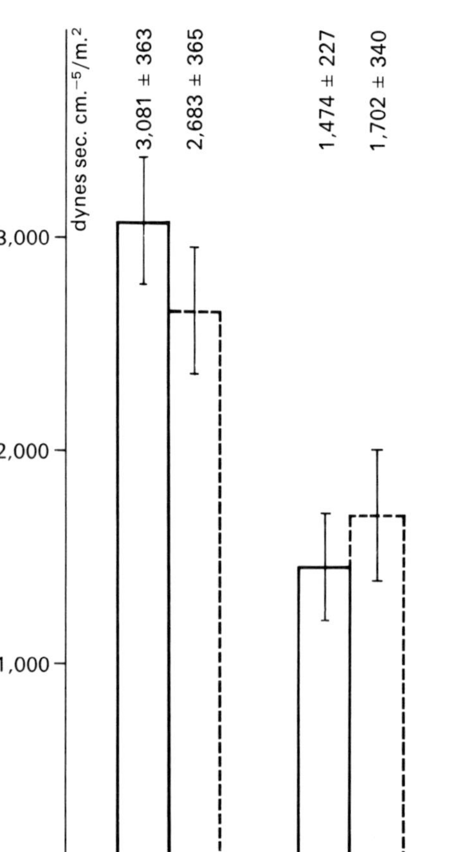

Fig. 2. Mean values ± S.E. for peripheral vascular resistance in six healthy volunteers before (———) and after (– – –) treatment with propranolol at rest (first pair of columns) and during exercise (second pair of columns).

* Difference significant

beats/min. (P<0.05). Although the absolute values recorded for cardiac output did not differ significantly (7.75±1.38 litres, as against 6.89±1.39 litres), the difference in the extent of the increase (0.71±0.88 litre) is statistically significant (P<0.05) (cf. also Figure 4). The mean arterial pressure attained

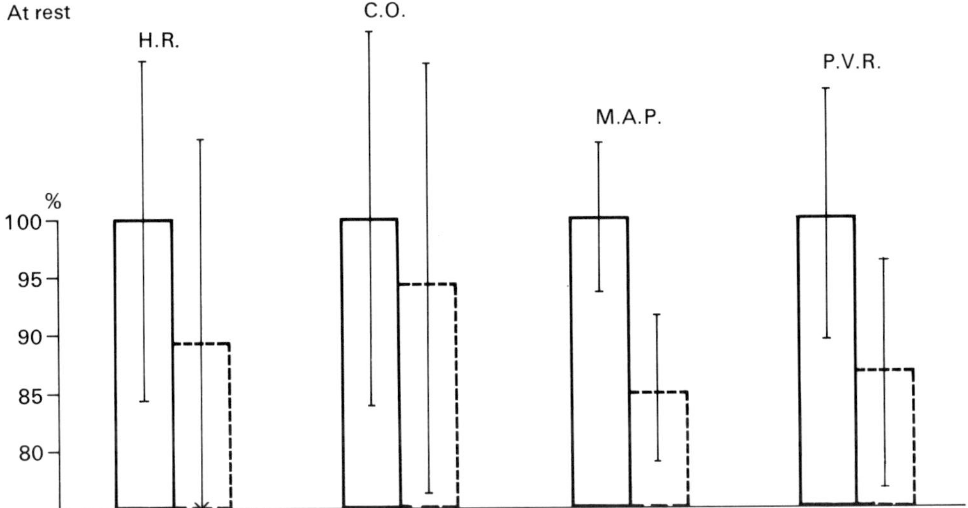

Fig. 3. Mean values for heart rate (H.R.), cardiac output (C.O.), mean arterial pressure (M.A.P.), and peripheral vascular resistance (P.V.R.) in six healthy volunteers at rest after two weeks of oral treatment with 40 mg. propranolol q.i.d. (–––), expressed as percentages of control values (——).

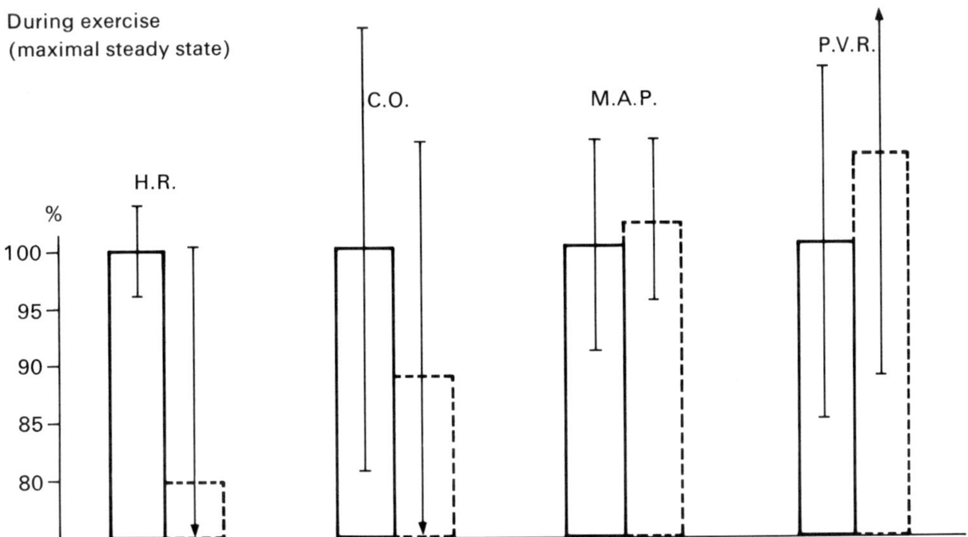

Fig. 4. Mean values for heart rate (H.R.), cardiac output (C.O.), mean arterial pressure (M.A.P.), and peripheral vascular resistance (P.V.R.) in six healthy volunteers during exercise after two weeks of oral treatment with 40 mg. propranolol q.i.d. (–––), expressed as percentages of the control values (——).

the same level following propranolol as in the control study (136 ± 10.65 mm. Hg, as compared with 139 ± 10.35, $P > 0.05$), with the result that peripheral vascular resistance during exercise following beta-blockade was *higher* than during exercise without beta-blockade (Figure 2). Although this difference ($1,702 \pm 340$, as against $1,474 \pm 227$ dynes sec. cm.$^{-5}$/m.^{2} was not in itself significant ($P > 0.05$), the difference between the exercise value and the control value – i.e. 626 ± 386 dynes sec. cm.$^{-5}$/m.^{2} (upon exercise the peripheral vascular resistance fell by 1,607 dynes sec. cm.$^{-5}$/m.^{2} without propranolol, but by only 981 following propranolol) – was in fact significant ($P < 0.05$) (cf. also Figure 4). During physical exercise the stroke volume was increased to a significantly greater extent following propranolol than without beta-blockade (46.83 ± 5.84, as against 41.33 ± 6.77 ml., $P < 0.05$) (Figure 1); at the same time, the rise in pulmonary artery pressure was significantly more marked during exercise following beta-blockade than during exercise without pro-pranolol (difference 3.5 mm. Hg, $P < 0.05$).

Also of interest is the behaviour of the plasma renin concentration (Figure 6): in the control study a highly significant increase, amounting to two to three times the initial value, was found immediately after exercise on the ergometer. The peak value, however, was not attained until some time after the blood pressure had increased to its maximal level. This rise in the plasma renin concentration was completely suppressed following treatment with propran-olol.

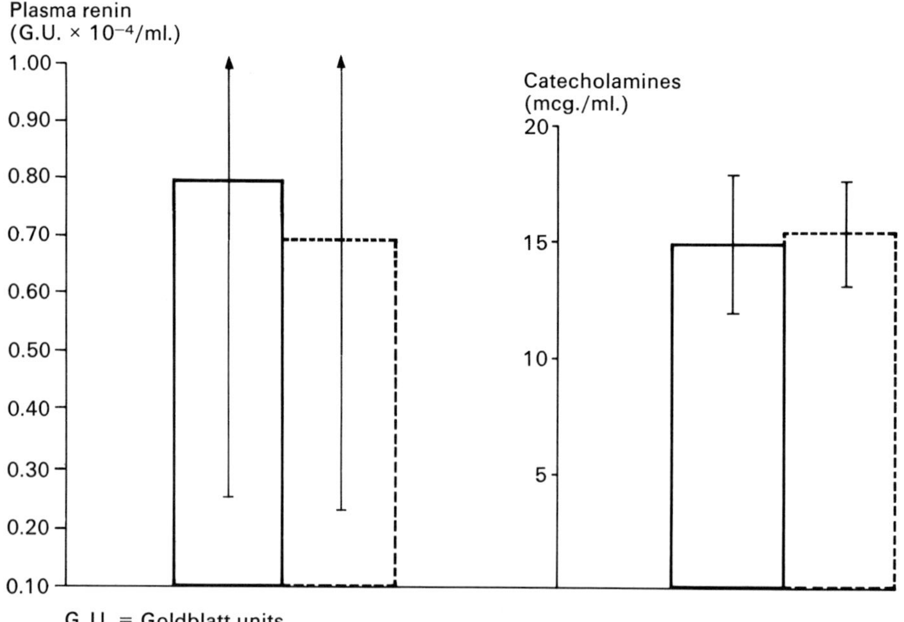

Fig. 5. Mean plasma renin concentration and urinary catecholamine concentration (0–5 hr) in six healthy volunteers at rest before (——) and after (– – –) two weeks of oral treatment with 40 mg. propranolol q.i.d.

The increase in urinary catecholamine excretion provoked by exercise was not significantly affected by propranolol (Figure 7).

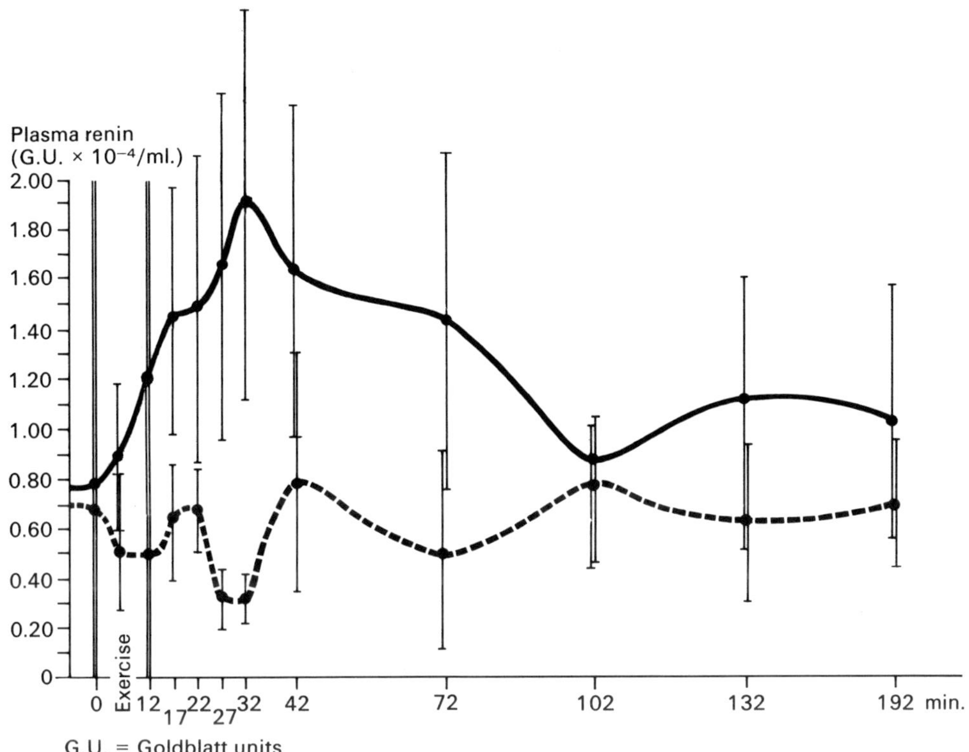

Fig. 6. Time course of mean values for plasma renin concentration in six healthy volunteers at rest, as well as during and after exercise, in a control study (——) and after two weeks of oral treatment with 40 mg. propranolol q.i.d. (– – –).

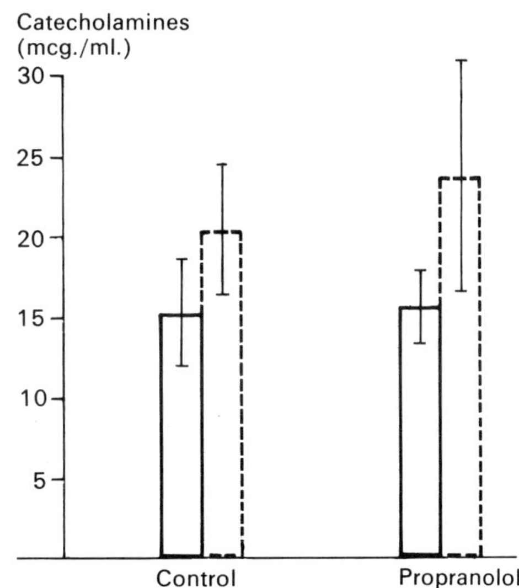

Fig. 7. Mean urinary catecholamine excretion (0–5 hr) in six healthy volunteers at rest (——) and after exercise on the bicycle ergometer (– – –) before (control) and after two weeks of oral treatment with 40 mg. propranolol q.i.d.

B. Hypertensives

The initial results obtained in the three hypertensive patients are shown in Table 1. These results are largely similar to those recorded in the healthy volunteers: whereas propranolol usually reduced cardiac output to a slight extent under resting conditions and to a somewhat larger extent during exercise, it did not prevent mean arterial pressure from increasing during exercise. The reason why it failed to prevent this increase is that peripheral vascular resistance during exercise, as compared with resting conditions, showed less of a decrease after propranolol. In these patients, as in the healthy volunteers, stroke volume increased following propranolol, both at rest (except in the case of E.K., in whom it was only increased after exercise) and during exercise, while the heart rate was reduced.

Discussion

Among the effects of propranolol observed by us in healthy subjects we consider the following to be worth noting:
1. Renin activity in peripheral blood was only slightly reduced under resting conditions. Although catecholamine production did not undergo any change at rest, the blood pressure decreased to a significant extent – not only via a diminution in cardiac output which, though significant, amounted to only about 12%, but also via a reduction in peripheral vascular resistance. Consequently, a clear-cut correlation between the effect of propranolol on blood pressure and its effect on renin seems to be lacking.

Table 1. Values obtained for heart rate (H.R.), cardiac output (C.O.), mean arterial pressure (M.A.P.), stroke volume (S.V.), and peripheral vascular resistance (P.V.R.) in three hypertensives at rest and during exercise before (upper rows of figures) and after (lower rows of figures in italics) two weeks of oral treatment with 40 mg. propranolol q.i.d.

		H.R.	C.O.	M.A.P.	S.V.	P.V.R.
E. K.	Rest	72	3.02	111	42	2,900
(labile hypertension)		*66*	*2.6*	*98*	*40*	*3,000*
	Exercise	170	5.05	112	29	1,900
		150	*5.1*	*131*	*34*	*2,000*
Z. F.	Rest	72	2.14	97.5	30	3,600
(labile hypertension)		*63*	*2.55*	*97*	*40*	*3,000*
	Exercise	170	6.3	152	37	1,950
		150	*6.2*	*162*	*41*	*2,100*
P. A.	Rest	69	2.85	131	41	3,600
(fixed essential		*60*	*2.80*	*120*	*46*	*3,400*
hypertension)						
	Exercise	159	7.1	166	43	1,819
		120	*6.6*	*190*	*55*	*2,300*

2. Activation of the sympathetic nervous system by physical exercise leads to the release of renin into the peripheral blood and to an increase in urinary catecholamine excretion. This release of renin was completely suppressed by propranolol. Nevertheless, the exercise-induced rise in blood pressure was more pronounced following beta-blockade with propranolol than in the control study. In response to blockade of the beta-receptors in the peripheral vessels, catecholamine-induced vasodilatation during exercise is inhibited and peripheral vascular resistance is higher than in subjects who have not received propranolol. Here, at all events, there is absolutely no correlation either between renin activity and blood pressure level or between renin activity and reactivity of the peripheral vascular system. In addition, one would have to consider the possibility that the apparently highly beneficial effect exerted by two weeks' propranolol treatment on blood pressure and peripheral vascular resistance under resting conditions might be cancelled out, or even turned into an undesirable effect, by exercise-induced activation of the sympathetic nervous system. This seems, in fact, to be the case in healthy subjects. Supposing it were to be found that similar conditions obtain in labile hypertensives (and our preliminary studies provide some justification for this supposition), then the question would arise as to whether beta-blockers – or, at least, those of the propranolol type administered in the dosage we used – can in fact continue to be recommended in this indication, i. e. in labile hypertension.

3. Although the increase in heart rate during exercise was greatly reduced by propranolol, the absolute values for cardiac output during exercise following beta-blockade were not significantly different from those recorded prior to treatment with propranolol. This absence of a significant change in cardiac output is no doubt to be accounted for by the appreciable increase in stroke volume observed during exercise after treatment with propranolol. If one bears in mind the rise in pulmonary artery pressure during exercise following beta-blockade, one is tempted to conclude that this increase in stroke volume might be due to activation of the heart's Frank-Starling mechanism.

Acknowledgment

This work was supported by the Austrian Fund for the Advancement of Scientific Research (Grants 2143 and 1783).

References

1 ASSAYKEEN, T.A., GANONG, W.F.: The sympathetic nervous system and renin secretion. In Martini, L., Ganong, W.F. (Editors): Frontiers in neuroendocrinology, 1971, p.67 (Oxford Univ. Press, New York/London/Toronto 1971)
2 DUSTAN, H.P., TARAZI, R.C., FROHLICH, E.D.: Functional correlates of plasma renin activity in hypertensive patients. Circulation *41*, 555 (1970)
3 EULER, U.S. VON, LISHAJKO, F.: Improved technique for the fluorimetric estimation of catecholamines. Acta physiol. scand. *51*, 348 (1961)

4 FOLKOW, B.: Role of the nervous system in the control of vascular tone. Circulation *21*, 760 (1960)

5 HAMET, P., KUCHEL, O., CUCHE, J.L., BOUCHER, R., GENEST, J.: Effect of propranolol on cyclic AMP excretion and plasma renin activity in labile essential hypertension. Canad. med. Ass. J. *109*, 1099 (1973)

6 LARAGH, J.H.: Vasoconstriction-volume analysis for understanding and treating hypertension: the use of renin and aldosterone profiles. Amer. J. Med. *55*, 261 (1973)

7 LARAGH, J.H., BAER, L., BRUNNER, H.R., BÜHLER, F.R., SEALEY, J.E., VAUGHAN, E.D., Jr.: Renin, angiotensin and aldosterone system in pathogenesis and management of hypertensive vascular disease. Amer. J. Med. *52*, 633 (1972)

8 SUTHERLAND, E.W., ROBISON, G.A.: The role of cyclic-3',5'-AMP in responses to catecholamines and other hormones. Pharmacol. Rev. *18*, 145 (1966)

9 SUTHERLAND, E.W., ROBISON, G.A., BUTCHER, R.W.: Some aspects of the biological role of adenosine 3',5'-monophosphate (cyclic AMP). Circulation *37*, 279 (1968)

10 TANIGAWA, H., ALLISON, D.J., ASSAYKEEN, T.A.: A comparison of the effects of various catecholamines on plasma renin activity alone and in the presence of adrenergic blocking agents. In Genest, J., Koiw, E. (Editors): Hypertension – 1972, p.37 (Springer, Berlin/Heidelberg/New York 1972)

11 WALDHÄUSL, W., LEWANDOWSKI, J.A.: Measurement of plasma renin concentration (PRC) using exogenous substrate and radioimmunoassay. Europ. J. clin. Invest. *3*, 1 (1973)

12 WINER, N., CHOKSHI, D.S., WALKENHORST, W.G.: Effects of cyclic AMP, sympathomimetic amines, and adrenergic receptor antagonists on renin secretion. Circulat. Res. *29*, 239 (1971)

Discussion

J. Brod: The muscular hyperaemia occurring in response to physical exercise, Dr. Hitzenberger, almost always involves – in addition to the metabolic component – an emotional component as well, which to a large extent is mediated beta-adrenergically. Since beta-blockade cuts out most of this component, I suggest that here you probably have the explanation for your findings. Would you agree with this suggestion?

G. Hitzenberger: Although this may perhaps be the explanation, I ought to point out that all the subjects on whom we performed these tests, including the hypertensives, had undergone a fairly prolonged period of preparatory training. Hence, the emotional component is not likely to have been responsible to any major extent for the differences in the results obtained before and after treatment – especially when you bear in mind that, though the medication was given on an ambulant basis, the test procedures were carried out in hospital, each person having first been subjected to the same preliminary training.

P. R. Imhof: Your observation, Dr. Hitzenberger, that the exercise-induced rise in blood pressure was more marked following treatment with propranolol than in the absence of beta-blockade does not tally with most of the findings published in the literature and relating to both acute and long-term studies conducted with a wide variety of beta-blockers. Your results may possibly be accounted for by the test procedure you employed, i. e. by the fact that you carried out your determinations in the lying position. In most of the other studies the measurements were made with the subjects sitting or in an upright position.

G. Hitzenberger: That's quite possible, Dr. Imhof. I should perhaps add that the statistically significant changes recorded when our subjects performed exercise after treatment with propranolol related, not to the absolute values, but to the differences before and after exercise. In other words, the decrease in peripheral resistance was less marked and the rise in blood pressure following exercise was more pronounced after treatment with propranolol than before; it is only these differences that were statistically significant.

G. Muiesan: I should like to present a few slides which may perhaps add something of interest to what Dr. Hitzenberger has already been telling us. Figure 1 shows that the excretion of noradrenaline in the urine of hypertensive patients can, of course, be stepped up by administering furosemide, but that, if you pre-treat the patients with oxprenolol (®Trasicor), you no longer observe such a marked increase in the urinary noradrenaline levels. From Figure 2, however, you can see that the same thing does not apply to adrenaline; consequently, noradrenaline and adrenaline should be measured, not together, but separately.
Similar findings were obtained when the arterial plasma concentrations of noradrenaline and adrenaline were measured. Figure 3 shows that with oxprenolol you abolish the increase in arterial plasma and urinary noradrenaline which would normally occur in response to the stimulus of a dose of furosemide; here, once again, however, as indicated in Figure 4, adrenaline behaves differently. As for plasma renin activity, this too is stimulated by furosemide in patients with high or normal basal renin values (Figure 5), whereas pre-treatment with oxprenolol significantly inhibits the furosemide-induced increase in plasma renin. Finally, if you measure these same parameters in patients with low basal renin values, you don't get an increase in the plasma renin values in response to furosemide, but you do observe an increase in plasma noradrenaline. May I therefore ask Dr. Hitzenberger whether or not he measured the urinary excretion of catecholamines in order to gain some idea of the index of sympathomimetic activity in the sympathetic nervous system.

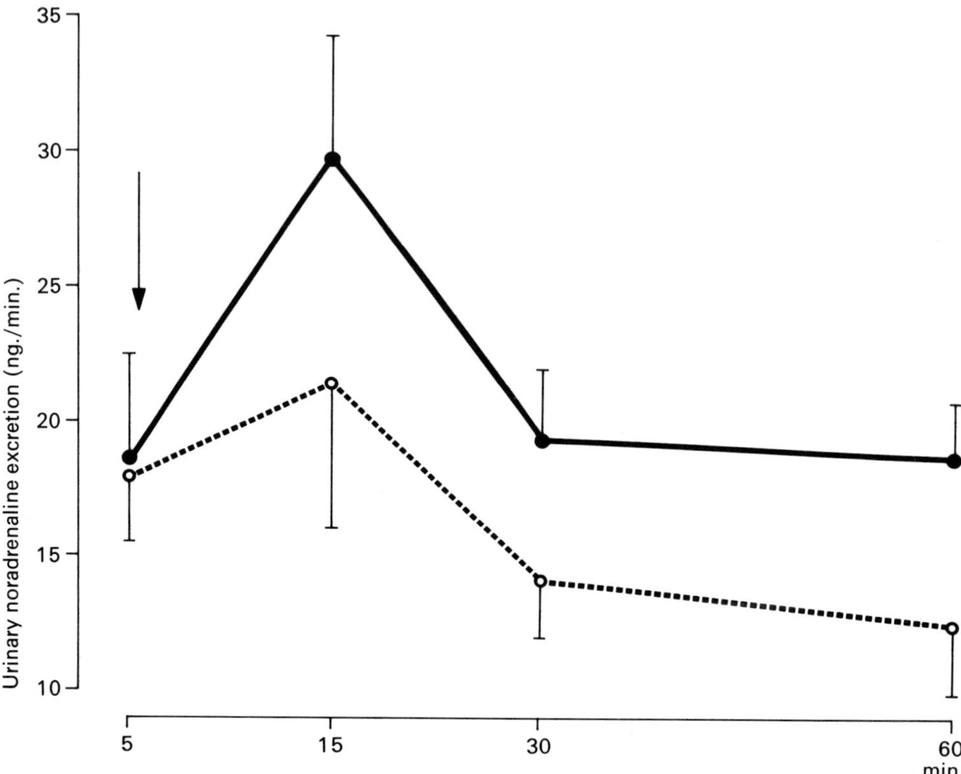

Fig. 1. Urinary excretion of noradrenaline (mean values ± S.E.) following intravenous administration of 40 mg. furosemide (↓) to hypertensive patients not pre-treated (●——●) and pre-treated (o·····o) with 10 mg. oxprenolol, likewise injected intravenously.

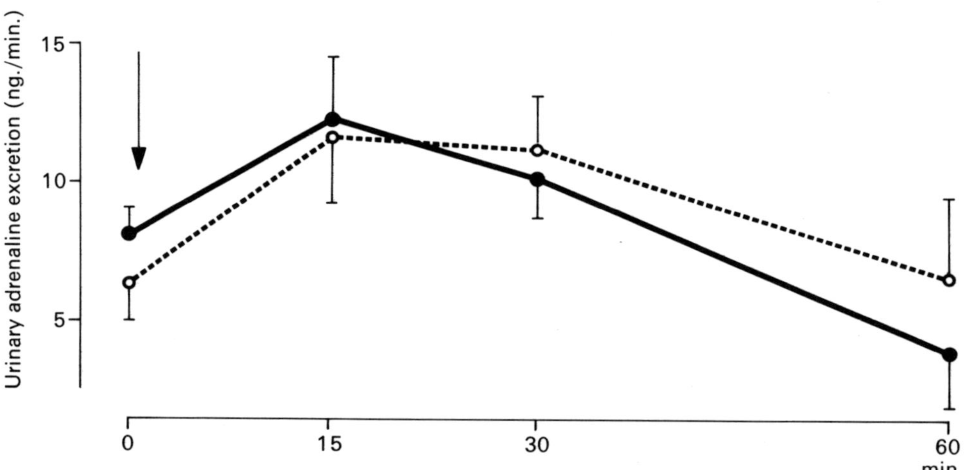

Fig. 2. Urinary excretion of adrenaline (mean values ± S.E.) following intravenous administration of 40 mg. furosemide (↓) to hypertensive patients not pre-treated (●——●) and pre-treated (o·····o) with 10 mg. oxprenolol, likewise injected intravenously.

G. Hitzenberger: Yes, we did. But we measured the catecholamine levels only in the urine, and not in the serum or plasma.

M. E. Carruthers: This question of exercise and beta-blockade seems to me to pose two problems in particular. The first is that, during the exercise, the person may cease to be able to gauge the intensity of his physical effort by reference to his pulse rate.

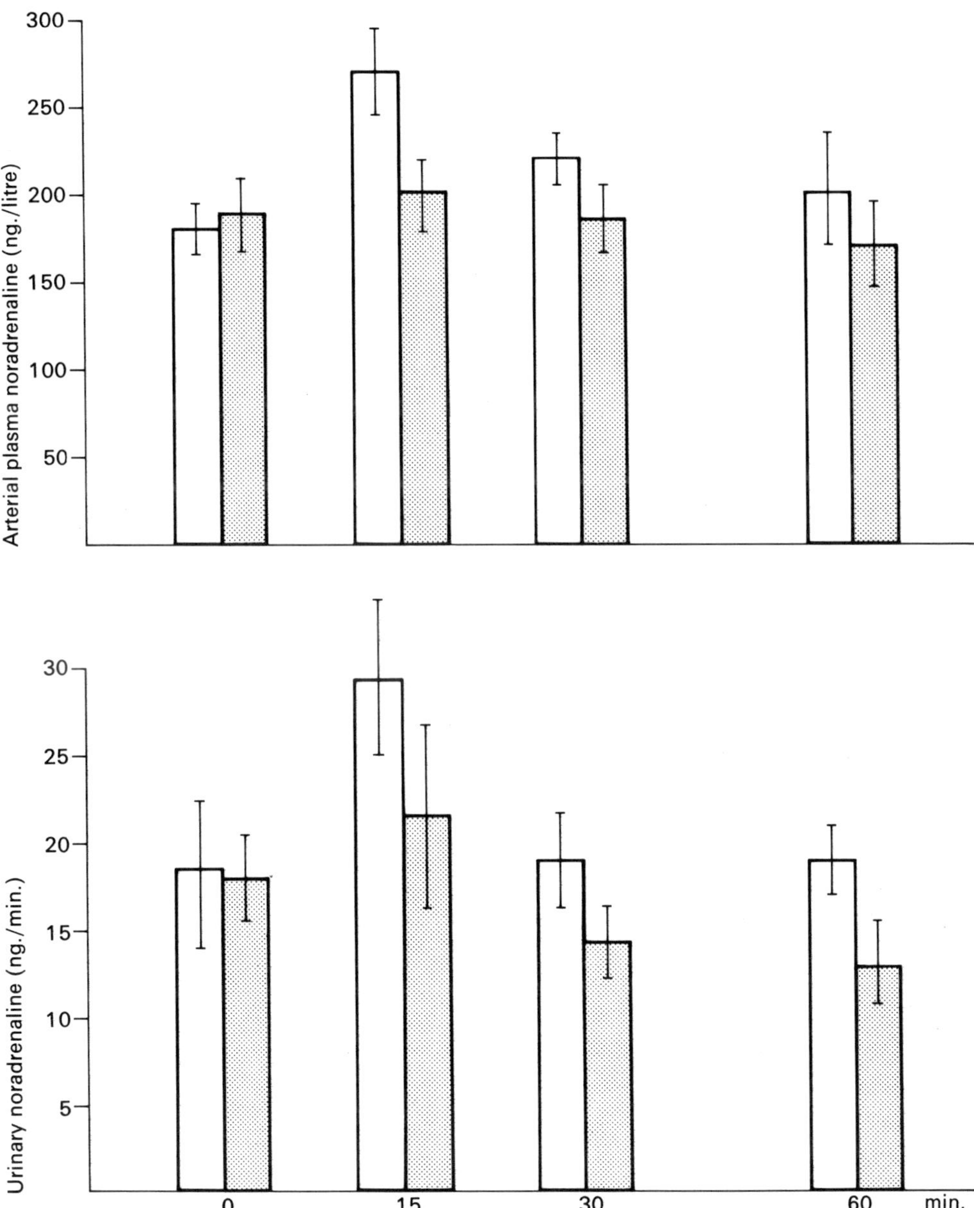

Fig. 3. Arterial plasma and urinary noradrenaline levels (mean values $\pm$ S.E.) following intravenous administration of 40 mg. furosemide to hypertensive patients not pre-treated (□) and pre-treated (▒) with 10 mg. oxprenolol, likewise injected intravenously.

Secondly, while working with a group at St. Bartholomew's Hospital in London, we found that in response to maximal exertion – which was sometimes so great as nearly to cause the subject to fall off the bicycle ergometer – very big increases could occur in the plasma catecholamine levels; for example, in the placebo group the plasma noradrenaline concentration was actually doubled. In a subject who is only relatively lightly blocked, I think that these very high catecholamine levels attained during extreme exertion might well be dangerous.

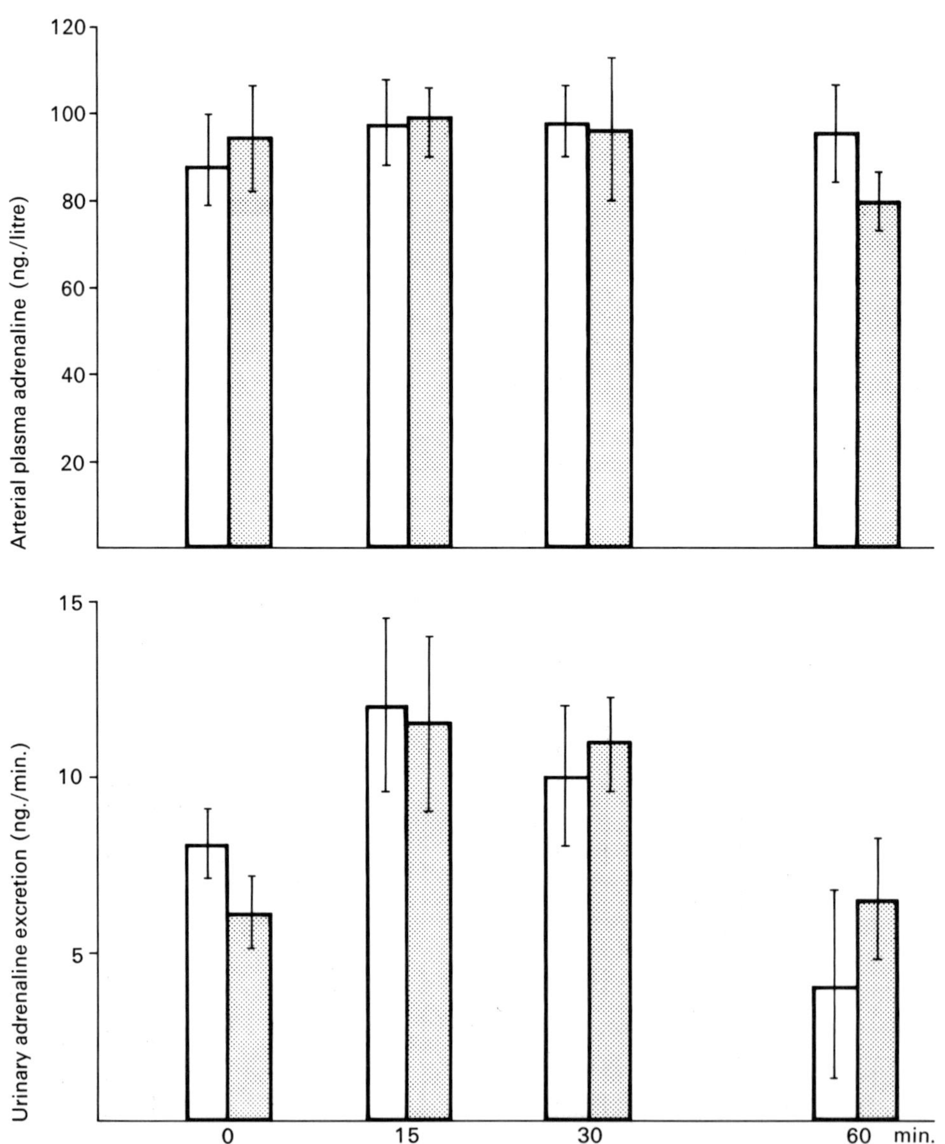

Fig. 4. Arterial plasma and urinary adrenaline levels (mean values ± S.E.) following intravenous administration of 40 mg. furosemide to hypertensive patients not pre-treated (□) and pre-treated (▨) with 10 mg. oxprenolol, likewise injected intravenously.

H. Bricaud: One question concerning your methodology, Dr. Hitzenberger – a question which may perhaps have a bearing on the interpretation of the results: were general consumption of oxygen and cardiac work the same in the exercise tests carried out after treatment with propranolol as in those performed before treatment with the beta-blocker?

G. Hitzenberger: Yes, they were.

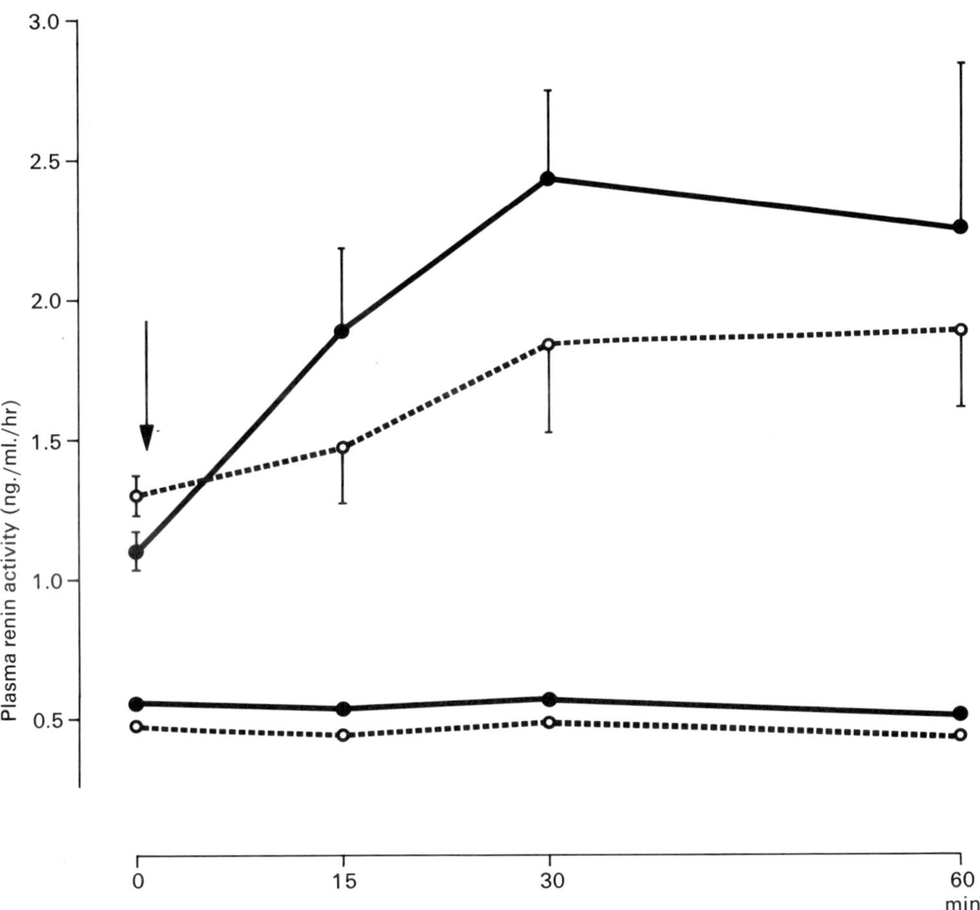

Fig. 5. Plasma renin activity (mean values ± S.E.) following intravenous administration of 40 mg. furosemide (↓) to hypertensive patients not pre-treated (●——●) and pre-treated (o------o) with 10 mg. oxprenolol, likewise injected intravenously. Note that furosemide failed to produce an increase in plasma renin activity in patients with low basal renin values.

The importance of plasma volume in the treatment of hypertension

by R. C. Tarazi, H. P. Dustan, and E. L. Bravo*

Introduction

Intravascular volume is one component in the mosaic of factors that determine arterial pressure and, as such, it may be of particular importance in the planning and follow-up of antihypertensive therapy. This importance derives from three observations: firstly, the frequency of subtle volume abnormalities in hypertension[57]; secondly, the demonstrated dependence, in many treated patients, of arterial pressure on blood volume variations[16]; and, thirdly, the identification, during the past few years, of some types of hypertension for which diuretic therapy (or a negative salt and water balance) seems to be specific[4, 17, 57].

Although significant correlations between blood volume and pressure have been described, this does not mean that the two are necessarily related in a direct or causal manner. The relationship between these two parameters is certainly not as simple or straightforward as that existing between the pressure within a static container and the volume that distends it. Neither the container nor its content are static in the intact organism: on the contrary, there are multiple factors which interfere to maintain or restore equilibrium, and, moreover, the volume measured in man is not restricted to the vascular segment in which the pressure is determined. The arterial segment of the vascular tree is thought to contain approximately 20% of the total blood volume; the capillary bed accounts for 5%, and the largest part is present on the venous side of the circulation[27]. It is not known whether these ratios are disturbed in hypertension; the total intravascular volume that we determine can therefore at best be only an index of arterial volume. Despite these limitations, however, blood volume studies have proved of great practical value in hypertension.

Whether the basic factor responsible for the various volume-pressure relationships discussed below is intravascular volume, total extracellular fluid volume[11], some connection between the two[55], or an alteration in sodium balance[65] is not known. These relationships do at all events indicate that there are certain aspects of body sodium and water which have a quantitative haemodynamic significance in hypertensive patients[18]. Furthermore, they also suggest that plasma volume is at the very least a good index of these important factors.

---

* Research Division, Cleveland Clinic Foundation and Cleveland Clinic Educational Foundation, Cleveland, Ohio, U.S.A.

Methodological considerations

The following factors have proved to be of special importance in the study of plasma and total blood volume in hypertension:

1. Volume determinations: We have not found any systematic derangement in the ratio of total body to large vessel haematocrit among different types of hypertension (HOFFMAN and TARAZI, unpublished data); moreover, the red-cell mass is usually normal in most hypertensive patients[13]. Total blood volume can, therefore, be calculated from the simultaneously determined plasma volume (radioactive iodinated human serum albumin) and haematocrit, using the correction factors developed by CHAPLIN et al.[9,55]. Plasma volume is measured in the morning after an overnight fast and a 30–45 minute rest in the supine position; its determination from one sample obtained after a ten-minute equilibration period was found to be as accurate as multiple samplings and extrapolation back to zero time[58]. As differences in body weight markedly influence values calculated in ml./kg.[10,42,58], the results are better expressed in relation to height (ml./cm.); this minimises differences due to obesity or age[10,58]. Where the analyses include both men and women, the values should be expressed as a percent of normal, because men have significantly ($P < 0.001$) higher plasma and total blood volumes than women (18.7 and 31.0 ml./cm., respectively, for men and 15.3 and 24.0 ml./cm. for women)[36,43,57].

2. Volume-pressure relationships: To avoid inaccuracies due to transient blood pressure fluctuations, weekly averages of four daily readings taken in the hospital (or two daily readings at home) were used for the study of these relationships. In our experience, a single blood pressure recording made during a particular test may sometimes differ widely from the weekly averages and may prevent any meaningful correlation with other indices[62].

Plasma volume in the choice of therapy

Although empirical treatment often proves successful in lowering high blood pressure, a rational approach to therapy is preferable for obvious reasons[33,60]: not only is it likely to result in the use of more specific forms of medication which will achieve better pressure control, but it is also especially indicated in the more resistant cases. The main difficulty, of course, lies in defining the pressor mechanism associated with each type of hypertension. In a sense the term itself, "pressor mechanism", is a misnomer because – except in phaeochromocytoma – the physiological abnormalities associated with hypertension have not been shown to be causal[15]. Moreover, in each type of hypertension, the mechanisms altered are usually multiple. In most cases, then, it is a fallacy to think in terms of a "single" pressor mechanism, because hypertension is a multifactorial response of the body to various stimuli. Nevertheless, it may still be possible to recognise certain factors as being particularly predominant in some instances.

1. Volume-dependent hypertension: Accumulated experience from many centres has made it possible gradually to outline the different characteristics of many types of hypertension[32, 33] and to advocate treatments specifically aimed at combating the predominant factors involved in each particular type – whether humoral, haemodynamic, or volume-related. Utilisation of plasma volume for the choice of therapy is based on its wide spectrum of variations in hypertension and on the fact that, whereas some patients become normotensive with diuretics alone, others displaying similar clinical characteristics show very little response to such drugs[17]. Research on volume-pressure relationships during the past few years has led to the identification of certain types of hypertension for which diuretic treatment (or a negative salt and water balance) seems to be specific. These are primary aldosteronism[4], hypervolaemic essential hypertension[12, 57], and a subset of the hypertension associated with chronic renal-parenchymal disease[13, 14, 61, 68]. In all three, plasma volume does not show the contraction usually associated with essential or renovascular hypertension[44, 57, 58, 64, 67], and in all three adequate blood pressure control is related to the achievement of a sustained plasma and extracellular fluid reduction.

Primary aldosteronism may be associated with some hypervolaemia[49, 57], but this is not found in all patients[2, 62]. In our experience, the main feature of

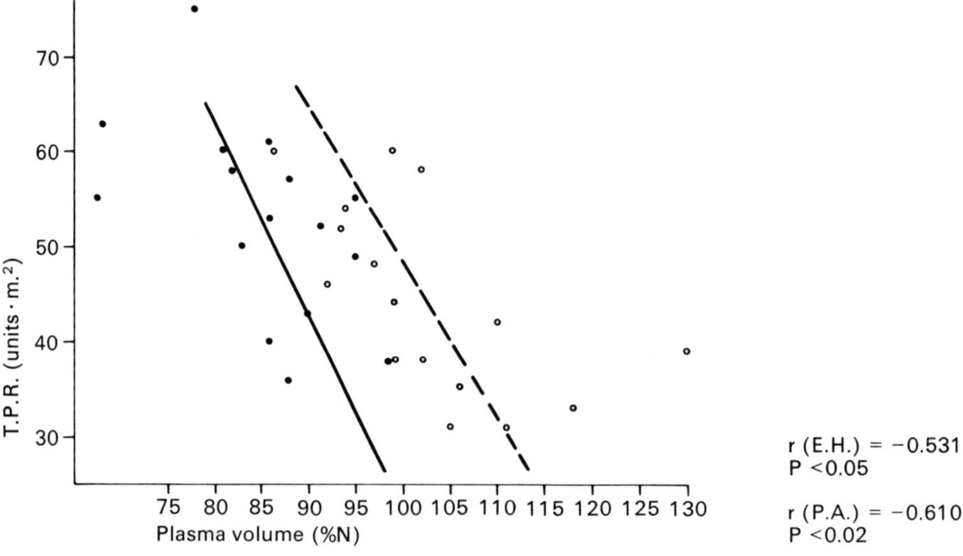

Fig. 1. Correlation between plasma volume and total peripheral resistance in 16 patients with essential hypertension (E.H., ——•) and 16 with primary aldosteronism (P.A., —— o). In both conditions the inverse relationship was significant; the slopes of the regression equations defining the two lines were not significantly different from each other, but the intercepts were. It is apparent that, for each particular level of peripheral resistance, plasma volume was greater in primary aldosteronism. T.P.R. indicates total peripheral resistance expressed in units · square metre. Plasma volume is expressed as percent of normal (%N) to allow inclusion of both men and women. (Reproduced from Tarazi et al.[62] by permission of the New England Journal of Medicine)

primary aldosteronism is a plasma volume inappropriate to the level of arterial pressure or total peripheral resistance (Figure 1). Hypervolaemia was frequently observed in patients with modest hypertension, whereas those with very high blood pressure tended to have a normal plasma volume; but there was no statistically significant correlation between arterial pressure and plasma volume[57, 62]. Despite this, good control of arterial pressure during medical treatment did prove to be dependent on the maintenance of reduced extracellular fluid volume or sodium stores[6]. We found that arterial pressure reduction correlated significantly with plasma volume contraction (Figure 2)[4], regardless of the means used to produce salt and water depletion. The therapeutic success of spironolactone is, therefore, only a reflection of its diuretic properties; low doses (100–200 mg. daily) will restore the blood potassium

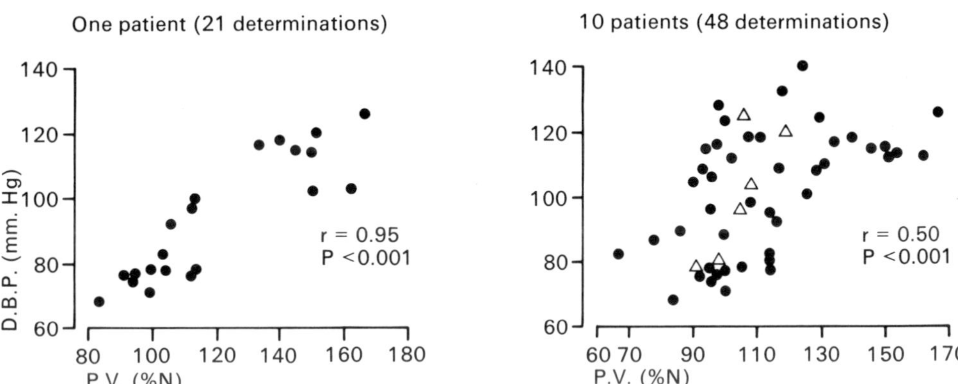

Fig. 2. Relationship between diastolic blood pressure (D.B.P.) and plasma volume (P.V., expressed as percent of normal) during long-term treatment of primary aldosteronism with spironolactone alone or in combination with either thiazide diuretics or low dietary sodium. The data were obtained over a period of six weeks to six years in patients with adrenal hyperplasia (Δ) or adrenal tumours (•). (Reproduced from Bravo et al.[4] by permission of The American Heart Association, Inc.)

Table 1. Results of surgery in relation to pre-operative haemodynamic indices in 11 patients operated upon for adrenal tumours. (Adapted from: Tarazi et al.[62])

Post-operative blood pressure*	Number of patients	Haemodynamic indices**				
		Cardiac index (litres/min./m.^{2})		M.R.L.V.E.*** (ml./sec./m.^{2})		Plasma volume (ml./cm.)
		>3.3	<3.3	>170	<170	>10%N****
Blood pressure normal without treatment	6	5	1	5	1	3
Blood pressure reduced, but treatment still needed	5	0	5	1	4	0

* Follow-up: three months to two years
** The values given represent the upper limit of normal (>10%) for this laboratory
*** M.R.L.V.E. = mean rate of left-ventricular ejection
**** %N = percent of normal value

levels to normal before bringing the blood pressure under control. The addition of conventional doses of hydrochlorothiazide will then rapidly lower the blood pressure as the plasma volume is reduced[4]. This therapeutic approach has the advantage of minimising the side effects of spironolactone.

In a group of 11 patients who were operated upon for adrenal tumours, better surgical results were obtained in those who before surgery had a higher plasma volume and cardiac output (Table 1)[62]. These patients differed neither in age nor in the known duration of their hypertension; kidney biopsy undertaken at the time of surgery revealed nephrosclerosis as frequently among those with excellent results as in the others. However, a higher plasma volume implies a lower peripheral resistance, and, hence, presumably less vascular disease. Persistence of hypertension in some patients might also be due to coincidental essential hypertension. Whatever the case, if continued studies uphold these results, determinations of plasma volume may help in selecting those patients most likely to be cured by removal of the tumour. The response of the arterial pressure to spironolactone was not helpful in this respect, because a response occurred in all patients (BRAVO, TARAZI, and DUSTAN, unpublished data).

"Hypervolaemic essential hypertensives" are a group of patients with essential hypertension and expanded plasma volume in the absence of any signs of cardiac or renal decompensation[57]; as is to be expected, their plasma renin activity tends to be low[19]. Our previous studies have yielded no evidence of primary aldosteronism or recognisable mineralocorticoid excess in any patients belonging to this group; not yet clarified is the relationship between these patients and the essential hypertensives with increased extracellular fluid volume described by JOSE et al.[30] or those with increased exchangeable sodium reported by WOODS et al.[73]. Hypervolaemic essential hypertension tends to be resistant to anti-adrenergic drugs, but responds dramatically to diuretics[12, 38].

In patients with end-stage kidney disease, two types of hypertension seem to exist[68, 69]. The first, which occurs only in a minority of patients, is "renin-dependent" and cannot be controlled by volume contraction[66, 68, 69]. The second is "volume-dependent" and shows a remarkable direct association between pressure and intravascular volume[14], extracellular fluid volume[3], or exchangeable sodium[37]. This is the only type present after bilateral nephrectomy (renoprival hypertension) and is also the more frequent in uraemic hypertensives prior to removal of the kidneys. Here, it is usually associated with a variable degree of activation of the renal pressor system, since the blood pressure is higher, at the same level of volume expansion, before than after bilateral nephrectomy[14]. Differentiation of the two types, and hence the choice of therapy, depends on a study of the volume-pressure relationships and on the determination of plasma renin activity under controlled conditions[50, 68, 69].

Not so widely recognised is the fact that patients with renal-parenchymal disease who are not yet severely azotaemic often have a salt and water dependent hypertension. The importance of extracellular fluid volume, or some function of it, in the maintenance of this hypertension is revealed by the direct correlation between arterial pressure and blood volume which can be demonstrated even in untreated patients[23, 57]. Here, the effectiveness of pressure con-

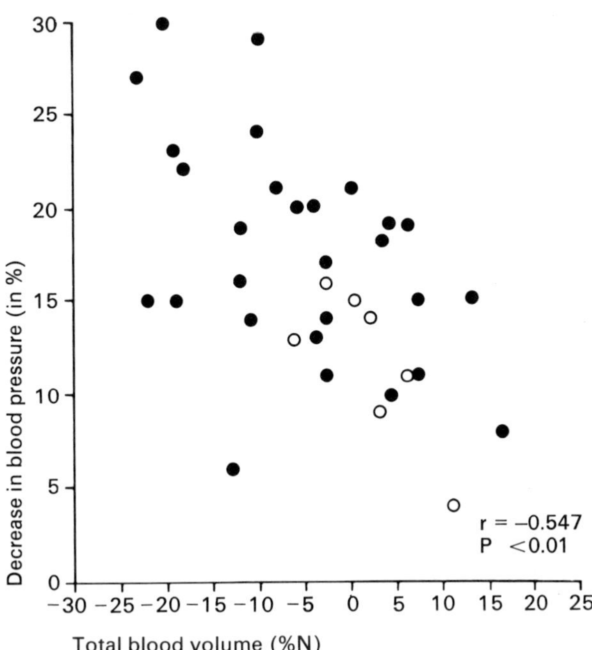

Fig. 3. Depressor response to trimetaphan (0.045 mg./kg. body weight) related to total blood volume expressed as per-cent of normal; the data from the seven normotensive sub-jects (o) fitted well with the data from the 28 essential hy-pertensives (●). The correla-tion for the 28 hypertensives (r = −0.465, P < 0.02) was im-proved by addition of the normotensives (r = −0.547, P < 0.01). (Reproduced from Tarazi and Dustan[54] by per-mission of Clinical Science)

trol depends on the choice of a diuretic capable of maintaining adequate volume control[17].

It is certainly *not* claimed that these three forms of hypertension are the only types responsive to diuretic treatment; on the contrary, they account for but a small fraction of the hypertensive patients responding to salt and water depletion. What they do represent, however, are typical instances in which determination of plasma volume can lead to a specific diagnosis and to a specific choice of treatment.

2. *Low-volume hypertension:* Of theoretical, but unproven, significance is the possibility that patients with a low blood volume may be more sensitive to neural-blocking agents – a possibility suggested by the inverse relationship found between intravascular volume and the magnitude of the immediate depressor response to trimetaphan (Figure 3)[54]. This relationship may also help to explain the resistance of hypervolaemic hypertensives to neural-blockers. The effect of prolonged therapy with neural-blocking agents, however, will depend on many other factors as well, one of the more important being the influence which such treatment exerts on secondary alterations in intravascular volume and on the development of a new set of volume-pressure relationships. These are discussed below.

Plasma volume in the follow-up of treatment

1. A recent study[16] has shown that in patients treated with *sympathetic-inhibiting drugs* ("beta-blockers" excepted) arterial pressure becomes directly related

to blood volume (Figure 4). This dependence of pressure on volume – which reflects an alteration in autonomic nervous control of haemodynamic functions[25, 51, 72] – is particularly significant inasmuch as drugs that suppress sympathetic vasomotor activity are capable of increasing plasma volume[29, 45, 46, 71] and thus of vitiating their own effectiveness. In experiments of brief duration, this response can occur without a change in body weight[71], and it probably results from a transfer of fluid from the interstitial to the intravascular compartment. Other studies of longer duration have shown increases in extracellular fluid volume, accompanied by weight gain, indicating fluid retention[29, 45, 46].

As volume expands, so therapeutic effectiveness is lost. This "false tolerance" is a well-recognised phenomenon. DUSTAN et al.[16] have studied it from the quantitative aspect and have noted individual variations in the effects which different drugs exert on intravascular volume. They found, for example, that neural-blockers do not produce volume expansion in all patients and that amounts of a diuretic that are usually sufficient to prevent fluid retention sometimes fail to do so. Two important corollaries emerge from these findings. Firstly, where neural-blocking agents administered as treatment for hypertension cease to elicit a response, one should not uselessly chop and change from one agent to another, but preferably make a determined attempt at achieving adequate diuresis and plasma volume contraction. The mere administration of a diuretic does not necessarily offer a guarantee that fluid retention or an inappropriate intravascular volume will be prevented or corrected; to obtain a guide as to how to adjust the treatment, one should also determine the plasma or blood volume (Table 2). Secondly, if and when the loss of effectiveness of an

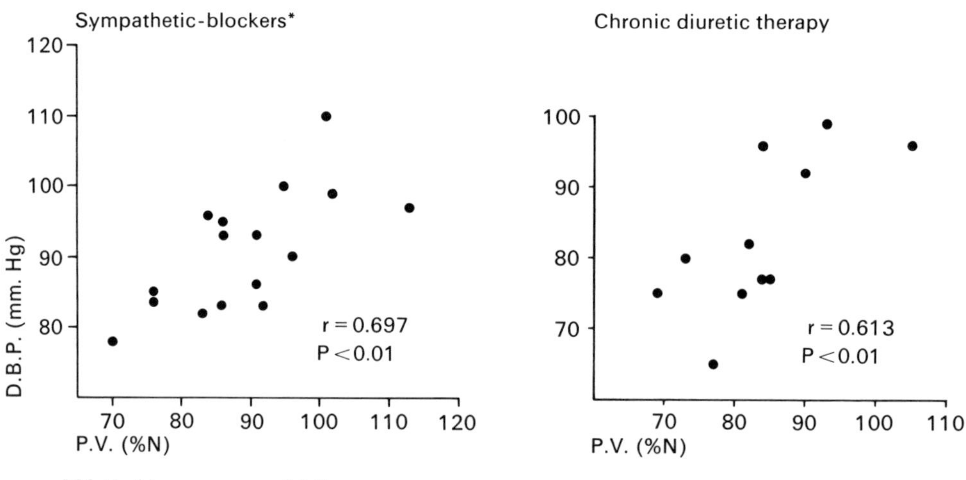

Fig. 4. A positive correlation between diastolic blood pressure (D.B.P.) and plasma volume (P.V., expressed as percent of normal) was likewise found among patients given either sympathetic-blockers or chronic diuretic therapy. (Reproduced from DUSTAN et al.[12] by permission of the American Journal of Cardiology and from DUSTAN et al.[16] by permission of the New England Journal of Medicine)

Table 2. Plasma volume and blood pressure response to sympathetic-blockers. The patients in Groups I and II were treated with sympathetic-blockers and diuretics; those in Group III received guanethidine or methyldopa alone for periods ranging from six months to six years. (Adapted from: DUSTAN et al.[16])

| | Blood pressure (mm. Hg) | | Plasma volume (%N) | Blood pressure (mm. Hg) | Plasma volume (%N) |
	Control	Treatment	Treatment	After addition of diuretics to the regimen	
Group I	212/128	167/105	109	146/93	93
Group II	169/109	126/83	85	–	–
Group III	175/111	136/85	83	–	–

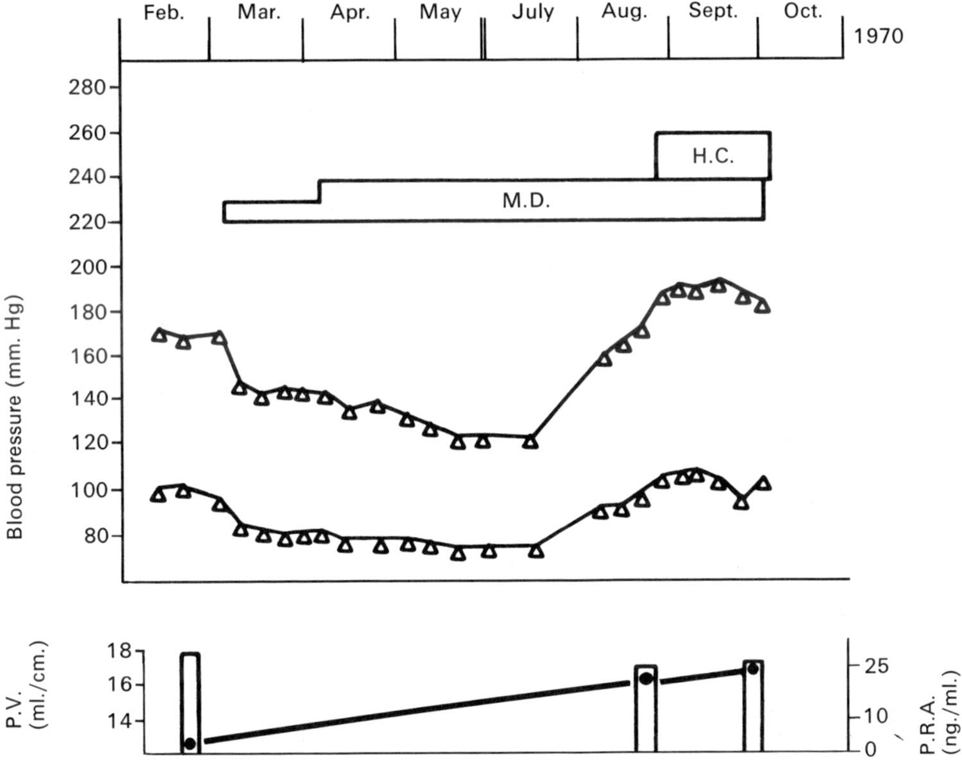

Fig. 5. Variations in blood pressure, plasma volume (P.V., □), and plasma renin activity (P.R.A., ●) coinciding with the development of renal arterial narrowing in a patient with essential hypertension. Initially, this patient had no arterial stenosis, and treatment with methyldopa (M.D.) led to a good blood pressure response; when pressure began to rise, this was not due to plasma volume expansion, but plasma renin activity was increased up to 25 ng./ml. Before these results became available, hydrochlorothiazide (H.C.) had been added, but had exerted no effect on the raised arterial pressure. A repeat renal arteriography at this point demonstrated a recently developed marked atherosclerotic narrowing of the right main renal artery.

antihypertensive agent or agents is unaccompanied by any demonstrable increase in plasma volume, interference by another pressor mechanism should be suspected. We, for instance, have encountered two patients with essential hypertension who became resistant to treatment even though contraction of plasma volume was maintained; in both cases, further investigation disclosed recently developed and markedly stenotic atherosclerotic lesions in a main renal artery (Figure 5).

2. The antihypertensive action of *diuretics* seems to be linked to the interplay of two factors throughout the duration of treatment, one of which is a mild to moderate volume contraction and the other an inadequate cardiovascular compensation for this contraction[52]. The complex alterations in plasma volume and haemodynamics that are associated with diuretic therapy are outlined in Figure 6. Although earlier studies failed to demonstrate any reduction in

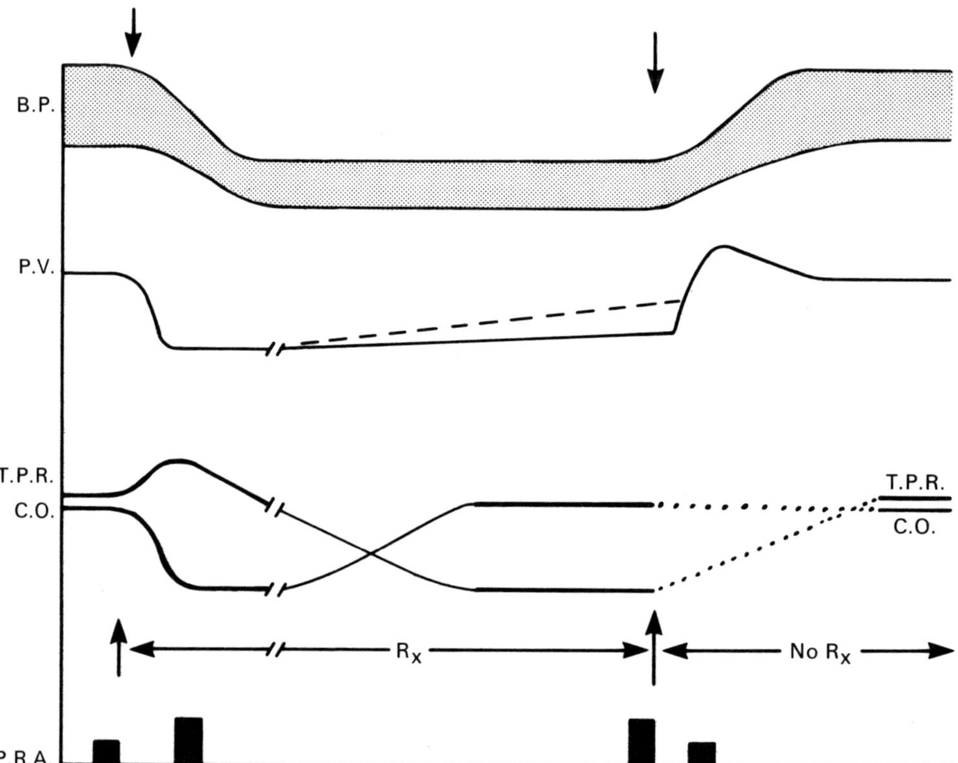

Fig. 6. Diagram showing changes in blood pressure (B.P.), plasma volume (P.V.), total peripheral resistance (T.P.R.), cardiac output (C.O.), and peripheral plasma renin activity (P.R.A.) during various stages of diuretic therapy (Rx) and in the first few weeks following its withdrawal (No Rx). The magnitude of the changes depicted is diagrammatic, not quantitative; thin and dotted lines (C.O. and T.P.R.) represent theoretical (non-documented) links between generally accepted (thicker lines) concepts. Different stages of plasma volume changes during treatment indicate varying degrees of volume restoration in different patients. (Reproduced from Tarazi[52] by permission of Grune & Stratton)

plasma volume in response to long-term treatment, recent evidence suggests that a sustained volume reduction does in fact occur during prolonged therapy[28, 34, 56] and that plasma renin activity remains elevated[56]. The magnitude of this volume reduction may vary in different patients and its initial degree might not be maintained in all cases; hence the different levels of statistical significance of the blood volume changes noted in various series.

More relevant than an arbitrary statistical significance, however, are the consistency of the plasma volume contractions reported[34] and the evident biological importance of even seemingly minor changes in volume. In our study[56], that importance was reflected in the persistent elevation of plasma renin activity during treatment with a diuretic and its prompt reduction within the first week of the drug's withdrawal, especially in view of the fact that the serum sodium concentration showed no appreciable alteration. The actual degree of fluid depletion also has an important bearing on blood pressure control: not all hypertensives respond to volume depletion, but, in those who do, the arterial pressure during diuretic treatment is found to be directly related to intravascular volume[12] (cf. Figure 4). Spironolactone showed no difference as compared with the thiazide diuretic in its effect on plasma volume and renin activity (Table 3). It therefore seems no longer possible to discount persistent fluid depletion as a significant factor contributing to the decrease in arterial pressure during chronic diuretic therapy. Whether the significant index here is the intravascular[12, 28, 34, 56] or the interstitial volume[73], is still a moot point; however, under ordinary conditions diuretics did not appear to influence the volume partition ratio of plasma to interstitial fluid[56].

On the other hand, recognition of the importance of volume reduction is not tantamount to implying that it alone is responsible for the lowering of arterial pressure by diuretics. An intact nervous system should be able to compensate

Table 3. Results of treatment for essential hypertension in 12 patients receiving hydrochlorothiazide and 14 receiving spironolactone. The duration of medication in the two groups ranged from one to 12 months (median > 4 months). All differences between the control and treatment values were significant (P < 0.001) in both groups.

Treatment (number of patients)	Arterial pressure (mm. Hg)	Plasma volume (ml./cm.)	Renin activity (ng./ml.)	Serum Na (mEq./litre)	Serum K (mEq./litre)
Hydrochlorothiazide (12)					
Before	175/106	18.2	0.7	141	4.0
During	151/94	16.1	3.2	139	3.6
S.E.D.*	2.8	0.4	1.1	0.5	0.02
Spironolactone (14)					
Before	165/109	18.0	0.8	142	3.9
During	144/97	16.2	3.7	137	4.6
S.E.D.*	1.6	0.28	0.37	0.6	0.06

*S.E.D. = standard error of difference; in the case of arterial pressure, this represents the S.E.D. for the diastolic pressure only.

for this volume loss. Logically, therefore, sodium depletion – either with thiazides, spironolactone, mercurials, low-sodium diets, or with other agents – must somehow interfere with autonomic nervous function to account for the sequence of haemodynamic events produced by these various measures. In fact, diuretics have been shown to reduce the vasoconstrictive effect of sympathetic nerve stimulation and to diminish the contractile response to noradrenaline[1, 8, 21]. Whatever the actual mechanism underlying this subtle autonomic dysfunction, its results are: firstly, a degree of venoconstriction that is insufficient to compensate for a diminished intravascular volume – hence a lowered cardiac output; and, secondly, an inadequate increase in peripheral resistance to compensate for the fall in output – hence a reduction in arterial pressure. This inadequacy of the increase in peripheral resistance is not necessarily a late manifestation[52].

Thus, just as volume depletion has been found to play a role throughout diuretic therapy (early and late), so interference with vascular adaptation to the reduced volume can also be detected right from the first days of effective diuresis. In our view, therefore, the decrease in elevated arterial pressure which diuretic agents produce is not achieved initially by one mechanism alone and then later by another, but is rather the result of an interplay between the same two mechanisms at all stages of therapy. The importance of volume contraction at all phases implies that the effectiveness of diuretic therapy must be judged in terms of the volume depletion achieved. Measurement of plasma volume may accordingly help in evaluating responsiveness, or resistance, to the agents used.

Plasma volume and beta-adrenergic blockade

Many reports have documented the effectiveness of beta-blockers in the treatment of hypertension[24, 41, 74], but relatively few studies – all of them dealing only with propranolol[31, 39, 47, 59] – have been published concerning their possible influence on plasma volume. Yet their action in that respect appears quite exceptional in comparison with common experience that practically all other antihypertensive agents, except diuretics, lead to fluid retention and plasma volume expansion.

1. Beta-blockers as monotherapy: Both acute[31] and long-term[59] administration of propranolol were found to be associated with a reduction in plasma volume. Following intravenous propranolol, plasma volume was decreased by an average of 13.4% in normotensive subjects and borderline hypertensives. With long-term oral treatment, the results were less clear-cut. TARAZI et al.[59] reported a significant reduction (more than 8%) in over half their patients; in the group of patients as a whole, plasma volume was contracted by an average of 12.5% after three to nine months of treatment, the variations ranging from +4% to –24%. However, the antihypertensive effect of propranolol was not dependent on the volume reduction, since there was no demonstrable rela-

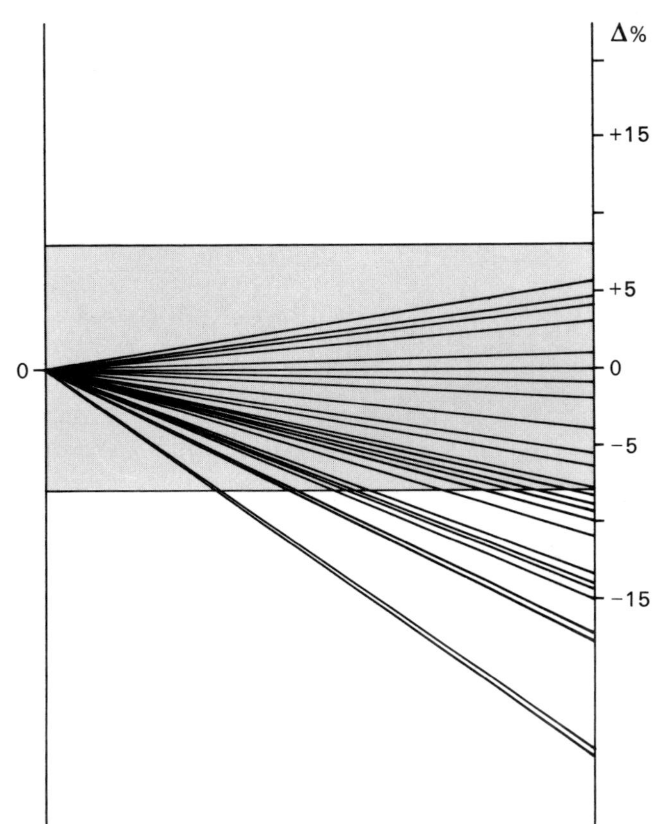

Fig. 7. Changes in plasma volume with oral propranolol therapy in 30 patients, expressed as percent of control values. The area between the two horizontal lines represents variations that are within the limits of error for the method. The group illustrated here includes data from 14 patients previously reported [59].

tionship between volume and pressure changes. This finding was also borne out by the results of acute studies, in which intravenous propranolol reduced the plasma volume but did not usually alter the arterial pressure [31, 41]. On the other hand, SEDERBERG-OLSEN and IBSEN [47] and PARVING and GYNTELBERG [39] claim to have observed no change in plasma volume in response to propranolol therapy.

Because of the apparent discrepancy between the results obtained in relatively small series, we reviewed our findings in a larger group comprising 28 hypertensives and two patients with hyper-beta-adrenergic syndrome. The alterations in plasma volume were evenly divided (Figure 7). Plasma volume was reduced by more than 8% (−8.5% to −24%) in half of the patients; the variations in the other half (range: +4.4% to −6.5%) were small and possibly within the limits of error for the method. Significant plasma volume expansion was encountered in repeated determinations in only two patients, both after more than one year of therapy. This expansion in volume was difficult to interpret with confidence, because both patients had impressive increases in body weight with no clinical, haemodynamic, or radiological signs of cardiac decompensation; both had normal renal function and no proteinuria.

The absence of plasma volume expansion in response to beta-adrenergic blockade and the fact that in a sizeable number of patients there was actually a

113

decrease in plasma volume contrast strikingly with the increase produced by other neural-blocking agents. The mechanism or mechanisms underlying this difference are not known for certain. In our experience, the changes in plasma volume elicited by propranolol did not correlate with either cardiac output or total peripheral resistance[59]; similar results have been obtained by Julius et al.[31]. We therefore suggested that the volume contraction provoked by beta-adrenergic blockade might possibly be due to unopposed alpha-adrenergic activity.

This tentative explanation has the advantage of providing one common denominator for the contraction of plasma volume produced by propranolol, by infusions of noradrenaline[22], or by sympathetic stimulation[70], as well as for its expansion by ganglionic[45], sympathetic[29, 71], or alpha-adrenergic[48] blockade. In this connection, it is interesting to note that we found no changes in plasma volume in a group of eight mild hypertensives treated for one month with a cardioselective beta-blocker (practolol)[63]. This explanation may also help to account for the variety of volume responses reported. Since propranolol apparently has no direct sympathomimetic or vascular action[35], the magnitude of its effects on volume must presumably depend on the relative functional importance of the alpha and beta-adrenergic activity present in each case. Hence, it is not surprising to find varying degrees of volume alteration in different patients or at different times during treatment. If our proposed explanation were true, one would expect propranolol to bring about a decrease in the ratio of plasma to interstitial fluid volume. Sederberg-Olsen and Ibsen[47], who determined both extracellular fluid and plasma volume, report that this ratio was indeed lower during treatment, but in the nine patients treated by them the results were not statistically significant. Apart from theoretical considerations, the volume aspects of propranolol therapy also have some special practical importance, as exemplified, for instance, by the lack of an expansion in intravascular volume – or even by an actual reduction in intravascular volume – during treatment. This may explain why false tolerance to beta-blockers has been neither encountered in our experience nor reported by others, although it is a common occurrence with neural-blocking drugs.

Of obvious relevance to the volume changes occurring during propranolol treatment are the drug's possible effects on cardiac compensation. Epstein and Braunwald[20] found altered patterns of sodium excretion in patients treated with 60–240 mg. propranolol daily; however, actual sodium retention occurred only in cases of evident cardiac failure. It is therefore all the more significant that plasma volume expansion was so unusual in our patients. The precipitation of cardiac decompensation by beta-adrenergic blockade is always a possible hazard in patients with overloaded or damaged hearts; but this complication has been most infrequent in our own experience[53] and in that of others[7, 40]. Prichard[40] observed no significant increase in body weight after several months of treatment with high dosages. The rarity of heart failure may be related to the reduction in systemic pressure – which diminishes costly cardiac work – and to the careful choice of patients without advanced heart disease.

Table 4. Results of treatment for essential hypertension in 12 patients receiving long-term diuretic medication to which propranolol was subsequently added. (Adapted from: BRAVO et al.[5])

	Plasma volume (%N)	Plasma renin activity (ng./ml.)	Blood pressure at home (mm. Hg)
Control values	102 ± 4	2.0 ± 0.5	171/109
Diuretic therapy	93 ± 5	9.5 ± 3.5	146/93
Propranolol added	94 ± 5	7.6 ± 3.7	134/84

The values listed are expressed as the mean ± S.E. Plasma renin activity was determined in samples obtained from patients while they were resting in the supine position. The blood pressure levels represent the weekly average calculated from two daily readings.

2. Beta-blockers in association with other drugs: When propranolol is associated with other antihypertensive agents, plasma volume may be influenced in a manner quite different from the changes referred to above. Thus, for example, in patients treated with a combination of propranolol and vasodilators – whether hydralazine[16] or minoxidil[26] – plasma volume increased, i. e. beta-blockade apparently did not counteract the fluid retention provoked by these drugs. The volume expansion produced by minoxidil may be quite impressive; surprisingly enough, however, in our experience this does not necessarily lead to loss of pressure control.

BRAVO et al.[5] recently reported an impressive and rapid lowering of arterial pressure in essential hypertensives when propranolol was added to chronic diuretic therapy – and this despite the fact that plasma renin activity remained elevated (Table 4). Average values for plasma volume showed no significant change, but this lack of change in mean values obscured an important difference as compared with results previously obtained by us using propranolol alone. In eight out of a total of 20 patients treated to date with combined diuretic and beta-blockade therapy, plasma volume was expanded by more than 8%; in only one was it reduced by more than 8%. It seems hardly likely that this volume expansion was due to cardiac decompensation, because the patients were after all receiving a diuretic and their blood pressure levels had been practically normalised; though certain theoretical explanations may be advanced to account for this phenomenon, no factual basis for it has yet been documented.

Acknowledgments

The authors' studies referred to in this paper were supported in part by Grants 6835 from the National Heart-Lung Institute and 70960 from the American Heart Association.

1 Aoki, V. S., Brody, M. J.: The effect of thiazide on the sympathetic nervous system of hypertensive rats. Arch. int. Pharmacodyn. *177*, 423 (1969)

2 Biglieri, E. G., Slaton, E. P., Jr., Kronfield, S. J., Deck, J. B.: Primary aldosteronism with unusual secretory patterns. J. clin. Endocr. *27*, 715 (1967)

3 Blumberg, A., Nelp, W. B., Hegstrom, R. M., Scribner, B. H.: Extracellular volume in patients with chronic renal disease treated for hypertension by sodium restriction. Lancet *ii*, 69 (1967)

4 Bravo, E. L., Dustan, H. P., Tarazi, R. C.: Spironolactone as a nonspecific treatment for primary aldosteronism. Circulation *48*, 491 (1973)

5 Bravo, E. L., Dustan, H. P., Tarazi, R. C.: β-Adrenergic blockade in essential hypertensives (EH) with chronic hyperreninemia. Circulation *48*, Suppl. IV: 72 (1973); abstract of paper

6 Brown, J. J., Davies, D. L., Ferriss, J. B., Fraser, R., Haywood, E., Lever, A. F., Robertson, J. I. S.: Comparison of surgery and prolonged spironolactone therapy in patients with hypertension, aldosterone excess, and low plasma renin. Brit. med. J. *ii*, 729 (1972)

7 Bühler, F. R., Laragh, J. H., Baer, L., Vaughan, E. D., Jr., Brunner, H. R.: Propranolol inhibition of renin secretion. A specific approach to diagnosis and treatment of renin-dependent hypertensive diseases. New Engl. J. Med. *287*, 1209 (1972)

8 Champlain, J. de, Krakoff, L., Axelrod, J.: Interrelationships of sodium intake, hypertension, and norepinephrine storage in the rat. Circulat. Res. *24/25*, Suppl. I: 75 (1969)

9 Chaplin, H., Jr., Mollison, P. L., Vetter, H.: The body/venous hematocrit ratio: its constancy over a wide hematocrit range. J. clin. Invest. *32*, 1309 (1953)

10 Chien, S., Usami, S., Simmons, R. L., McAllister, F. F., Gregersen, M. I.: Blood volume and age: repeated measurements on normal men after 17 years. J. appl. Physiol. *21*, 583 (1966)

11 Davidov, M., Gavrilovich, L., Mroczek, W., Finnerty, F. A., Jr.: Relation of extracellular fluid volume to arterial pressure during drug-induced saluresis. Circulation *40*, 349 (1969)

12 Dustan, H. P., Bravo, E. L., Tarazi, R. C.: Volume-dependent essential and steroid hypertension. Amer. J. Cardiol. *31*, 606 (1973)

13 Dustan, H. P., Frohlich, E. D., Tarazi, R. C.: Pressor mechanisms. In Frohlich, E. D. (Editor): Pathophysiology, p. 41 (Lippincott, Philadelphia/Toronto 1972)

14 Dustan, H. P., Page, I. H.: Some factors in renal and renoprival hypertension. J. Lab. clin. Med. *64*, 948 (1964)

15 Dustan, H. P., Tarazi, R. C., Bravo, E. L.: Physiologic characteristics of hypertension. Amer. J. Med. *52*, 610 (1972)

16 Dustan, H. P., Tarazi, R. C., Bravo, E. L.: Dependence of arterial pressure on intravascular volume in treated hypertensive patients. New Engl. J. Med. *286*, 861 (1972)

17 Dustan, H. P., Tarazi, R. C., Bravo, E. L.: Diuretic and diet treatment of hypertension. Arch. intern. Med. *133*, 1007 (1974)

18 Dustan, H. P., Tarazi, R. C., Bravo, E. L., Dart, R. A.: Plasma and extracellular fluid volumes in hypertension. Circulat. Res. *32/33*, Suppl. I: 73 (1973)

19 Dustan, H. P., Tarazi, R. C., Frohlich, E. D.: Functional correlates of plasma renin activity in hypertensive patients. Circulation *41*, 555 (1970)

20 Epstein, S. E., Braunwald, E.: Effects of beta adrenergic blockade on patterns of urinary sodium excretion. Studies in normal subjects and in patients with heart disease. Ann. intern. Med. *65*, 20 (1966)

21 Feisal, K. A., Eckstein, J. W., Horsley, A. W., Keasling, H. H.: Effects of chloro-

thiazide on forearm vascular responses to norepinephrine. J. appl. Physiol. *16,* 549 (1961)

22 Finnerty, F.A., Jr., Buchholz, J.A., Guillaudeu, R.L.: Blood volumes and plasma protein during levarterenol-induced hypertension. J. clin. Invest. *37,* 425 (1958)

23 Frohlich, E.D., Tarazi, R.C., Dustan, H.P.: Hemodynamic and functional mechanisms of two renal hypertensions: arterial and pyelonephritis. Amer. J. med. Sci. *261,* 189 (1971)

24 Frohlich, E.D., Tarazi, R.C., Dustan, H.P., Page, I.H.: The paradox of beta-adrenergic blockade in hypertension. Circulation *37,* 417 (1968)

25 Frye, R.L., Braunwald, E.: Studies on Starling's law of the heart. I. The circulatory response to acute hypervolemia and its modification by ganglionic blockade. J. clin. Invest. *39,* 1043 (1960)

26 Gilmore, E., Weil, J., Chidsey, C.: Treatment of essential hypertension with a new vasodilator in combination with beta-adrenergic blockade. New Engl. J. Med. *282,* 521 (1970)

27 Guyton, A.C.: Textbook of medical physiology, 4th Ed. (Saunders, Philadelphia/London/Toronto 1971)

28 Hansen, J.: Hydrochlorothiazide in the treatment of hypertension. The effects on blood volume, exchangeable sodium and blood pressure. Acta med. scand. *183,* 317 (1968)

29 Hansen, J.: Alpha-methyl-dopa (Aldomet) in the treatment of hypertension. The effects on blood volume, exchangeable sodium, body weight and blood pressure. Acta med. scand. *183,* 323 (1968)

30 Jose, A., Crout, J.R., Kaplan, N.M.: Suppressed plasma renin activity in essential hypertension, roles of plasma volume, blood pressure, and sympathetic nervous system. Ann. intern. Med. *72,* 9 (1970)

31 Julius, S., Pascual, A.V., Abbrecht, P.H., London, R.: Effect of beta-adrenergic blockade on plasma volume in human subjects. Proc. Soc. exp. Biol. (N.Y.) *140,* 982 (1972)

32 Laragh, J.H.: Biochemical profiling and the natural history of hypertensive diseases: low-renin essential hypertension, a benign condition. Circulation *44,* 971 (1971)

33 Laragh, J.H., Baer, L., Brunner, H.R., Bühler, F.R., Sealey, J.E., Vaughan, E.D., Jr.: Renin, angiotensin and aldosterone system in pathogenesis and management of hypertensive vascular disease. Amer. J. Med. *52,* 633 (1972)

34 Leth, A.: Changes in plasma and extracellular fluid volumes in patients with essential hypertension during long-term treatment with hydrochlorothiazide. Circulation *42,* 479 (1970)

35 Lucchesi, B.R., Whitsitt, L.S.: The pharmacology of beta-adrenergic blocking agents. Progr. cardiovasc. Dis. *11,* 410 (1969)

36 Moore, F.D., Olesen, K.H., McMurrey, J.D., Parker, H.V., Ball, M.R., Boyden, C.M.: The body cell mass and its supporting environment. Body composition in health and disease (Saunders, Philadelphia/London 1963)

37 Onesti, G., Swartz, C., Ramirez, O., Brest, A.N.: Bilateral nephrectomy for control of hypertension in uremia. Trans. Amer. Soc. artif. intern. Org. *14,* 361 (1968)

38 Parijs, J., Joossens, J.V., Van der Linden, L., Verstreken, G., Amery, A.K.P.C.: Moderate sodium restriction and diuretics in the treatment of hypertension. Amer. Heart J. *85,* 22 (1973)

39 Parving, H.-H., Gyntelberg, F.: Albumin transcapillary escape rate and plasma volume during long-term beta-adrenergic blockade in essential hypertension. Scand. J. clin. Lab. Invest. *32,* 105 (1973)

40 Prichard, B.N.C.: Propranolol as an antihypertensive agent. Amer. Heart J. *79,* 128 (1970)

41 Prichard, B. N. C., Gillam, P. M. S.: Treatment of hypertension with propranolol. Brit. med. J. *i*, 7 (1969)

42 Rettori, O., Mejía, R. H., Fernández, L. A.: Factors of variation in blood volume determinations in the rat. Acta physiol. lat. amer. *14*, 221 (1964)

43 Retzlaff, J. A., Tauxe, W. N., Kiely, J. M., Stroebel, C. F.: Erythrocyte volume, plasma volume, and lean body mass in adult men and women. Blood *33*, 649 (1969)

44 Rochlin, D. B., Shohl, T., Blakemore, W. S.: Blood volume changes associated with essential hypertension. Surg. Gynec. Obstet. *111*, 569 (1960)

45 Rønnov-Jessen, V.: Blood-volume and tolerance to pentolinium in the treatment of hypertension. Lancet *ii*, 669 (1960)

46 Rønnov-Jessen, V., Hansen, J.: Blood volume and exchangeable sodium during treatment of hypertension with guanethidine and hydrochlorothiazide. Acta med. scand. *186*, 255 (1969)

47 Sederberg-Olsen, P., Ibsen, H.: Plasma volume and extracellular fluid volume during long-term treatment with propranolol in essential hypertension. Clin. Sci. *43*, 165 (1972)

48 Skillman, J. J., Eltringham, W. K., Zollinger, R. M., Jr., Lauler, D. P., Moore, F. D.: Phenoxybenzamine-induced vasodilatation: a stimulus to increased plasma volume with reduced central venous pressure and aldosterone hypersecretion in man. Surgery *64*, 368 (1968)

49 Slaton, P. E., Biglieri, E. G.: Hypertension and hyperaldosteronism of renal and adrenal origin. Amer. J. Med. *38*, 324 (1965)

50 Stokes, G. S., Mani, M. K., Stewart, J. H.: Relevance of salt, water and renin to hypertension in chronic renal failure. Brit. med. J. *iii*, 126 (1970)

51 Takagi, H., Dustan, H. P., Page, I. H.: Relationships among intravascular volume, total body sodium, arterial pressure, and vasomotor tone. Circulat. Res. *9*, 1233 (1961)

52 Tarazi, R. C.: Diuretic drugs: mechanisms of antihypertensive action. In Onesti, G., Kim, K. E., Moyer, J. H. (Editors): Hypertension, mechanisms and management, XXVIth Hahnemann Symp., p. 251 (Grune & Stratton, New York/London 1973)

53 Tarazi, R. C., Dustan, H. P.: Beta adrenergic blockade in hypertension. Practical and theoretical implications of long-term hemodynamic variations. Amer. J. Cardiol. *29*, 633 (1972)

54 Tarazi, R. C., Dustan, H. P.: Neurogenic participation in essential and renovascular hypertension assessed by acute ganglionic blockade: correlation with haemodynamic indices and intravascular volume. Clin. Sci. *44*, 197 (1973)

55 Tarazi, R. C., Dustan, H. P., Frohlich, E. D.: Relation of plasma to interstitial fluid volume in essential hypertension. Circulation *40*, 357 (1969)

56 Tarazi, R. C., Dustan, H. P., Frohlich, E. D.: Long-term thiazide therapy in essential hypertension. Circulation *41*, 709 (1970)

57 Tarazi, R. C., Dustan, H. P., Frohlich, E. D., Gifford, R. W., Jr., Hoffman, G. C.: Plasma volume and chronic hypertension. Relationship to arterial pressure levels in different hypertensive diseases. Arch. intern. Med. *125*, 835 (1970)

58 Tarazi, R. C., Frohlich, E. D., Dustan, H. P.: Plasma volume in men with essential hypertension. New Engl. J. Med. *278*, 762 (1968)

59 Tarazi, R. C., Frohlich, E. D., Dustan, H. P.: Plasma volume changes with long-term beta-adrenergic blockade. Amer. Heart J. *82*, 770 (1971)

60 Tarazi, R. C., Gifford, R. W., Jr.: Drug treatment of hypertension. In Donoso, E. (Editor): Current cardiovascular topics, Vol. II: Drugs in cardiology (printing)

61 Tarazi, R. C., Hirano, J., Dustan, H. P., Masson, G. M. C.: Rôle du sodium dans l'hypertension artérielle. Path.-Biol. *16*, 547 (1968)

62 Tarazi, R. C., Ibrahim, M. M., Bravo, E. L., Dustan, H. P.: Hemodynamic characteristics of primary aldosteronism. New Engl. J. Med. *289*, 1330 (1973)

63 Tarazi, R.C., Savard, Y., Dustan, H.P., Bravo, E.L.: Cardioselective beta adrenergic blockade in hypertension. Clin. Pharmacol.Ther. *13*, 154 (1972); abstract of paper

64 Tibblin, G., Bergentz, S.E., Bjure, J., Wilhelmsen, L.: Hematocrit, plasma protein, plasma volume, and viscosity in early hypertensive disease. Amer. Heart J. *72*, 165 (1966)

65 Tobian, L., Jr.: A viewpoint concerning the enigma of hypertension. Amer.J. Med. *52*, 595 (1972)

66 Toussaint, C., Verniory, A., Vereerstraeten, P., Kinnaert, P., Buchin, R., Geertruyden, J. van: Indications de la néphrectomie bilatérale dans l'hypertension du mal de Bright au stade terminal. Actualités néphrol. Hôp. Necker *1970*, 113

67 Ulrych, M.: Plasma volume decrease and elevated Evans Blue disappearance rate in essential hypertension. Clin. Sci. molec. Med. *45*, 173 (1973)

68 Vertes, V., Canziano, J.L., Berman, L.B., Gould, A.: Hypertension in end-stage renal disease. New Engl. J. Med. *280*, 978 (1969)

69 Weidmann, P., Maxwell, M.H., Lupu, A.N., Lewin, A.J., Massry, S.G.: Plasma renin activity and blood pressure in terminal renal failure. New Engl. J. Med. *285*, 757 (1971)

70 Weil, J.V., Battock, D.J., Grover, R.F., Chidsey, C.A.: Venoconstriction in man upon ascent to high altitude: studies on potential mechanisms. Fed. Proc. *28*, 1160 (1969)

71 Weil, J.C., Chidsey, C.A.: Plasma volume expansion resulting from interference with adrenergic function in normal man. Circulation *37*, 54 (1968)

72 Werkö, L., Sannerstedt, R., Varnauskas, E., Bojs, G.: Experimentally induced hypervolemia during ganglionic blockade: the effect on cardiac function in patients with mitral disease. Scand. J. clin. Lab. Invest. *14*, 289 (1962)

73 Woods, J.W., Liddle, G.W., Stant, E.G., Jr., Michelakis, A.M., Brill, A.B.: Effect of an adrenal inhibitor in hypertensive patients with suppressed renin. Arch. intern. Med. *123*, 366 (1969)

74 Zacharias, F.J., Cowen, K.J.: Controlled trial of propranolol in hypertension. Brit. med. J. *i*, 471 (1970)

Discussion

F. R. Bühler: You suggest, Dr. Tarazi, that the fact that volume does not increase in response to beta-blockade might be due to venous pooling. If this is so, don't you think it surprising that, whereas venous pooling usually causes an orthostatic response, patients being treated with a beta-blocker do not experience a fall in blood pressure when they get up? Another question I'd like to ask is why does the volume increase in the non-responders? Perhaps it's the way in which beta-blockers affect aldosterone that might provide an answer to both these questions. Treatment with diuretics produces a decrease in volume, and this too could conceivably be due to the fact that diuretics amplify the effect of aldosterone.

R. C. Tarazi: Allow me first of all to correct your statement that the non-responders show an increase in plasma volume. They don't. Both the non-responders and the responders have a decrease in plasma volume. This decrease exhibits no relationship to the degree of control of the arterial pressure. Secondly, I wasn't trying to suggest that the decrease in plasma volume is the result of pooling. We ascribe this decrease in plasma volume to a venoconstriction which is probably due to the fact that beta-blockade permits uninhibited alpha-constriction of the veins. It is along these lines that we would account for the plasma volume changes; but this doesn't help us much to understand what happens to arterial pressure – which is a far more complex issue. Incidentally, we have found, in our patients at least, that the relationship between renin and aldosterone remains pretty much the same, irrespective of whether or not they are receiving propranolol. This might be due to the fact that we have not observed the hyperpotassaemia which you and others have encountered with propranolol. Whatever it is that happens in response to diuretics, we prefer to explain it on a haemodynamic basis; in other words, our explanation would be that a marked reduction in peripheral resistance occurs together with a refilling of the vascular system.

C. R. B. Joyce: I entirely agree with Dr. Tarazi that it is very important to distinguish the characteristics of responders and non-responders. One possible very simple explanation for failure to respond may be that, particularly in long-term trials, at least some of the patients simply do not comply with the therapeutic regime. As I'm sure everybody knows, even in quite short-term trials it's not uncommon to find as many as 50% of patients failing to adhere strictly to the prescribed regime; and in trials lasting for more than three months, I think one could fairly confidently predict quite an appreciable degree of departure from the regime. I wonder what measures Dr. Tarazi and his colleagues took to check this point.

R. C. Tarazi: In our research division, Dr. Joyce, we possibly have an advantage inasmuch as the patients collaborate closely and directly with us and are perhaps better motivated than is often the case in clinical trials. We treat them with a firm but very friendly hand; they record their blood pressure twice a day at home, lying and standing. We do in fact feel fairly confident that they are obeying our instructions. Despite this, we don't usually carry out tests on a patient under treatment with propranolol without first having given him an isoprenaline infusion to make sure that he has adequate beta-blockade. I must admit, however, that we don't measure the propranolol levels in the plasma, and maybe we should.

A. Zanchetti: Since Leonetti, Marganti, Terzoli, and I have also been studying some of the interrelationships between propranolol and diuretics with regard to plasma renin activity, I'd like now to present two slides by way of example. Here, in Figure 1, we have a patient whose plasma renin activity was measured in the lying and then in the standing position, and again in the lying position before and after an intravenous injection of 40 mg. furosemide. We then increased the dose of propranolol stepwise and also measured the plasma propranolol level. In this figure you can see that all

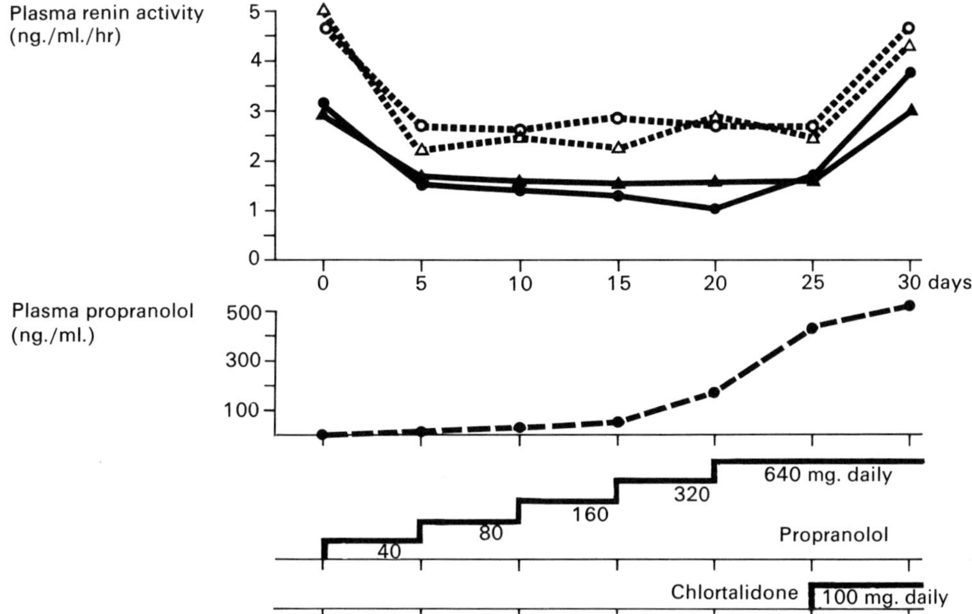

Fig. 1. Plasma renin activity in the standing (o·····o) and lying (●——●) positions, as well as in the lying position before (▲——▲) and after (△·····△) an intravenous injection of 40 mg. furosemide in a hypertensive patient given propranolol first and then propranolol plus chlortalidone.

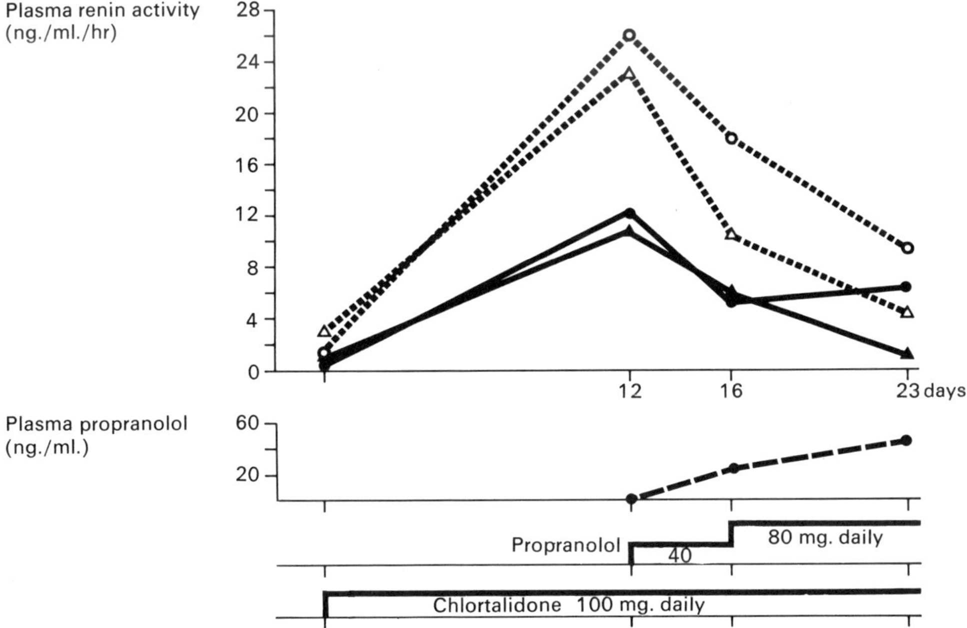

Fig. 2. Plasma renin activity in the standing (o·····o) and lying (●——●) positions, as well as in the lying position before (▲——▲) and after (△·····△) an intravenous injection of 40 mg. furosemide in a hypertensive patient receiving chlortalidone first and then chlortalidone plus propranolol.

the values for plasma renin activity went down; when we added 100 mg. chlortalidone, the values went up. Thus, propranolol markedly depressed the rise in renin occurring in response to a single injection of furosemide, but could not prevent the sharp increase produced by chronic administration of chlortalidone. In Figure 2 the reverse procedure was adopted. Here, the patient was given the diuretic first and propranolol afterwards, but was tested in the same way. After we had been giving this patient chlortalidone for a few days, we got a dramatic increase in plasma renin activity, both in the lying and standing position and following furosemide. When we then started administering propranolol, even in rather low doses, the plasma renin values showed quite a marked decrease, which became even more pronounced as higher plasma propranolol levels were reached. So with propranolol it is indeed possible to reduce the effect of chronically administered diuretics, though the inhibition is not as complete as following a single intravenous injection of the diuretic given in an acute experiment.

K. H. WINTERHALTER: From data I shall be presenting later in my own paper it will become apparent that, at least *in vitro*, propranolol produces very pronounced shrinkage of the erythrocytes; this probably means that the haematocrit is also lowered *in vivo*. I should therefore like to ask you, Dr. TARAZI, whether your technique enables you to distinguish between plasma shrinking and total blood volume shrinking.

R. C. TARAZI: We determine plasma volume directly, so that the shrinking of red cells which you have observed would not affect that determination. We inject I^{131} or I^{125}, spin down the blood, and determine radioactivity only in the plasma without any red cells. As regards the effect of administering propranolol, we have observed a very rapid decrease in plasma volume, and I think JULIUS et al.* have reported the same type of observation. This decrease is associated with an immediate increase in the haematocrit. The long-term figures we have on the haematocrit are, however, unfortunately not as reliable as those we have on plasma volume.

S. H. TAYLOR: Do you think, Dr. TARAZI, that measurement of the plasma volume tells us anything about the mechanism of hypertension at all and, in particular, that it is of any relevance to the therapeutic correction of hypertension? I'm sure that intra-arterial blood volume is relevant in this connection, but I strongly suspect that in the case of plasma volume you may be measuring a variable which is simply changing together with other circulatory processes that are taking place.

R. C. TARAZI: In the paper I presented, we have tried to distinguish between arterial volume – which is the more direct variable involved – and plasma volume; the latter is an index of both intravascular volume and sodium balance. I'm not aware of any study showing that the relationship between arterial volume (which accounts for 20% of the total plasma volume) and total plasma volume is different from normal in hypertension, even in hypertension due to phaeochromocytoma. Plasma volume is thus, I think, a convenient index of "volume"; whether the basic factor responsible for the various volume-pressure relationships is intravascular volume, total extracellular fluid volume, or a change in sodium balance, is not known. Plasma volume is at all events easier to measure. As for your question, Dr. TAYLOR, whether blood volume and plasma volume have anything to do with hypertension, I should say that of course they do. Under physiological conditions, in which so many other mechanisms interfere to minimise the effect of volume changes, the impact of volume variations is of course blunted, but if, for example, you depress sympathetic nervous activity, intravascular volume becomes an important determinant of arterial pressure. In a patient treated with guanethidine, for instance, blood pressure becomes a direct function of volume. So I think that volume definitely does have something to tell us about hypertension.

---

* JULIUS, S., PASCUAL, A. V., ABBRECHT, P. H., LONDON, R.: Effect of beta-adrenergic blockade on plasma volume in human subjects. Proc. Soc. exp. Biol. (N.Y.) *140*, 982 (1972)

Clinical experience with beta-blockers in the treatment of hypertension in the United Kingdom

by J. G. LEWIS*

This paper will present the views of a middle-of-the-road, ordinary clinician who, though not engaged in research on hypertension, sees many hypertensive patients in a large London hospital, as well as in domiciliary practice, and who has an interest in pharmacotherapeutics.

Twenty-five years ago there were no antihypertensive agents worthy of the name and, during the intervening years, many drugs have shown great initial promise in the treatment of hypertension, only to fall out of favour because of shortcomings encountered in practice. Although the ideal antihypertensive has yet to be discovered, the advantages and limitations of the beta-blockers as antihypertensives can be properly appreciated only by reference to those drugs which preceded them and to those that are available today.

We now have at our disposal a wide variety of drugs which lower elevated blood pressure to a greater or a lesser extent. Among the weaker ones are the Rauwolfia alkaloids, the thiazide diuretics, and, in my opinion, methyldopa, which I do not find very effective in the more severe grades of hypertension. Nowadays reserpine is used in hospital practice on only a limited scale in the United Kingdom because of the problem of depression. It is the thiazides that are perhaps the most frequently prescribed type of antihypertensive: for example, over 70% of the hypertensive patients seen at the Hammersmith Hospital are taking thiazides[6], often, of course, in combination with other drugs. The disadvantages of the thiazides – including the fact that they are liable to cause diabetes, gout, and sometimes potassium depletion – are well known. But it should also be borne in mind that they may give rise to postural hypotension, especially in the elderly; moreover, in cases where they aggravate nocturia, which is quite a common feature of hypertensive disease, they interfere with something that is highly conducive to the relief of hypertension, namely, unbroken sleep.

Of the potent antihypertensives, the ganglion-blockers, which heralded the arrival of effective chemotherapy for this condition, are now employed relatively infrequently, owing to the disadvantages of unselective blockade. The guanidine derivatives, on the other hand, still find regular use despite a comparatively high incidence of side effects, of which the most troublesome are postural hypotension, interference with sexual function in the male, and in some cases diarrhoea.

Methyldopa is widely prescribed in the United Kingdom, especially by general practitioners, who probably appreciate the fact that, in doses which will

---

produce a modest, if sometimes inadequate, reduction in blood pressure, it is often easier to handle than guanethidine or betanidine. But methyldopa, too, has its problems. Particularly troublesome may be lethargy, sedation, and "heavy legs"; contrary to some earlier beliefs, postural hypotension can also occur with methyldopa.

Hydralazine, which still commands wide respect in other countries, fell into early disfavour in the United Kingdom, largely because it had not been properly used and because it proved unsatisfactory as monotherapy. Until recently hydralazine was prescribed by relatively few British clinicians, who – since it also has the effect of increasing renal blood flow – had largely confined its use to the management of hypertension occurring during pregnancy or in the presence of renal failure. Now, however, the interesting possibilities opened up by combined treatment with beta-blockers and hydralazine, which exert a mutually complementary pharmacological action, are attracting increasing attention.

After this rapid review, which merely serves to corroborate the conclusion that the ideal antihypertensive drug has yet to be found, let us now turn to the beta-blockers, whose antihypertensive activity was first described by PRICHARD[12] in 1964 following a trial with pronetalol.

Allow me to begin with one or two basic facts about the overall usage of beta-blockers in the United Kingdom. In 1973 something approaching three million prescriptions were issued for the three products available in this category, i.e. propranolol, practolol, and oxprenolol (®Trasicor). Of these, more than half were in respect of angina pectoris or ischaemic heart disease; only one-sixth, or about half a million of the prescriptions, were made out for patients with hypertension. The fact remains, however, that in the United Kingdom beta-blockers as a whole have come into widespread use for the treatment of hypertension, both in hospital and in general practice, and that the rate of increase in their usage has been, and still is, very high. This phenomenon is partly related to the unsatisfactory nature of the alternative drugs and particularly to the desire to find an antihypertensive agent displaying only minimal side effects that can be used in asymptomatic hypertension or in patients with very few symptoms. The increasing trend towards use of the beta-blockers must therefore reflect the acceptability of this group of compounds in terms of efficacy, tolerability, and safety in their established indications. Reviews published by authors specially interested in hypertension also confirm the increasing utilisation of the beta-blockers.

Although at the present time beta-blockers account for only one in about every 25 prescriptions for hypertension *per se*, this is already double the figure of a year ago, and it seems that we are only just beginning to learn what potential such compounds may have in this field. I myself have prescribed beta-blockers extensively for a number of years, but, since this experience has not been quantified, it may be helpful at this point to review the results obtained by others whose findings have been documented.

The effectiveness of propranolol in lowering elevated blood pressure, which was originally established by PRICHARD and GILLAM[14], was subsequently con-

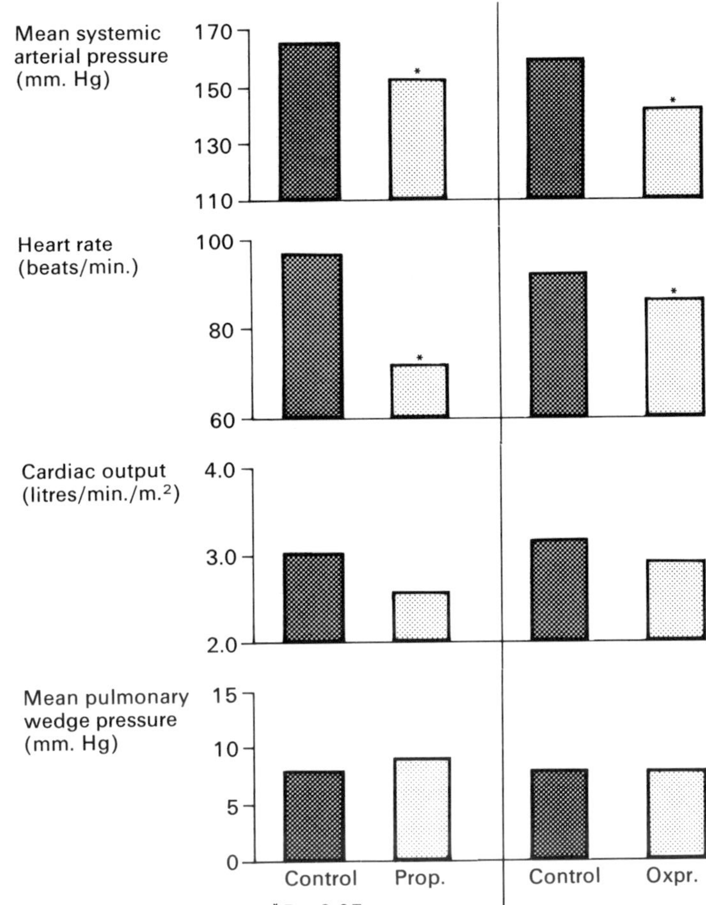

Fig. 1. Comparison between the haemodynamic effects of prolonged oral treatment with propranolol (Prop.) and oxprenolol (Oxpr.) in hypertensive patients. (Adapted from: Taylor et al.[18])

firmed by Zacharias et al.[24] and others. Percentages of patients brought under "good control", some of whom received a diuretic in addition to the beta-blocker, ranged from 66 to 88%. Practolol has been found by several investigators (e.g. Prichard et al.[13] and, more recently, Wood et al.[22]) to be less effective than propranolol. Oxprenolol, on the other hand, has been shown in several studies (e.g. Simpson and Waal-Manning[16], Taylor et al.[18], Marshall and Barritt[9]) to exhibit roughly the same antihypertensive potency as propranolol (Figure 1). Marshall and Barritt, for example, report that, out of 20 patients with mild to moderate hypertension, all 17 who completed the trial achieved a diastolic pressure of 95 mm. Hg or less on oxprenolol alone, roughly half of them requiring a daily dosage of only 160 mg. Some patients, however, needed 480 mg. daily or even more.

Comparable results have also been obtained in general hospital practice, as illustrated by a typical series of 91 patients, 74 of whom showed very satisfactory blood pressure reductions in response to a combination of oxprenolol and cyclopenthiazide[5]. In general practice, too, moderate to severe hypertension has been found to respond well to this same combination[4].

Table 1. Effects of practolol and oxprenolol on parameters of airway resistance in 15 asthmatic patients. None of the differences as compared with the control values was statistically significant.

	Control	Practolol (100 mg. t.i.d)	Control	Oxprenolol (40 mg. t.i.d)
Airway resistance	3.22	3.50	3.43	3.44
Thoracic gas volume	4.16	3.97	4.01	4.08
Forced expiratory volume	2.96	3.08	3.04	2.97
Forced vital capacity	3.91	4.02	3.99	3.95

With regard to tolerability, it has been the general impression that beta-blockers produce fewer side effects than other potent antihypertensive drugs[9, 14]. Experience with oxprenolol may be worth documenting in this connection. Of the 91 patients participating in the hospital study already mentioned[5], 12 were withdrawn because of side effects; of these 12, one had heart failure and one, a known asthmatic, developed bronchospasm. In over 500 cases reported from general practice, by contrast, the withdrawal rate was only 6%, and neither overt heart failure nor bronchospasm were encountered at all – a fact presumably reflecting a careful choice of patients. The most common problems appear to be a variety of central nervous effects, such as dizziness and tiredness, as well as gastro-intestinal upsets. Noticeably absent, however, are postural hypotension and any interference with sexual function.

Interesting light on the question of bronchospasm as a side effect is shed by a study* in which orally administered practolol and oxprenolol were compared in 15 known asthmatic patients who were kept under careful supervision and in whom various parameters reflecting airway resistance were investigated. No significant deviation from control values occurred in any patient on either drug (Table 1). The findings reported in this study must of course be interpreted with reserve, since only moderate doses were used and the medication given over only a three-day period. It is nevertheless probably safe to conclude that only a small number of asthmatics are sensitive to beta-blockers. But, as the practical significance of so-called cardioselectivity also appears uncertain, the only proper advice would seem to be that any beta-blocker should be prescribed for an asthmatic with caution, if at all.

A rarer side effect which appears to occur with practolol[2, 15] is systemic lupus erythematosus (S.L.E.), and positive tests for antinuclear factor may occur in about 10% of patients. The development of an overt clinical syndrome is, of course, less frequent. So far as is known, this problem does not arise with oxprenolol, and in the literature on propranolol – a drug which has been available for almost ten years – only one case has apparently been reported. Incidentally,

---

* A comparison between the effects of practolol and oxprenolol on airway resistance in asthmatic patients. Unpublished report to CIBA LABORATORIES LIMITED, Horsham, Sussex.

there might possibly be some relationship between the occurrence of S.L.E. as a side effect of practolol and the fact that the latter seems to produce a higher incidence of drug rash than other beta-blockers[23].

Another interesting phenomenon which has been reported with beta-blockers, including oxprenolol, is the occasional manifestation of a paradoxical hypertensive reaction[8]. This reaction, which is apparently often associated with weight gain and presumably with fluid retention, does not appear to occur in cases where a diuretic has been given concomitantly. In any event it must be uncommon.

The response rate of hypertensives to beta-blockers alone lies within the range of 50–90% and possibly depends among other things on the initial height of the blood pressure. These agents, however, have also proved most valuable when combined with other drugs. The combined use of a beta-blocker and a diuretic has already been commented upon, and there is clear evidence that in such combinations the diuretic has its usual effect of enhancing the antihypertensive response – an effect resulting in the attainment of a correspondingly lower blood pressure, in a reduction of the dosage in which the beta-blocker requires to be given, or both.

Likewise of interest in this context is evidence obtained with other types of combination. In a small, but carefully designed, trial carried out at the Hammersmith Hospital, PEARSON et al.[11] studied the possible interaction of oxprenolol and guanethidine and concluded that the two drugs produced an additive but not a synergistic effect on the blood pressure. Since this combination proved acceptable in terms of tolerability, guanethidine is clearly one of the more potent drugs that can be added to a beta-blocker in the small percentage of cases not responding adequately to the latter when prescribed either alone or with a diuretic.

Methyldopa remains the most widely used antihypertensive in the United Kingdom, although, as has frequently been pointed out, it produces a by no means insignificant number of side effects, including lethargy and tiredness in particular. Sometimes it is only when the drug is discontinued that the patient or the doctor realises that such side effects have occurred. By contrast, at least 20% of patients receiving beta-blockers report a new sense of well-being[3]. Perhaps it is therefore worthwhile to compare and contrast these two types of drug.

Two papers have come to my attention[20, 21] which show that monotherapy with beta-blockers is more effective than monotherapy with methyldopa. In addition, there is a study published by MURPHY et al.[10] which appears to indicate that the two drugs can be given together, if necessary, and that both control of blood pressure and tolerability are better with this combined treatment than with methyldopa alone.

The combination of a beta-blocker with a vasodilator, such as an alpha-blocker or hydralazine, has also yielded interesting results – as illustrated by reports on the use of a beta-blocker plus hydralazine (®Apresoline)[1, 7], as well as by a recent study in which TAYLOR[17] employed a beta-blocker in combination with phentolamine (®Regitine). The chief reason for the interest in this form of combined therapy lies in the fact that the two types of drug possess pharmacological properties which in many respects are complementary.

In acute studies involving 12 patients with severe hypertensive disease (diastolic pressures greater than 120 mm. Hg), TAYLOR showed that, although neither oxprenolol nor phentolamine alone – administered intravenously – produced a satisfactory lowering of the blood pressure, they did reduce it to normal levels in all cases (both during exercise and at rest) when employed in combination. Moreover, haemodynamic studies showed a reduction in left-ventricular end-diastolic pressure to normal, as well as a normal cardiac output. Six of the patients were also given continuous oral treatment for six months with progressively increasing doses of oxprenolol alone (up to 480 mg. daily), which produced a moderate decrease in blood pressure. Here again, however, the addition of oral phentolamine to the regimen, in a dose of 20 mg. three times daily, resulted in an immediate reduction of the blood pressure to normal levels (Figure 2), both at rest and while walking. No postural or other side effects were reported, and no habituation occurred in the 6–18 months during which the patients received this combined treatment. It was concluded that combined selective beta and alpha-blockade with these two drugs, i.e. with oxprenolol and phentolamine, offered distinct theoretical advantages over other methods of treatment and that it might find general application, particularly in patients uncontrolled by other means.

I should now like to offer a few personal comments in an attempt to put the beta-blockers into context. In doing so, I shall take it for granted that the important question of whether *any* drug treatment is needed by a specific patient would be given proper consideration, and that such other measures as weight reduction, which can make a valuable contribution to the overall plan of treatment, would also be taken.

Although beta-blockers may certainly be used in cases of mild hypertension, I prefer to start with a thiazide alone, unless, of course, thiazides are contra-indicated. If this treatment does not provide an acceptable degree of control, by which I mean the attainment of a diastolic pressure under 95 mm. Hg, I then resort either to a beta-blocker or to methyldopa.

For moderate hypertension, I have tended in recent years to use a diuretic in combination with a beta-blocker. From the evidence of others, this should lead to success in something like 90% of cases. I myself have probably not managed to achieve such a high success rate as this, but I think that – since a maximal response may take several weeks to attain – I may possibly not have persevered long enough. Be that as it may, if the blood pressure does fail to respond adequately, it is now clear that one can add modest doses of methyldopa or, say, guanethidine; in such cases, the possibility of resorting to a combination with phentolamine or hydralazine would also appear to hold out promising prospects for the future.

In patients with severe or refractory hypertension, as evidenced by papilloedema, retinal haemorrhages and exudates, or left-ventricular failure, my preference is to prescribe a quick-acting sympathetic-inhibitor, such as betanidine, together with a diuretic. In my view, any degree of urgency rules out

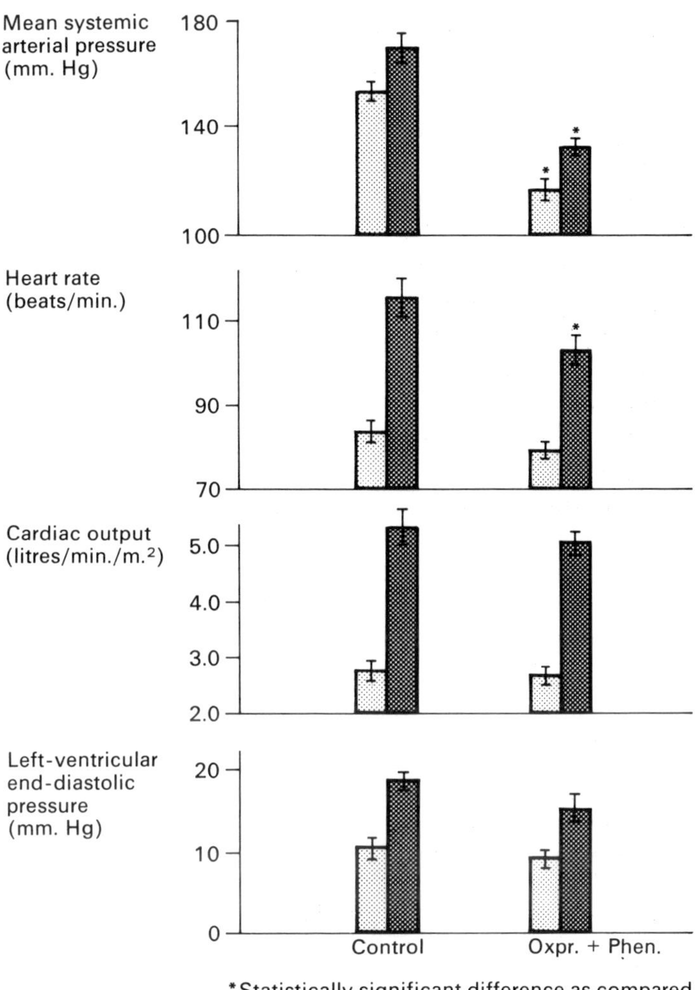

Fig. 2. Haemodynamic effects of beta plus alpha blockade in six hypertensive patients at rest (lighter columns) and while walking (darker columns). The beta-blocker employed was oxprenolol (Oxpr.) and the alpha-blocker phentolamine (Phen.). (Adapted from: Taylor[17])

initial treatment with a beta-blocker, although I am aware that beta-blockers do occasionally exert a rapid antihypertensive effect. One particularly worrying category of patient is the Negro suffering from hypertension; in such patients I have been using a potent guanidine derivative – even at the risk of provoking postural hypotension – in preference to prescribing the more mildly acting methyldopa. In cases where uraemia is present, on the other hand, I have deliberately chosen to employ methyldopa, since it has been alleged to cause little deterioration in renal function. Incidentally, under some, perhaps exceptional, circumstances, it seems that the beta-blockers may exacerbate renal failure; they should therefore be used with caution if and when renal impairment is already moderately advanced.

The particular advantage of the beta-blockers would appear to be their relative lack of side effects, including especially postural hypotension and interference with sexual function; provided due care is exercised, heart failure and bronchospasm, too, should pose very few problems. Where angina pectoris co-exists with hypertension, there would also seem to be a special indication for the use of beta-blockers.

Finally, reference should also be made to the speculative aspect of the protection which beta-blockade might afford the cardiovascular system against the potentially adverse influence of excess catecholamines, an aspect which has been dealt with by TURNER[19], among others. The hypothesis has been advanced that this protective effect – or some other action – might improve the prognosis in hypertension, presumably by reducing the incidence of myocardial infarction. The interaction between beta-blockers and the plasma renin levels, and the significance of the latter, likewise need to be defined. Though a great deal of work is now being done in these areas with a view to clarifying such questions, the answers are unlikely to be quickly forthcoming. Meanwhile, however, the beta-blockers are obviously of great interest and of considerable promise in this field.

References

1 ÄNISHÄNSLIN, W., PESTALOZZI-KERPEL, J., DUBACH, U.C., IMHOF, P.R., TURRI, M.: Antihypertensive therapy with adrenergic beta-receptor blockers and vasodilators. Europ. J. clin. Pharmacol. *4*, 177 (1972)

2 ASSEM, E.S.K., BANKS, R.A.: Practolol-induced drug eruption. Proc. roy. Soc. Med. *66*, 179 (1973)

3 FORREST, W.A.: Oxprenolol and a thiazide diuretic together in the treatment of essential hypertension – a large general practice study. Brit. J. clin. Pract. *27*, 331 (1973)

4 FORREST, W.A.: Treatment of moderately severe essential hypertension in general practice with a combination of cyclopenthiazide with potassium chloride (Navidrex K) and oxprenolol (Trasicor 80 mg): a report on 554 patients. J. int. med. Res. *2*, 7 (1974)

5 FORREST, W.A.: Hospital experience with a combination of cyclopenthiazide with potassium chloride (Navidrex-K) and oxprenolol (Trasicor 80 mg). In preparation

6 GEORGE, C.F., BRECKENRIDGE, A.M., DOLLERY, C.T.: Comparison of the potassium-retaining effects of amiloride and spironolactone in hypertensive patients with thiazide-induced hypokalaemia. Lancet *ii*, 1288 (1973)

7 HANSSON, L., OLANDER, R., ÅBERG, H., MALMCRONA, R., WESTERLUND, A.: Treatment of hypertension with propranolol and hydralazine. Acta med. scand. *190*, 531 (1971)

8 KELLETT, R.J.: The treatment of hypertension with oxprenolol. In Burley, D.M., et al. (Editors): New perspectives in beta-blockade, Int. Symp., Scanticon, Aarhus, Denmark, 1972, p.215 (Metropolis Press, London 1973)

9 MARSHALL, A.J., BARRITT, D.W.: Oxprenolol in hypertension. Brit. J. clin. Pract. *27*, 337 (1973)

10 MURPHY, J.E., STANDEN, S.M., FORREST, W.A.: The addition of oxprenolol to hypertensive patients treated with methyldopa – a general practice study. J. int. med. Res. *2*, 1 (1974)

11 PEARSON, R.M., BULPITT, C., GEORGE, C.F., HOLE, D., BRECKENRIDGE, A.: A trial of the combination of guanethidine and oxprenolol in hypertension. Scot. med. J. *19*, 45 (1974); abstract of paper

12 PRICHARD, B.N.C.: Hypotensive action of pronethalol. Brit. med. J. *i*, 1227 (1964)

13 PRICHARD, B. N. C., BOAKES, A. J., DAY, G.: Practolol in the treatment of hypertension. Postgrad. med. J. *47*, Suppl. (Jan.): 84 (1971)

14 PRICHARD, B. N. C., GILLAM, P. M. S.: Treatment of hypertension with propranolol. Brit. med. J. *i*, 7 (1969)

15 RAFTERY, E. B., DENMAN, A. M.: Systemic lupus erythematosus syndrome induced by practolol. Brit. med. J. *ii*, 452 (1973)

16 SIMPSON, F. O., WAAL-MANNING, H. J.: Hypertension and beta-adrenergic blockade. Symp. Beta adrenergic blocking agents, Sydney, Australia, 1970, p. 59

17 TAYLOR, S. H.: Combined alpha and beta-receptor blockade in the treatment of severe hypertension. Scot. med. J. *19*, 47 (1974); abstract of paper

18 TAYLOR, S. H., MAJID, P. A., SAXTON, C., STOKER, J. B.: Comparison of the circulatory effects of oxprenolol and propranolol in hypertensive patients. In Simpson, F. O. (Editor): Beta-adrenergic receptor blocking drugs, Proc. Symp., Auckland 1970, p. 47

19 TURNER, R. L.: Discussion, loc. cit. [8], p. 230

20 VAN COLLER, P. E.: Clinical experience with Visken (prindolol) in essential hypertension: its special comparative action to Aldomet (methyldopa). J. int. med. Res. *1*, 561 (1973)

21 VEDIN, J. A., WILHELMSSON, C.-E., WERKÖ, L.: Comparative study of alprenolol and methyldopa in previously untreated essential hypertension. Brit. Heart J. *35*, 1285 (1973)

22 WOOD, R. A., FORRESTER, T. M., JOHNSTON, A. W., PALMER, K. V.: Management of hypertension: a trial of practolol (Eraldin). Clin. Trials J. *10*, 53 (1973)

23 ZACHARIAS, F. J.: Discussion, loc. cit. [8], p. 238

24 ZACHARIAS, F. J., COWEN, K. J., PRESTT, J., VICKERS, J., WALL, B. G.: Propranolol in hypertension: a study of long-term therapy, 1964–1970. Amer. Heart J. *83*, 755 (1972)

Discussion

J. Brod: Why, Dr. Lewis, didn't you mention in your paper clonidine as one of the antihypertensive drugs available in Great Britain? Though it is only of moderate potency, I have found it a good drug. One of the snags of guanethidine, of course, is that it may produce marked orthostatic effects – which, incidentally, are also a problem in patients treated with methyldopa. I agree with you that the use of reserpine is gradually diminishing, one reason for this being the possibility of mental depression as a side effect. On the other hand, I have never stopped using hydralazine, which now seems to be coming back into favour again. Hydralazine is an excellent drug, but treatment has to be started with a very low dosage, and the doses have to be raised very gradually; that's the whole secret of success with hydralazine.

J. G. Lewis: I should perhaps also have mentioned clonidine in my paper, but I omitted to do so because it is prescribed in Great Britain on only a modest scale. Clonidine is, of course, another of the imidazoline derivatives and, as such, it bears a resemblance to histamine. It certainly works, though nobody quite seems to know how. It has a central effect, as well as an effect on the sympathetic nervous system or on sympathetically mediated tone, but its side effects are very similar to those of methyldopa. One problem involved in its use is the rebound phenomenon, which can prove rather worrying. If a patient decides to stop taking the drug, there is a great risk of his developing a phaeochromocytoma-type reaction.

R. C. Tarazi: May I at this point draw attention to an investigational drug that is now being used more and more frequently in patients with severe hypertension, namely, to minoxidil, which is a piperidine derivative. We ourselves, as well as other groups*, have found it very effective in patients with severe hypertension, including those with advanced renal disease and renal failure. This effect is achieved, however, at the price of pronounced fluid accumulation – which means that a diuretic usually has to be given at the same time. But, then, it's very difficult in any case to manage severe hypertension without using diuretics.

J. G. Lewis: Minoxidil is one of several vasodilators which I could have mentioned in addition to hydralazine; the list also includes, for example, phentolamine and prazosin. Though minoxidil is not yet available in Britain, I gather that it also has some literally hair-raising side effects: apparently it causes the hair to come off the head and to grow instead on the face. But maybe hirsutism is not too great a price to have to pay for the control of previously refractory hypertension.

R. C. Tarazi: It makes hair grow on the face, but it doesn't take hair off the head!

G. Muiesan: In your paper, Dr. Lewis, you seemed to suggest that beta-blockade interferes with catecholamine secretion. But following treatment with oxprenolol (®Trasicor), for example, the plasma catecholamine levels may become even higher.

J. G. Lewis: Yes, this is an error in thinking into which one can easily slip. As adrenergic-receptor blockers, these drugs have in fact a competitive effect.

W. Schweizer: Wouldn't someone like to comment on the most intriguing combination of phentolamine plus a beta-blocker, to which Dr. Lewis referred in his paper?

A. Zanchetti: I was a little surprised at the very good effect that Dr. Taylor got with this type of combination. I have nothing in principle against the idea of combining an alpha-blocker with a beta-blocker in the treatment of hypertension, but I was merely rather surprised that he managed to get such a nice potentiation by giving

---

* Gilmore, E., Weil, J., Chidsey, C.: Treatment of essential hypertension with a new vasodilator in combination with beta-adrenergic blockade. New Engl. J. Med. *282*, 521 (1970)

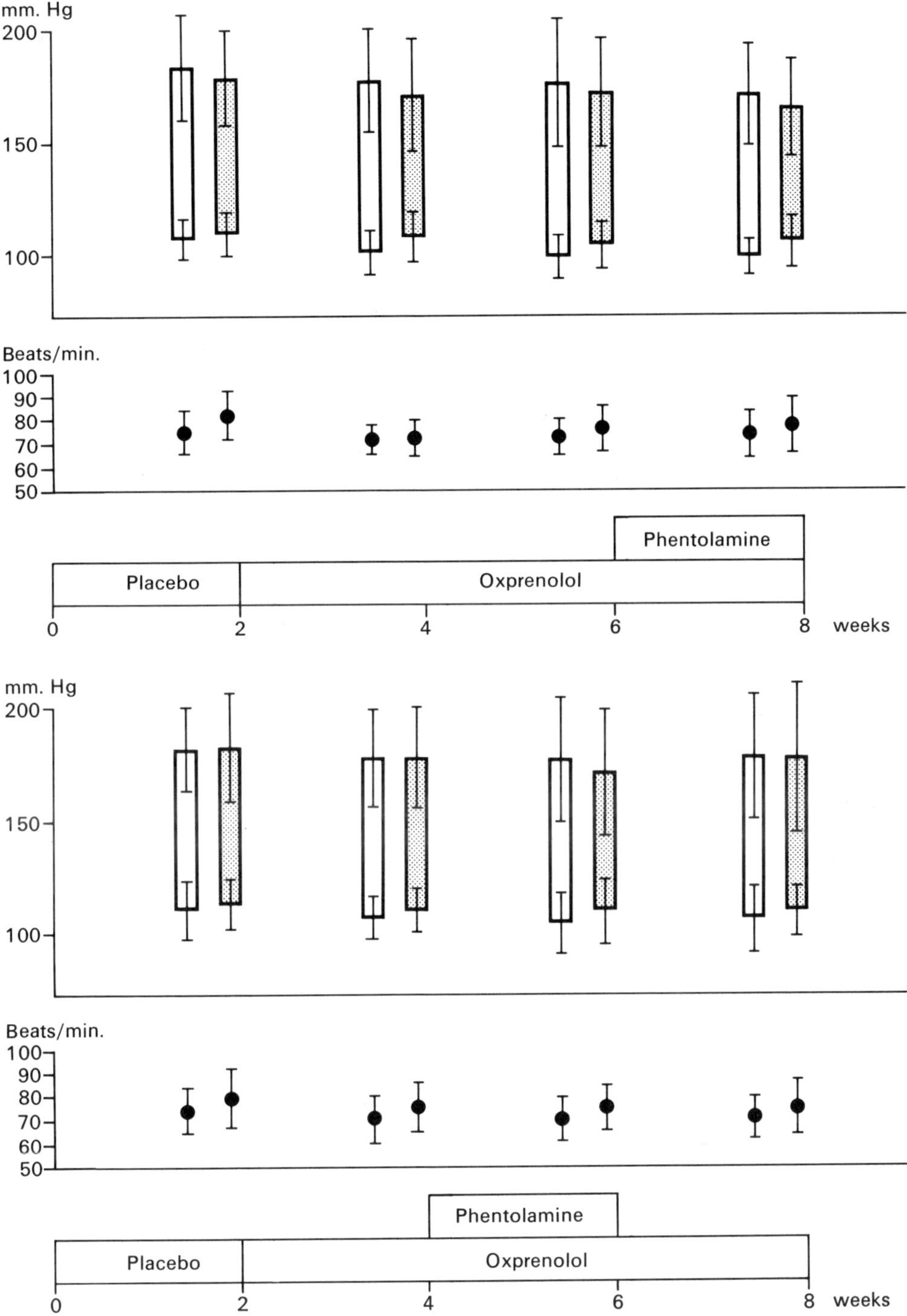

Fig. 1. *Above:* blood pressure and heart rate (mean values ± S.E.M.) in the lying (□) and standing (▨) positions in 40 patients suffering from mild to moderate hypertension and treated with placebo for two weeks, oxprenolol (240 mg. daily) for four weeks, and then oxprenolol (240 mg. daily) plus phentolamine (60 mg. daily) for two weeks. *Below:* findings obtained in 35 similar patients treated in the same way, except that phentolamine was added to the regimen from Week 4 to Week 6.

phentolamine three times a day. Phentolamine, of course, has a short duration of action.

S. H. Taylor: Phentolamine has a short half-life, both biologically and pharmacologically. Following its administration the blood pressure falls for some two to four hours and then rises again. With a thrice-daily regimen, however, we found that we could in fact keep the blood pressure largely under control.

K. D. Bock: We have tried this combination of a beta-blocker with an alpha-blocker in seven patients. They were all receiving metoprolol, to which we added phenoxybenzamine (20 mg. four times daily by mouth) in six cases and phentolamine (20 mg. three times daily by mouth) in the seventh patient. Though we did get an additional effect, it was not very intensive, nor was it very regular. We think that phenoxybenzamine poses problems of absorption. In one further case in which we resorted to intravenous administration of phentolamine, we did observe a transient but distinct additional reduction in blood pressure over and above the reduction produced by the beta-blocker.

P. R. Imhof: To supplement what Dr. Lewis had to say about Dr. Taylor's study on the antihypertensive effect of combined treatment with an alpha-blocker and a beta-blocker, I'd like to mention some further observations that have been made with this type of combination. A Swiss group of general practitioners studied the effect of combined treatment with oxprenolol and phentolamine in a fairly large series of patients suffering from mild to moderate hypertension. Shown in the upper part of Figure 1 are the mean values for blood pressure and heart rate in the lying and standing positions in 40 patients treated with placebo for two weeks, oxprenolol (240 mg. daily) for four weeks, and then oxprenolol (240 mg. daily) plus phentolamine (60 mg. daily) for two weeks. The lower part of the figure indicates the findings obtained in another group of 35 patients who were treated in a similar way, except that phentolamine was added to the regimen from Week 4 to Week 6 instead of from Week 6 to Week 8. In both groups the lowest blood pressures were recorded during combined treatment with oxprenolol and phentolamine; the differences, however, were not very marked.
Dr. Taylor had pre-treated his patients for several months with oxprenolol in a dosage of 480 mg. daily before he added the phentolamine, and this possibly may explain why his results were better.

F. R. Bühler: So far as phentolamine is concerned, does it really act in hypertension by blocking the alpha-receptors, or is its effect perhaps due to some other vasodilating property that has nothing to do with its action on the alpha-receptors?

S. H. Taylor: There is no doubt that these drugs are general vasodilators. Drugs with an imidazoline ring produce general vasodilatation by acting on the smooth muscle of the arteries and veins. But there is also good pharmacological evidence to indicate that, in addition, phentolamine selectively blocks the alpha-receptors in the skin and in the kidney in particular.

J. G. Lewis: Dr. Bock has mentioned phenoxybenzamine. Beilin and Juel-Jensen* did a similar study at Oxford and also got poor results. This is why I was so interested in Dr. Taylor's work with phentolamine, especially since these two alpha-blockers are chemically different.

---

* Beilin, L. J., Juel-Jensen, B. E.: Alpha and beta adrenergic blockade in hypertension. Lancet *i*, 979 (1972)

Use of beta-blockers in general practice in the United Kingdom

by D.W. GAU*

I have been asked to provide some insight into the use of beta-blockers in hypertension, based on the standpoint and experience of a British general practitioner. I should like to preface my remarks by stressing that perhaps not all the views I express may be typical and that I cannot claim my experience to be adequately documented in the research sense. On the other hand, since hypertension is a disease which in the United Kingdom is certainly treated far more often by general practitioners than by specialists or research workers, the comments I have to offer may well prove of interest.

I practise in the small country town of Beaconsfield, which has about 15,000 inhabitants and is situated 26 miles from London, in the county of Buckinghamshire. In my group practice, a colleague and I have sole care of about 5,000 patients, with a slight preponderance of females and a higher proportion of younger patients than in the average British practice.

For various reasons it is possible to analyse both the nature of our consultations, and the drug treatment we prescribe, with some precision. So perhaps these statistics should be presented first, after which I can then comment on my own approach to hypertension and on the reasons underlying it.

During the past year, we have had 94 patients – 67 females and 27 males – under active treatment for hypertension. This represents about 2% of the total practice population, but about 13% of the patients for whom, during the same period, we were prescribing drug treatment for subacute or chronic conditions. As we all know, hypertension is a common disease.

The drugs prescribed for these 94 patients are shown in Table 1. It should be noted that more than one drug is often administered, so that the total exceeds 100%.

As can be seen from this table, current prescribing, across our practice, shows an above-average use of diuretics, reserpine, and beta-blockers. However, this pattern is not representative of the way in which we are now treating *new* patients, because the 94 cases surveyed include patients who have been under treatment for a considerable time, often for years. Naturally, those who are satisfactorily controlled on a given regime are left on it. At present, the beta-blockers – prescribed with or without a thiazide diuretic – are my drugs of first choice, and I have virtually stopped using methyldopa altogether because of its side effects.

In commenting specifically on this group of drugs, I should like to tell you briefly about my first experience with them. In 1961/62 I was an intern work-

---

* "Upper Burgess", 11 Stratton Road, Beaconsfield, Buckinghamshire, England.

Table 1. Drugs prescribed for the treatment of hypertension in a group of 94 patients seen in general practice.

Drug or drug group	Patients		Average for the United Kingdom (%)
	Number	% of group (N = 94)	
Diuretics	62	67	39
Methyldopa	38	40	42
Reserpine	31	33	7
Beta-blockers	10	11	5
Sympathetic-inhibitors	7		
Apresoline	4	15	18
Clonidine	3		

ing in an intensive care ward of a surgical thoracic unit. One day, while I was doing internal cardiac massage on a heart with ventricular fibrillation, a research assistant gave the patient some pronetalol intravenously. The effect was staggering. The small, coarsely fibrillating heart suddenly became a dilated sac, quivering like a jelly, for which nothing further could be done. This has given me a very healthy respect for the beta-blockers, especially when administered intravenously.

I myself use either oxprenolol (®Trasicor) or propranolol. The average dose of, say, oxprenolol with a diuretic is probably in the region of 240 mg., but, if need be, I am willing to increase the dose gradually to up to about 1.6 g., although such a high dose involves the taking of a lot of tablets. Before one can say that the treatment is *not* working, it is necessary to wait for six to eight weeks; though some patients do of course fail to respond, such failures are encountered only in a minority of cases. Under these circumstances, I then reduce the dose of the beta-blocker to about 640 mg. and add ®Apresoline (up to 100 mg. daily) in divided doses, increasing the dosage to 150 mg. if necessary. Only two patients have not responded to this regime, and in these two cases a small dose of clonidine subsequently brought the blood pressure under control.

I used to give beta-blockers four times a day, but I now prefer a twice-daily dosage. I also have two patients who are satisfactorily controlled on once-a-day therapy. Incidentally, I make a point of taking the blood pressure as late in the day as possible; otherwise, diurnal variations in the blood pressure may lull one into a false sense of security.

Since I started prescribing beta-blockers for hypertension, three patients have had to discontinue them because of side effects. These side effects – in the form of vivid dreams, malaise, tiredness, and constipation – generally occurred in response to higher dosages, but not always. None of my hypertensive patients has had bronchospasm or heart failure while undergoing treatment with a beta-blocker.

In common, I think, with most doctors in the United Kingdom, I am at present treating more cases of mild hypertension than in the past, the chief reason being that it has now become a great deal easier to avoid troublesome side effects. At the moment I treat anybody under 65 years of age with a diastolic pressure of over 100 mm. Hg; for this purpose, I take as my criterion the first casual reading, since I feel this to be more relevant to the real life situation.

In a high percentage of patients it seems to me that essential hypertension is a psychosomatic disease, and I was therefore interested to read in a recent paper by Cobb and Rose[1] that hypertension had been found to occur significantly more frequently in a group of air-traffic controllers, who work under sustained stress, than in a matched control group. Bearing in mind the strong possibility of a relationship between stress and hypertension – and in view of the known action of beta-blockers in inhibiting the influence of the sympathetic nervous system, coupled with the demonstrated fact that beta-blockers exert an anti-anxiety effect[2] – I think they should be considered the drugs of choice for hypertension, unless there are special grounds for not employing them.

For this same reason, I also prescribe beta-blockers for normotensive patients who are facing short-term stressful situations which give rise to manifest anxiety, so that they are thus spared the undesirable sedation produced by the conventional tranquillisers. In addition, I have nine patients with angina pectoris on beta-blockers, as well as three patients suffering from paroxysmal tachycardia – due in two cases to thyrotoxicosis – whom I successfully treated with beta-blockers pending definitive diagnosis and treatment.

In conclusion, it is my opinion that the beta-blockers have a major part to play in the management of hypertension and that they are also destined to assume a role of growing importance in the treatment of short-term stress situations seen in general practice.

References

1 Cobb, S., Rose, R. M.: Hypertension, peptic ulcer, and diabetes in air traffic controllers. J. Amer. med. Ass. *224*, 489 (1973)
2 McMillin, W. P.: Trasicor in the treatment of emotional stress. In Burley, D. M., et al. (Editors): New perspectives in beta-blockade, Int. Symp., Scanticon, Aarhus, Denmark, 1972, p.313 (Metropolis Press, London 1973)

Discussion

F. Gross: I was interested to learn from your paper, Dr. Gau, that in Great Britain the percentage of prescriptions for beta-blockers in general practice is about 5%. I assume that this figure relates to the actual numbers of prescriptions. On the European continent – excluding the socialist countries – the corresponding figure was still less than 1% in 1973. In this respect, however, the countries on the continent of Europe are still far better off than the U.S.A., where I would like to remind you that hypertension is not yet an accepted indication for beta-blockers and where there is only one such drug on the market, i. e. propranolol.

J. Brod: You said, Dr. Gau, that the beta-blockers may exert an anxiolytic effect. I wonder whether you were implying that they also have a central action on the nervous system. I can quite see that they serve to combat the peripheral repercussions of anxiety, such as the rise in cardiac output and the increase in smooth-muscle tone that are partially mediated by beta-adrenergic activity. But I wonder what happens in the case of the blood pressure, because when you give a beta-blocker the alpha-adrenergic vasoconstriction remains unopposed.

P. Kielholz: We in Basle have also found that beta-blockade has a good anxiolytic effect, and it is our impression that this effect is not solely attributable to relief of the physical signs and symptoms of the anxiety. I suspect that there must also be a central component contributing to this anxiolytic effect. It has been our experience – and I want to revert to this later in the proceedings – that in response to a beta-blocker anxiety is also effectively relieved in depressed patients, but that the beta-blocker has no influence on the depression itself. Since I believe that beta-blockers must evidently also exert a central effect, may I ask what, if anything, is known about such an effect?

D. W. Gau: This question interests me too. What about reports, for example, of patients experiencing vivid dreams while under treatment with beta-blockers? Is it possible that there may be differences between one beta-blocker and another in this respect, because I understand that some of them don't in fact cross the blood-brain barrier?

H. Brunner: In animal experiments it is possible, under suitable conditions, to obtain evidence that beta-blockers do indeed exert a central effect. Rats, for example, can be conditioned to display anticipatory anxiety which leads to a rise in temperature, and this can be prevented with beta-blockers. In another type of experiment, animals pre-treated with a monamine-oxidase inhibitor are given an injection of reserpine; the injection provokes very strong excitation, which can also be inhibited by beta-blockers. Beta-blockers thus certainly do have cerebral effects which can be demonstrated in animal experiments by reference to changes in behaviour.

P. R. Imhof: I should merely like to add that acute anxiety states induced by infusions of lactic acid in patients suffering from anxiety neuroses can be prevented by beta-blockers.

E. Sowton: Bad dreams in patients receiving beta-blockers are certainly a side effect that has come to our notice. They seem to be more common with pindolol than with the other beta-blockers, and I have encountered them only very occasionally after propranolol or oxprenolol (®Trasicor).

B. N. C. Prichard: Evidence in favour of the idea that the anxiolytic effect of beta-blockers is indeed a peripheral one is perhaps afforded by the finding that practolol proves quite effective as an anti-anxiety drug in small doses. Since practolol shows an extremely low degree of penetration into the central nervous system, it could be supposed that the concentrations reached in the brain would probably be far too weak to account for the drug's anxiolytic effect. SCALES and COSGROVE* have studied various animal species and have shown that the concentrations in the brain are low. If you do an autoradiograph on the mouse, for instance, it looks as though – in contrast to propranolol and other beta-blockers – there is no practolol getting into the C.N.S. at all.

---

* SCALES, B., COSGROVE, M.B.: The metabolism and distribution of the selective adrenergic beta blocking agent, practolol. J. Pharmacol. exp. Ther. *175*, 338 (1970)

Monitored release studies with Trasicor

by D. M. Burley*

Before describing the results obtained in monitored release studies on oxprenolol (®Trasicor), I should first like to say a few words about monitored release studies in general and to explain how they came to be so described, their scope, and their limitations.

Since the Second World War, and more particularly during the last 15 years, which span the time I have worked in the pharmaceutical industry, the science of clinical pharmacology and clinical trials methodology has forged ahead and become more and more sophisticated – a trend that has been fostered by the advent in most developed countries of government regulatory bodies operating in the drug domain.

It is now necessary, not only to know how well a drug works and what its principal adverse effects are, but also to investigate its mode of absorption and excretion, its distribution in body fluids, its metabolites, its concentration in various organs, its effect on biochemical parameters, its interactions with other drugs, and so on. As a result, clinical trials tend to be conducted on fewer and fewer patients but in greater and greater depth, so that it is not uncommon to seek permission to market a drug in Great Britain with only 200–300 patients ever having received it. Furthermore, the clinical trials will often have been carried out under atypical conditions, in hospital rather than in general practice, in specialised units rather than in general units, and in "classic" cases rather than in the general run of patients. The results, while yielding a great deal of information about drug handling and metabolic effects, give a most imperfect picture of how the drug will fare in general practice in the hands of non-specialist doctors – or of the true incidence of adverse effects, including particularly the unusual ones. A few simple examples will suffice to illustrate this point: dosages of hypnotics are invariably higher in hospital trials than in out-patient or general practice trials; classic angina pectoris, with a normal E.C.G. at rest, but characteristic S-T segment changes and chest pain on exercise, is very different from the "angina" diagnosed in general practice, quite apart from the fact that the concept of what constitutes angina pectoris also differs from country to country; finally, before it can be stated with confidence that the incidence of a particular adverse effect is, say, 0.1%, it is necessary for statistical reasons to treat 3,000 patients with the drug.

Because of these disadvantages of formal clinical trials, a number of people have suggested that at the time when a drug is first marketed there should be some system whereby sample patients from a large number of widely scattered

---

140   * CIBA LABORATORIES LIMITED, Horsham, Sussex, England.

practices are carefully documented in order to obtain a large volume of data on such simple problems as response, optimum dosage, side reactions, and toxicity. Five years ago VERE[11] introduced the concept of "monitored release". His suggestion was that 1,000 practitioners first using a drug should submit carefully recorded data on the first ten patients who received it, at the same time collecting similar data on ten other matched patients receiving the best available previous treatment. Data on 20,000 subjects would thus be obtained. The idea behind this proposal was that adverse incidents in the group of patients treated with the new drug could so be compared with the spontaneous frequency of adverse incidents in a matched population[2,9]. By such methods one could at least satisfy some of the requirements of an orthodox trial (Table 1).

VERE intended that data from monitored release studies should be collected centrally by an independent coordinator, using a variety of sophisticated aids such as a biochemical laboratory and computer facilities. In practice no such system has yet become available, and on the few occasions when the Committee on Safety of Medicines has required monitored release information in Great Britain, notably with ®Symmetrel, this was obtained by sending out reply-paid cards to doctors using the drug.

It was against this background that the Medical Division of CIBA LABORATORIES in Horsham decided four years ago to conduct monitored release studies on most of its new drugs; a report on the first of these studies, dealing with rifampicin[1], was presented in 1971. For practical reasons, we found straight away that it was necessary to modify many of the criteria postulated by VERE for the ideal study. We obviously lacked the wherewithal to collect the very large case numbers (20,000) that he had suggested. Also, there were severe limitations on the quantity of information that busy general practitioners could be expected to provide. For the same logistic reasons, one could rarely assemble data on a matched population. However, I think we have shown that with good organisation (and here I must pay tribute to my colleague Dr. W.A. FOR-

Table 1. Main features of an orthodox clinical trial as compared with a monitored release study.

Orthodox trial	Monitored release
1. Random sampling Matched comparison	Matched pairs Large population Random allocation
2. Eliminates concurrent variation	Rapid collection over wide area
3. Independent coordinator	Group of trained personnel with expert coordinator

4. Patient cooperation
   Toxicity detection
   Plasma levels
	Central biochemical
	laboratories and computer
	facilities

Initials of patient		Sex	M	F	Age	
		Occupation				

Is pain angina of effort?	☐	Or chest pain with anxiety?	☐

Other symptoms   + + + + + +   + + + + + +

Breathlessness				Palpitations			
Fatigue				Dizziness			

Treatment notes:

	Date	Progress	Dosage (see overleaf)
Start			
Follow-up			

Optimum dose (see overleaf):

Optimum daily dose ☐ Trasicor 40 mg. tablets   Achieved in ☐ weeks

Overall assessment of therapeutic response (for key see overleaf):

Excellent ☐   Good ☐   Fair ☐   Nil ☐   Worse ☐

Grade of angina:	Before	On Trasicor
0 Symptom-free		
1 Hurrying, especially hills/stairs		
2 Walking on level		
3 Walking slowly		
4 At rest		

Other effects (key overleaf):

Date	Symptoms/signs	Action taken

Comments	Name and address (capitals)
	Dr.

 Fig. 1. Example of the record cards used in monitored release studies with Trasicor.

REST, whose special preserve these studies have now become), it is possible to collect very valuable data serving as a basis for many decisions relevant to medicine and marketing. I mention the latter because it is obviously of value to a pharmaceutical company – which, after all, has an interest in selling its products – to involve several hundred doctors in a study of a new drug! But at the same time I am also convinced that these studies are clinically justifiable, since any such undertaking which ultimately enables the doctor to know how best to use the drugs in his armamentarium is bound to make for the better practice of medicine.

A note of warning must nevertheless be sounded, since uncontrolled studies, which these mostly are, have an inherent tendency to produce favourable responses and to give a falsely low estimate of the incidence of adverse effects. This, of course, results from optimism and enthusiasm on the part of doctors participating in the study, who may thus become biased and tend to discard poor responses and bad reactions. We have gone out of our way to try and ensure that doctors send in record cards on *all* their cases, whatever the response; but, despite this, we are probably only partially successful in obtaining an unbiased picture. However, all the data supplied to us are processed in our "Data-Handling and Statistics" department, and incomplete cards are sent back for the necessary additions to be made.

Monitored release "kits", as we call them, are of convenient size and easy to distribute and handle. The record cards are simple in design – consisting usually of a single sheet with boxes to tick or numbers to enter on one side, and simple instructions on the other – and their format is such that they fit into the National Health Service record envelopes. The doctor is not expected to fill in more than four or five of these record cards (cf. Figure 1).

I shall now outline for you the results we have obtained in our monitored release studies with oxprenolol. These studies were carried out in two categories of patient: those suffering from angina pectoris and those with hypertension.

Angina pectoris

The total number of patients involved in the studies was 5,492, of whom 5,238 were treated in general practice and 254 in hospital practice. These studies have been the subject of three separate reports[3,4,6], but for the purpose of this paper it is convenient to review them together.

The diagnosis in the patients concerned was angina of effort, and attempts were made to exclude those with chest pain at rest and anxiety. In other words, the aim was to study cases in which chest pain of characteristic distribution developed after a degree of exertion that was predictable and repeatable, but to exclude those patients who were more likely to be suffering from Da Costa's syndrome. Despite all attempts at differentiation, however, it seems probable that 4–5% of the general practice patients did in fact fall into this latter category.

In the general practice studies, 4,403 cases were treated with oxprenolol alone, whereas in the remaining 835 patients oxprenolol was used to supplement treatment with long-acting nitrates, which was the commonest form of long-

Table 2. Response rates obtained with oxprenolol in angina pectoris in three studies: the first conducted in general practice, the second in hospital, and the third a study in which oxprenolol was added to long-term nitrate medication.

Oxprenolol in angina pectoris	Study	Response rate
	General practice study	89%
	Hospital study	90%
	Nitrate study	85%

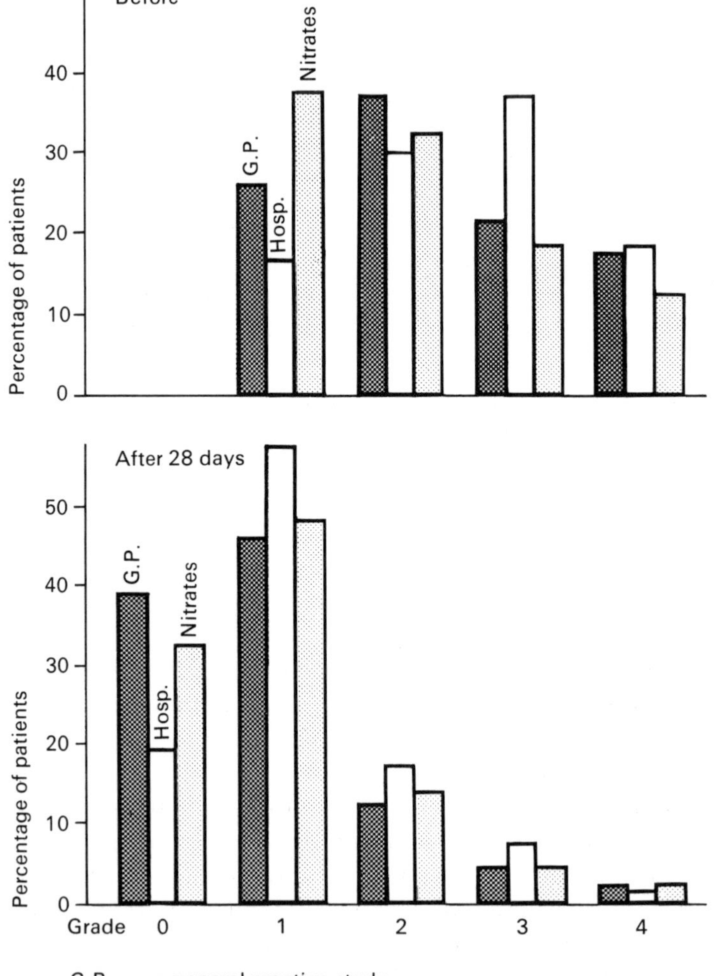

Fig. 2. Anginal grade before and after treatment with oxprenolol in the three studies (see text and Table 2).

term therapy for angina pectoris at that time. As can be seen from Table 2, the response to treatment was remarkably similar in all three groups: 89% of the cases in the study with oxprenolol alone in general practice responded well, as compared with 90% in the hospital study involving much smaller numbers, and 85% in the study in which oxprenolol was added to long-term nitrate medication. The third percentage is interesting insofar as it suggests that long-acting nitrates may perhaps be no better than a placebo, since the response rate was no higher than in the other two groups of patients receiving oxprenolol alone. The degree of improvement observed was similar in all the trials (Figure 2). Valuable information was also obtained concerning dosage (Table 3): in the three studies the daily dose of oxprenolol required in the cases that responded satisfactorily was 120–200 mg. in 80%, 70%, and 84% of the patients, respectively; and in the remaining successfully treated cases it was 240–480 mg. daily.

Table 3. Successfully treated patients listed in percent according to the daily dosage of oxprenolol administered.

Daily dosage	120 mg.	160–200 mg.	>200 mg.
General practice study	53	27	20
Hospital study	54	16	30
Nitrate study	68	16	16

Table 4. Percentages of adverse reactions to oxprenolol in the three studies.

Adverse reactions	Severe	Total
General practice study	6.8	13.7
Hospital study	2.4	15
Nitrate study	4	16

Table 5. Types of adverse reaction occurring in 4,657 patients treated with oxprenolol.

Type of adverse reaction	Incidence	
Gastro-intestinal	5%	(2.4% severe)
Dizziness	3.5%	(1.3% severe)
Heart failure	0.7%	
Bronchospasm	1.1%	
Others	3.6%	(1.3% severe)

The findings were also very similar in respect of side effects. In the three studies, the percentage of patients experiencing adverse reactions was 13.7%, 15%, and 16%, respectively (Table 4); the majority of these side effects, however, were of a minor character and did not interfere with the continuance of therapy or detract appreciably from the therapeutic response. On the average, 6% of the patients had side effects sufficiently severe to cause inconvenience or to necessitate withdrawal of therapy. The commonest side effects were gastro-intestinal disturbances, next in order of frequency being dizziness and giddiness (Table 5). More severe reactions, such as heart failure, bradycardia, and aggravation of angina and respiratory symptoms, were relatively rare and certainly far less frequent than was feared likely when beta-blocking drugs were introduced into general medicine.

Certain incidental findings emerging from these monitored release studies may possibly be of interest. Breathlessness, for instance, proved to be a relatively common symptom in untreated cases of angina pectoris – a fact that is hardly ever mentioned in text-books. Also, in all three trials, up to 10% of the patients reported in response to oxprenolol a positive sense of well-being which did not seem to be merely related to the improvement in their angina.

Hypertension

In these two studies, upon which FORREST has also already reported[5, 7], 2,854 patients with diastolic pressures ranging from 95 to 140 mm. Hg were

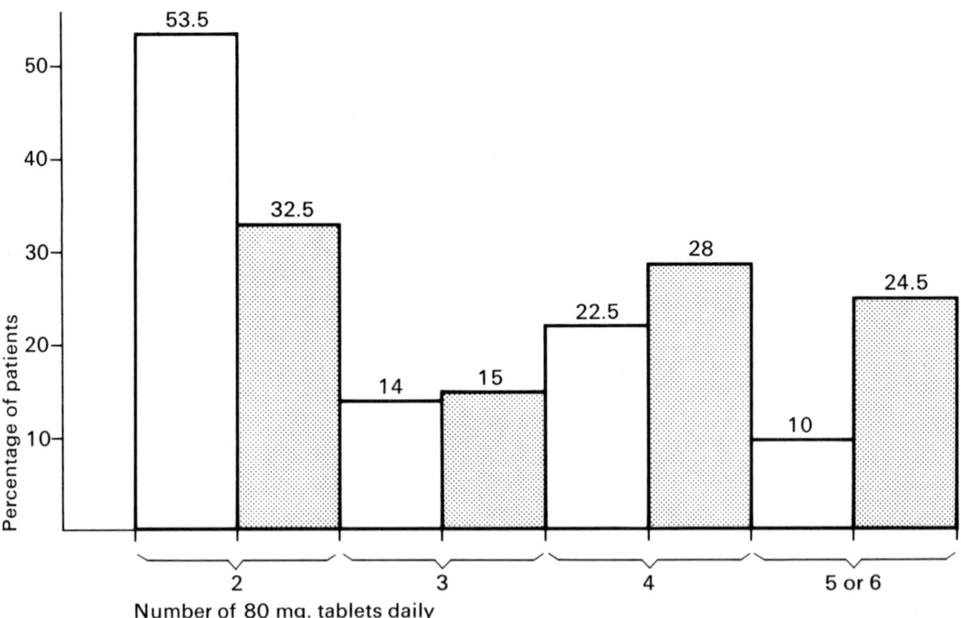

Fig. 3. Dosages of oxprenolol (expressed in tablets of 80 mg. daily) required in patients with initial diastolic pressures of 95–120 mm. Hg (□) and over 120 mm. Hg (▨).

treated with oxprenolol, together with cyclopenthiazide and potassium chloride (®Navidrex-K). For purposes of analysis, they were divided into those patients with pressures between 95 and 120 mm. Hg (average 108) and those with pressures between 121 and 140 mm. Hg (average 130). The dosage of the diuretic was kept constant at two tablets of cyclopenthiazide ($= 0.5$ mg.), whereas the dosage of oxprenolol was increased until an optimum response had been obtained. The patients in the higher blood pressure range required an average of 160 mg. more oxprenolol daily, but it will be noted from Figure 3 that in both groups a large number of patients responded to 160 mg. (80 mg. twice daily). The *mean* systolic and diastolic pressures for the milder hypertensive group were 184/108 falling to 154/90 mm. Hg, and for the more severe group 210/130 falling to 169/100 mm. Hg (Figure 4). In the first group, 8% of the patients were considered to be poorly controlled, i.e. their blood pressure either did not fall at all or remained above 100 mm. Hg (diastolic); in the second group, 31% were poorly controlled, i.e. had diastolic pressures that were still higher than 100 mm. Hg. These figures are comparable with those reported by other workers who have employed beta-blocking drugs for the management of hypertension[8, 10, 12].
The incidence of adverse reactions was similar in both studies (15% and 18.5%, respectively), but in only 4% and 6% of cases were these side effects considered severe enough to necessitate discontinuation of the therapy. The commonest were once again gastro-intestinal disturbances, dizziness, and gid-

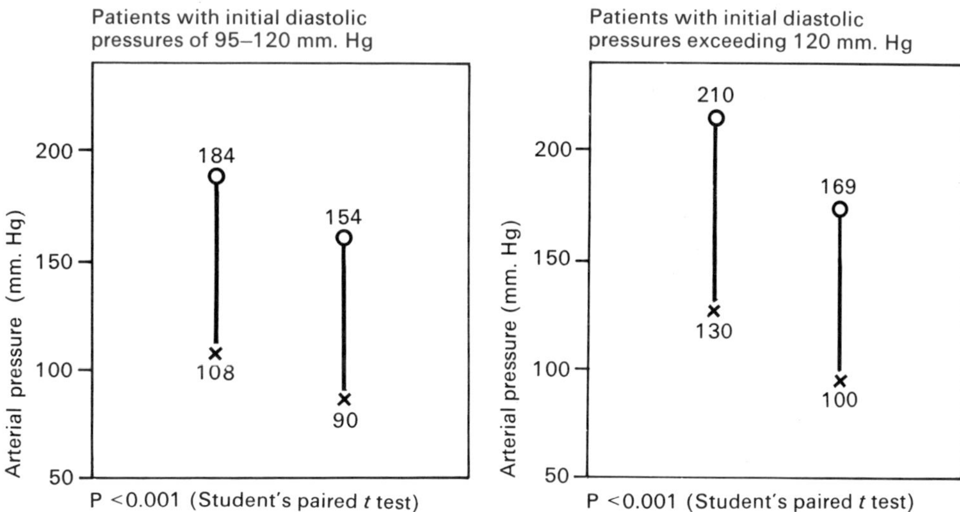

Fig. 4. *Left:* mean initial and final systolic (O) and diastolic (✗) pressures in 2,211 patients with mild to moderate hypertension after four weeks' treatment on optimum dosage (see text). Although the total number of cases studied was 2,300, 89 patients had to be withdrawn from the trial owing to the occurrence of adverse effects; this explains why the data given relate to only 2,211 patients. *Right:* mean initial and final systolic (O) and diastolic (✗) pressures in 554 patients with moderate to severe hypertension.

diness; heart failure, bradycardia, and bronchospasm occurred in only a few isolated cases, their overall incidence amounting to less than 0.5%.

In conclusion, one can say that this group of monitored release studies, involving a total of 8,346 patients suffering either from angina pectoris or from hypertension, has demonstrated the value, safety, and acceptability of oxprenolol. The side effects documented were those common to all beta-blocking drugs; no unexpected adverse reactions were encountered; and fears concerning serious hazards such as heart failure and bronchospasm were largely dispelled.

References

1 BURLEY, D.M.: A monitored release study. Danish pneumol. Soc., 258th Meet., Scanticon, Aarhus, Denmark 1971, p.43
2 EDITORIAL: Adverse non-drug reactions. Brit. med. J. *i*, 798 (1969)
3 FORREST, W.A.: A total of 254 cases of angina pectoris treated with oxprenolol in hospital practice – a monitored release study. Brit. J. clin. Pract. *26*, 217 (1972)
4 FORREST, W.A.: A monitored release study: a clinical trial of oxprenolol in general practice. Practitioner *208*, 412 (1972)
5 FORREST, W.A.: Oxprenolol and a thiazide diuretic together in the treatment of essential hypertension – a large general practice study. Brit. J. clin. Pract. *27*, 331 (1973)
6 FORREST, W.A.: A subjective comparison between oxprenolol (Trasicor) and long-acting nitrates in the treatment of angina pectoris in general practice. J. int. med. Res. *1*, 253 (1973)
7 FORREST, W.A.: Treatment of moderately severe essential hypertension in general practice with a combination of cyclopenthiazide with potassium chloride (Navidrex K) and oxprenolol (Trasicor 80 mg.): a report of 554 patients. J. int. med. Res. *2*, 7 (1974)
8 PRICHARD, B.N.C., BOAKES, A.J., DAY, G.: Practolol in the treatment of hypertension. Postgrad. med. J. *47*, Suppl. (Jan.): 84 (1971)
9 REIDENBERG, M.M., LOWENTHAL, D.T.: Adverse nondrug reactions. New Engl. J. Med. *279*, 678 (1968)
10 SIMPSON, F.O., WAAL-MANNING, H.J.: Hypertension and beta-adrenergic blockade. Symp. Beta adrenergic blocking agents, Sydney, Australia, 1970, p.59
11 VERE, D.W.: Controlled trials to detect efficacy and toxicity: training to meet tomorrow's needs. In Harris, E.L., Fitzgerald, J.D. (Editors): The principles and practice of clinical trials, p.242 (Livingstone, Edinburgh/London 1970)
12 ZACHARIAS, F.J., COWEN, K.J., PRESTT, J., VICKERS, J., WALL, B.G.: Propranolol in hypertension: a study of long-term therapy, 1964–1970. Amer. Heart J. *83*, 755 (1972)

Discussion

P. E. Lucchelli: I should like to quote a few data confirming what Dr. Burley has said about the antihypertensive action of beta-blockers as reflected in the results of large-scale trials. In a monitored release trial carried out in Italy, more than 900 patients with angina pectoris who were already receiving a nitrite derivative were given additional treatment with oxprenolol (®Trasicor) in doses of up to 120 mg. daily. Of those patients – numbering almost 300 – who also had above-normal blood pressure levels before the start of oxprenolol medication, almost all showed a decrease in systolic and diastolic pressure after four to ten weeks of treatment with oxprenolol added to their previous nitrite therapy (Figure 1).

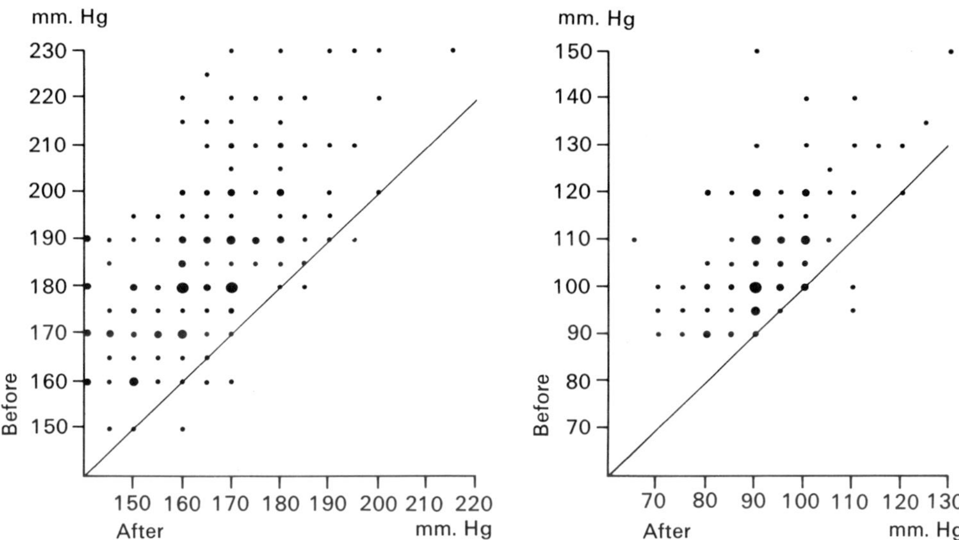

Fig. 1. Systolic pressures in 289 patients *(left)* and diastolic pressures in 270 patients *(right)* before and four to ten weeks after oxprenolol had been added to the treatment with a nitrite derivative which they had been receiving for the relief of angina pectoris. Since, for reasons of space, it would have been impossible to assign one dot to each of the 289 and 270 patients, respectively, the size of the dots plotted varies roughly proportionally to the number of patients in whom the values in question were recorded.

M. Motolese: The main purpose of the monitored release study to which Dr. Lucchelli has just referred was to assess the additional benefit which patients with angina pectoris might derive from oxprenolol when added – in daily doses of up to 120 mg. for a period of four to ten weeks – to the treatment with a nitrite derivative which they had previously been receiving.

In all, 935 patients were treated by a total of 78 general practitioners and practising cardiologists distributed throughout Italy. During check-ups performed at various intervals in the course of the treatment, any changes occurring in the frequency of attacks of chest pain, in the patient's nitroglycerin consumption, in his physical working capacity, or in the electrocardiogram were recorded. In addition, at the end of the trial, the physician was requested to give his overall assessment of the treatment, using a visual analogue scale.

As can be seen from Figure 2, in most cases the attacks of chest pain disappeared progressively or decreased in frequency during the treatment; similar improvements were

149

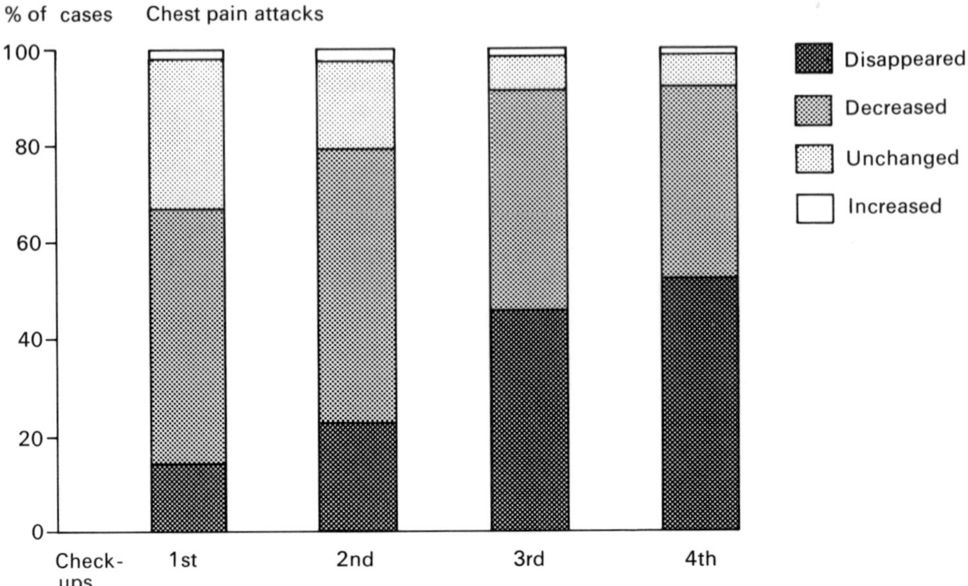

Fig. 2. Influence exerted by additional treatment with oxprenolol (given in daily doses of up to 120 mg. for a period of four to ten weeks) on attacks of chest pain in 935 patients with angina pectoris who had previously been receiving a nitrite derivative only.

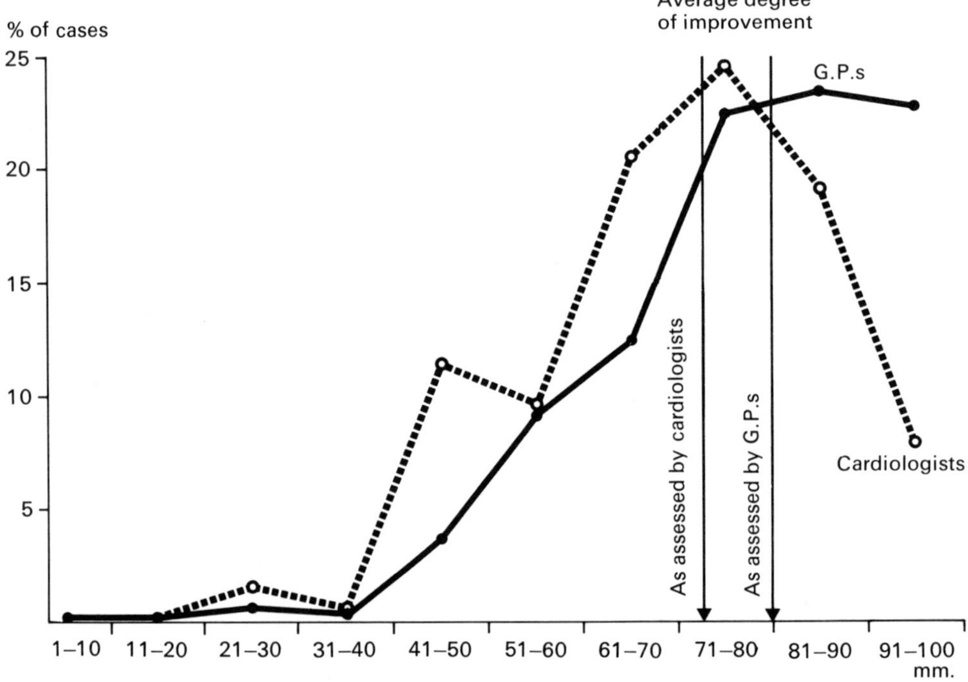

Fig. 3. General practitioners' and cardiologists' overall assessment of the response to treatment with oxprenolol plus a nitrite derivative in 935 patients with angina pectoris, based on a visual analogue scale in which 100 mm. represents the maximum degree of improvement.

also observed in the other criteria of assessment I have mentioned. It should be added, however, that in the case of the patients treated by cardiologists the final assessments made in respect of the various individual criteria were consistently less favourable than those of the general practitioners. The results of the overall assessment based on the visual analogue scale are shown in Figure 3, in which 100 mm. represents the maximum degree of improvement. Here again, the improvement ratings given by the cardiologists were on the whole lower – possibly because the patients seen by them were generally in a more serious condition.

Side effects occurred in approximately 12% of cases. These side effects, which usually appeared during the first week of medication, showed no direct relationship either to the dosage of oxprenolol employed or to the age of the patients. Oxprenolol had to be withdrawn in only 2% of cases. The results obtained in this study suggest that a combination of oxprenolol and a nitrite derivative constitutes a safe and effective form of treatment for angina pectoris.

E. SOWTON: I'd like to know, Dr. BURLEY, how you managed to exclude misdiagnoses. You admit that the diagnoses were probably wrong in 4–5% of the cases, but I should have thought that probably about 30–35% of the patients entered on the cards as having angina pectoris due to coronary artery disease were likely in fact to be not suffering from angina of this type.

D. M. BURLEY: You may well be right, Dr. SOWTON. We tried to exclude these cases by getting the doctors to write down some simple symptom complexes on the basis of which we thought we might be able to judge which diagnoses were likely to be wrong. But I'm quite prepared to accept that the proportion of misdiagnoses may have been higher than the 4–5% I quoted.

W. SOMERVILLE: Monitored release studies suffer from one great disadvantage. While they enable you to collect data quickly on large numbers of patients, they involve taking the first ten patients seen by the doctor, and in these first ten cases side effects may be overlooked or may be magnified; some of the virtues of the treatment under trial may be initially blurred. I am not sure, Dr. BURLEY, whether you took the first, or, preferably, the second or third ten.

D. M. BURLEY: I'm afraid we had no way of knowing whether the first, second, or even perhaps the third ten were taken. The original idea of the monitored release studies was that they should be a pre-condition for the releasing of a drug on to the general market. In other words, the drug would be made available to the doctor virtually on the condition that he would agree to document his results in the first ten patients. The primordial consideration was one of safety, but studies of this kind also provide a lot of other valuable information, and the data they yield can be regarded as affording a not unreasonable guide. A disadvantage, of course, is that these are uncontrolled studies, as I took care to point out at the beginning of my paper. Given the limited resources that are available, I quite honestly believe that it would be very difficult to get results on such a large scale by any other means. Moreover, as I tried to indicate in Table 1, in which I compared the criteria for a conventional study with the criteria for this sort of study, you do tend to overcome some of the problems, but, I agree, not all of them.

General discussion on the use of beta-blockers in hypertension

W. Schweizer: We are now going to have a general discussion on the use of beta-blockers in arterial hypertension, bearing in mind all the facts and findings presented in the course of this first session. I suggest that we try to concentrate our attention on two specific questions, the answers to which will, I hope, help us later when we come to consider guidelines for the general practitioner. The first of these specific questions is: Via what mechanisms do the beta-blockers exert their antihypertensive effect?

H. Brunner: Perhaps I can best answer this question by outlining – without attempting to assess their relative merits – the various hypotheses which have so far been put forward in the literature to account for the mechanisms by which the beta-blockers produce their antihypertensive effect.
The first hypothesis postulates a site of attack in the central nervous system. It is in experiments performed by Dollery's group on conscious rabbits* that this possibility of a central effect has been most thoroughly investigated. These authors succeeded in demonstrating that the prolonged fall in blood pressure occurring in the rabbit in response to intracerebroventricular injections of a beta-blocker (the drug used was propranolol) could only have been due to beta-blockade. Neither the D-isomer nor local anaesthetics elicited a comparable effect. Similar findings have been reported by Day and Roach**, who also observed protracted decreases in blood pressure in non-anaesthetised cats following intracerebroventricular injections.
In addition, Carter et al.*** found that in anaesthetised dogs propranolol – given in relatively high doses – produced a greater fall in blood pressure when injected into the cerebral ventricles, into the cisterna, or into the vertebral artery than when administered intravenously. In connection with all these experiments, however, it must be pointed out that a beta-blocker injected intracerebroventricularly reaches other centres in the brain in other concentrations than when it is simply introduced into the bloodstream. Studies of this kind therefore afford only limited proof of a central mechanism.
Another possible mechanism, which Prichard and Gillam**** in particular have discussed, might take the form of a resetting of the baroceptors, i.e. an increase in the sensitivity of the baroceptors.
A third mechanism, to which frequent reference has already been made at this symposium, concerns the reduction in cardiac output which beta-blockers provoke. It is probably true to say that in acute experiments all beta-blockers diminish cardiac output to quite a similar degree when administered intravenously. But it seems fairly certain that the various beta-blockers do differ from one another in this respect when given orally on a long-term basis – and this, in the last analysis, is the decisive point so far as antihypertensive therapy is concerned. Studies such as those undertaken by

---

* Reid, J.L., Lewis, P.J., Myers, M.G., Dollery, C.T.: Cardiovascular effects of intracerebroventricular d-, l- and dl-propranolol in the conscious rabbit. J. Pharmacol. exp. Ther. *188*, 394 (1974)
** Day, M.D., Roach, A.G.: β-Adrenergic receptors in the central nervous system of the cat concerned with control of arterial blood pressure and heart rate. Nature new Biol. (Lond.) *242*, 30 (1973)
*** Carter, J.K., Mitchell, H.W., Poyser, R.H.: Comparison of some haemodynamic changes between central and intravenous administration of ($\pm$)-propranolol in anaesthetized dogs. Proc. Brit. pharmacol. Soc., Southampton, 28th–29th March, 1974, p.34
**** Prichard, B.N.C., Gillam, P.M.S.: The use of propranolol (Inderal) in the treatment of hypertension. Brit. med. J. *ii*, 725 (1964)

the FREIS group* with timolol, or by JOHNSSON et al.** with alprenolol, appear to
have shown that chronic treatment is not associated with a decrease in cardiac output.
These observations, however, conflict with those reported by other authors, e. g. by
LUND-JOHANSEN*** with alprenolol, by HANSSON**** and TARAZI and DUSTAN*****
with propranolol, or by AMERY et al.****** with the beta-blocker ICI 66,082. I think
we must therefore assume that a decrease in cardiac output does in fact play some role in
the mechanism by which beta-blockers lower elevated blood pressure.
A fourth explanation for the antihypertensive action of beta-blockers may be a reduction in plasma volume. Dr. TARAZI has dealt at length with this mechanism in the
paper he presented here. Similar findings have also been published by JULIUS et
al.*******, who observed quite a pronounced decrease in plasma volume following
intravenous administration of propranolol.
Finally, the inhibitory effect of beta-blockade on the release of renin may offer a
fifth explanation. Here it would seem more likely that beta-blockers act, not so much
by lowering renin levels at rest, but rather by preventing them from rising in response
to an increase in sympathetic tone – an increase which may be due simply to the fact
of the patient's assuming an upright position or, alternatively, to treatment with a
diuretic or a vasodilator.

L. WERKÖ: The proceedings of this whole session on hypertension have demonstrated
what a lot we have now learned about the use of beta-blockers in hypertensive disease
and also about the mechanisms that have an important bearing on this disease. However, I also think these proceedings show that we have an insight into only a very small
part of the general mosaic of hypertensive cardiovascular disease and its treatment.
In order to further our knowledge, may I at this point make a plea for larger studies
in well-defined series of patients. What has so far been reported consists only of observations that may be suitable for use as a starting point for new and more extended
studies which, I hope, will eventually enable us to decide upon exactly what mechanisms the antihypertensive effects of the beta-blockers are based. I feel, for example,
that the haemodynamic studies undertaken so far merely provide a *point de départ* for
more detailed studies of the various vascular beds. Data seem to be slowly accumulating which indicate that, in the long run, treatment for hypertension with any drug
which decreases the blood pressure will also have the effect of diminishing peripheral
vascular resistance – as determined in the extremities, for example, at maximal vasodilatation. It is studies of this kind that we need before we can really say what the
mechanisms are by which the beta-blockers exert their antihypertensive effect. It must

---

* FRANCIOSA, J.A., FREIS, E.D., CONWAY, J.: Antihypertensive and haemodynamic
properties of the new beta adrenergic blocking agent timolol. Circulation *48*, 118
(1973)
** JOHNSSON, G., GUZMAN, M. DE, BERGMAN, H., SANNERSTEDT, R.: The haemodynamic effects of alprenolol and propranolol at rest and during exercise in hypertensive patients. Pharmacol. clin. *2*, 34 (1969)
*** LUND-JOHANSEN, P.: Hemodynamic changes at rest and during exercise in long-
term β-blocker therapy of essential hypertension. Acta med. scand. *195*, 117 (1974)
**** HANSSON, L.: Beta-adrenergic blockade in essential hypertension. Effects of pro-
pranolol on hemodynamic parameters and plasma renin activity. Acta med. scand.
194, Suppl. 550 (1973)
***** TARAZI, R.C., DUSTAN, H.P.: Beta adrenergic blockade in hypertension. Amer.
J. Cardiol. *29*, 633 (1972)
****** AMERY, A., BILLIET, L., JOOSSENS, J.V., MEEKERS, J., REYBROUCK, T., VAN
MIEGHEM, W.: Preliminary report on the haemodynamic response of hypertensive
patients treated with a beta blocker (ICI 66082). Acta clin. belg. *28*, 358 (1973)
******* JULIUS, S., PASCUAL, A.V., ABBRECHT, P.H., LONDON, R.: Effect of beta-
adrenergic blockade on plasma volume in human subjects. Proc. Soc. exp. Biol. (N.Y.)
140, 982 (1972)

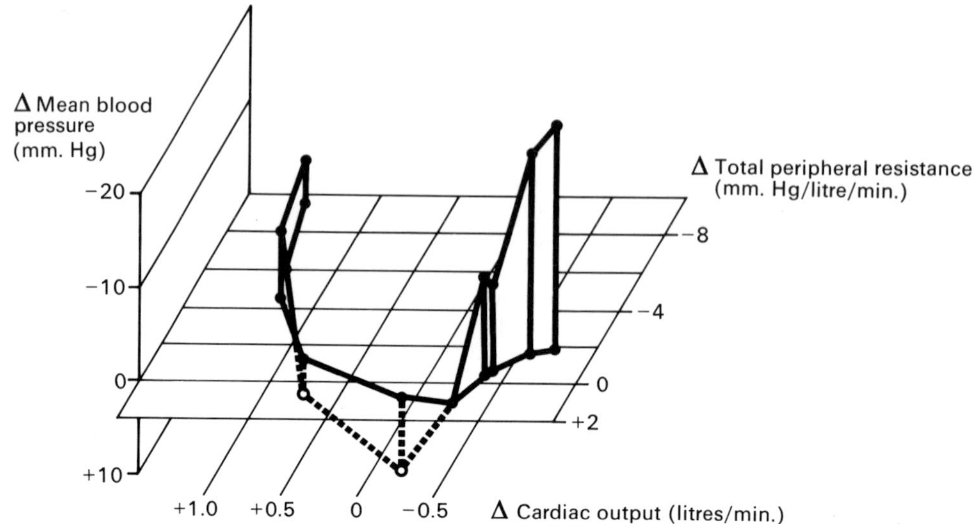

Fig. 1. Three-dimensional plot (based on haemodynamic data obtained in nine hypertensive patients treated with practolol for three weeks) showing the relationship between the changes occurring in blood pressure, total peripheral resistance, and cardiac output.

also be stressed that, so long as we don't know how regulation of blood pressure really works in the patient with hypertension, it is very difficult to say how treatment influences it. Another thing that is quite clear is that most of the observations reported have been made in patients studied in some sort of undefined basal state. Consequently, what we need, too, is research indicating how patients react to the stresses of daily life. Exercise has been used as one of these stresses that can be reproduced fairly well; I say fairly well, because you have of course to define what level of exercise the patient can tolerate and at what level you are really studying him. We are thus still only beginning to acquire knowledge as to how hypertension influences the vascular system and how we should treat it; we certainly haven't yet reached a point where we can make many dogmatic statements.

K. D. BOCK: Since only a few reports have been published on the haemodynamic effects of longer-term treatment with beta-blockers, I should like at this point to refer briefly to a study which Dr. M. ANLAUF has carried out at our clinic in Essen on patients with mild or moderate hypertension before and after three weeks of oral medication with 300–800 mg. practolol daily. These patients showed a moderate fall in blood pressure (determined by direct intra-arterial measurement), an increase in heart rate, no change in cardiac index, and a small but significant decrease in total peripheral resistance. The central venous pressure diminished slightly, but no alteration in blood volume occurred. These findings are similar to those reported by TARAZI et al.* and BODEM et al.**. No significant correlation was found between the fall in blood pressure and the change either in cardiac output or in total peripheral resistance. However, when the changes in blood pressure, cardiac output, and total peripheral

---

* TARAZI, R.C., SAVARD, Y., DUSTAN, H.P., BRAVO, E.L.: Cardioselective beta adrenergic blockade in hypertension. Clin. Pharmacol.Ther. *13*, 154 (1972); abstract of paper
** BODEM, G., BRAMMELL, H.L., WEIL, J.V., CHIDSEY, C.A.: Pharmacodynamic studies of beta adrenergic antagonism induced in man by propranolol and practolol. J. clin. Invest. *52*, 747 (1973)

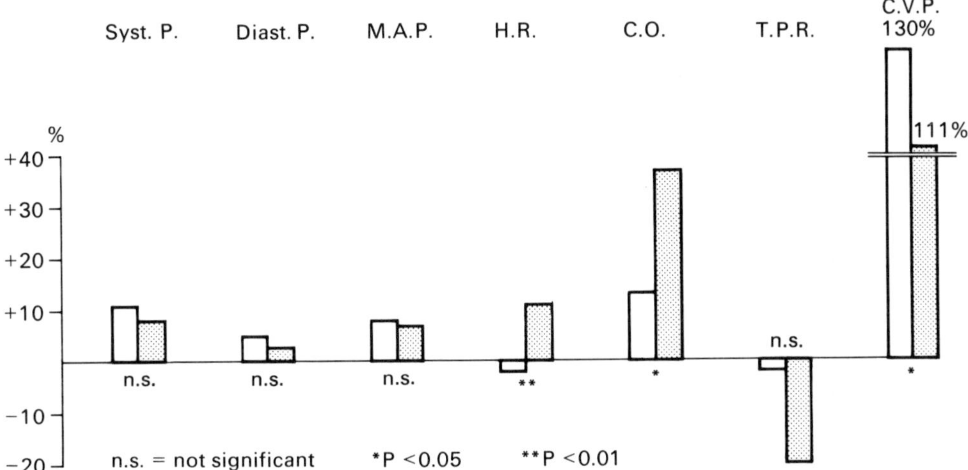

Fig. 2. Haemodynamic effects of plasma expansion (using sodium-free dextran solution) in ten hypertensive patients before (□) and after (▨) three weeks of treatment with practolol. Syst. P. = systolic pressure; Diast. P. = diastolic pressure; M.A.P. = mean arterial pressure; H.R. = heart rate; C.O. = cardiac output; T.P.R. = total peripheral resistance; and C.V.P. = central venous pressure.

resistance were plotted in a three-dimensional system (Figure 1), it appeared that a distinction could be drawn between three types of patient: firstly, patients in whom a slight increase or no change in blood pressure was accompanied by only small alterations in the other variables measured; secondly, patients who responded with a fall in blood pressure plus a decrease in total peripheral resistance and a concomitant increase in cardiac output; and, thirdly, patients who also showed a fall in blood pressure, but in whom there was a decrease in cardiac output and only a slight reduction in total peripheral resistance.

In order to obtain information on blood pressure regulation under the influence of practolol, the plasma volume of the ten patients was acutely expanded by rapidly infusing sodium-free dextran solution (in a quantity equivalent to 15% of the blood volume) before and after treatment with practolol. Figure 2 reveals that the response of the blood pressure to the infusion remained practically unchanged after three weeks of practolol medication, whereas heart rate and cardiac output showed a considerable increase; total peripheral resistance diminished, and central venous pressure showed a significantly less pronounced rise. It is curious to note that, following treatment with practolol, the increase in cardiac output provoked by plasma expansion caused no additional rise in blood pressure. The increase in heart rate and cardiac output occurring in response to plasma expansion after three weeks of practolol therapy may have been due to the drug's intrinsic sympathomimetic activity.

These results appear to confirm the conclusion reached by other authors that the changes which beta-blockers produce in cardiac output are not the only factor responsible for their antihypertensive activity. This activity seems to be the net result of an interplay between the peripheral – in the case of practolol, mainly cardiac – effects of the beta-blockers, their action on peripheral resistance (which is possibly a central one), and autoregulatory phenomena which help to determine the final extent of the reduction in blood pressure.

J. BROD: When discussing the haemodynamic mechanisms by which blood pressure is lowered, we have so far largely confined our attention to cardiac output and to total peripheral vascular resistance, which of course is only a calculated value. I should

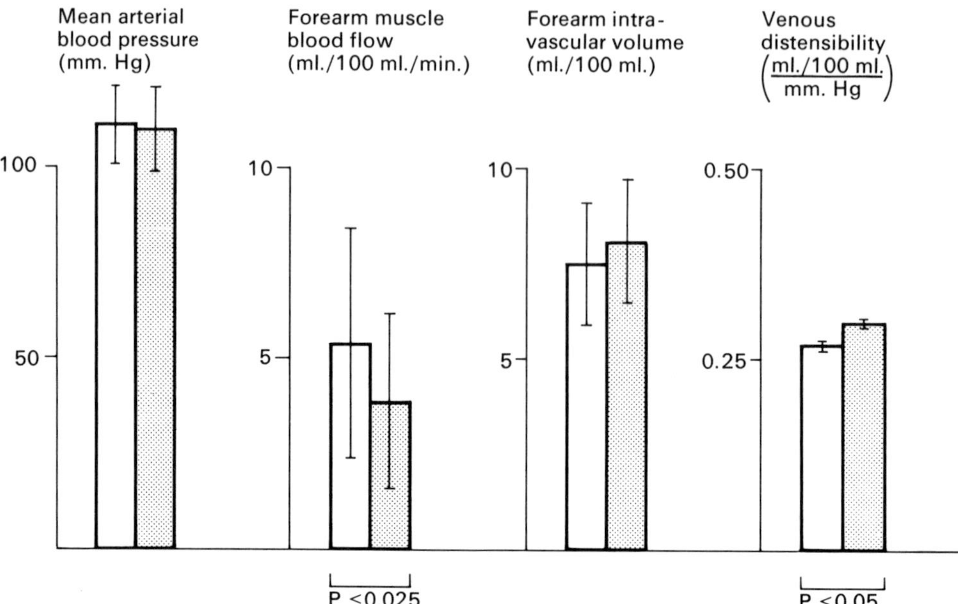

Fig. 3. Blood pressure, forearm blood flow and intravascular volume, and venous distensibility in seven healthy normotensive subjects (aged 17–22 years) at rest (□) and in response to propranolol administered intravenously in a dose of 2.5 mg. over a period of 2½ minutes (▨). (From: CACHOVAN, M.: Veränderungen der peripheren Durchblutung und venösen Distensibilität unter β-adrenergischer Rezeptoren-Stimulation und -Blockade. Paper presented at the Jahresvers. Dtsch. Ges. Angiol., Würzburg 1972)

therefore like now to present some additional data. If you examine the findings reproduced in Figure 3, you will see here that there is no change in the blood pressure in response to propranolol, but that blood flow through the forearm shows a significant drop. This is exactly the opposite of what happens as a result of beta-stimulation, e. g. under conditions of emotional stress. Note, on the other hand, that the volume of blood in the forearm – in contrast to the blood flow – does not change significantly. In the last two columns of Figure 3, we have data on venous distensibility. This increases. In other words, the mean systemic pressure (i. e. the pressure which would be established in the whole vascular system in the absence of heart action) obviously drops, and this may well be another contributory factor, in addition to the drug's direct effect, in the lowering of cardiac output.

Let me now show you a slide (Figure 4) dealing with data on haemodynamic parameters in essential hypertension. In the first two columns are data on cardiac output. The values exhibit a greater scatter in the hypertensives than in the normotensives, confirming what Dr. SANNERSTEDT has already said. From the second pair of columns it can be seen that blood flow through the forearm muscle is higher in essential hypertensives than in normotensives. In view of what I have just said about the effect of beta-blockers on peripheral blood flow, it seems conceivable that the rise in total peripheral vascular resistance occurring under these conditions may be due to closing of the vascular bed in the muscles, and not to some compensatory mechanism.

Figure 5 shows that in essential hypertension there are also marked changes on the venous side of the circulation. In the patients with essential hypertension, intravascular volume in the forearm, as well as venous distensibility, is considerably diminished. Hence, if a beta-blocker is given to a patient with essential hypertension, and if it in-

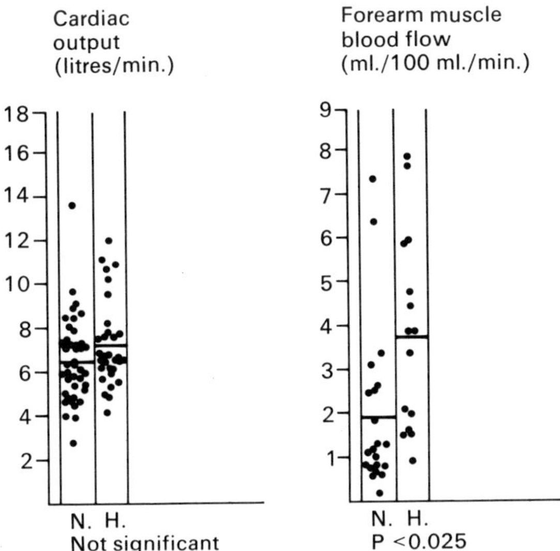

Fig. 4. Cardiac output and forearm muscle blood flow in normotensives (N.) and patients with essential hypertension (H.). [Adapted from: BROD, J.: Clinical significance of labile hypertension. In White, P. D. (Editor): Cardiovascular clinics, Vol. II, No. 3: International Cardiology, p. 17 (Davis, Philadelphia 1971)]

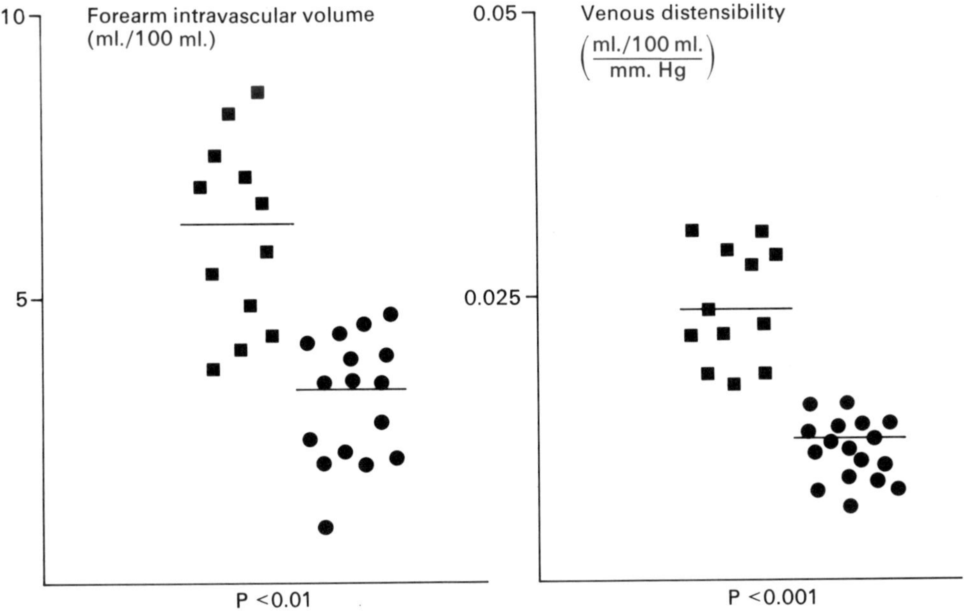

Fig. 5. Forearm intravascular volume and venous distensibility in normotensives (■) and in patients with essential hypertension (●). (From: BROD, J.: Zentrale und periphere Haemodynamik bei gesunden, essentiellen Hypertonikern und renalen Hypertonikern mit einigen Bemerkungen über den Einfluss von Training und körperlicher Belastung. Paper presented at the Jahresvers. Dtsch. Ges. Angiol., Würzburg 1972)

creases his venous distensibility, perhaps this too may contribute to the lowering of his cardiac output and to the normalisation of his blood pressure.

Presented in Figure 6 is a brief review of our latest data on hypertension in chronic renal disease. From the cardiac index values it is apparent that some of the patients with renal disease and hypertension have high cardiac outputs, though only in this initial stage. From the bottom row, it can be seen that the patients with high cardiac outputs are those who have the lowest venous distensibility. They have also the lowest amount of blood in the forearm. Again, it would not be unreasonable to assume that, if we give a beta-blocker to this type of patient, we shall increase his venous distensibility and that this in turn might favourably affect his cardiac output and blood pressure.

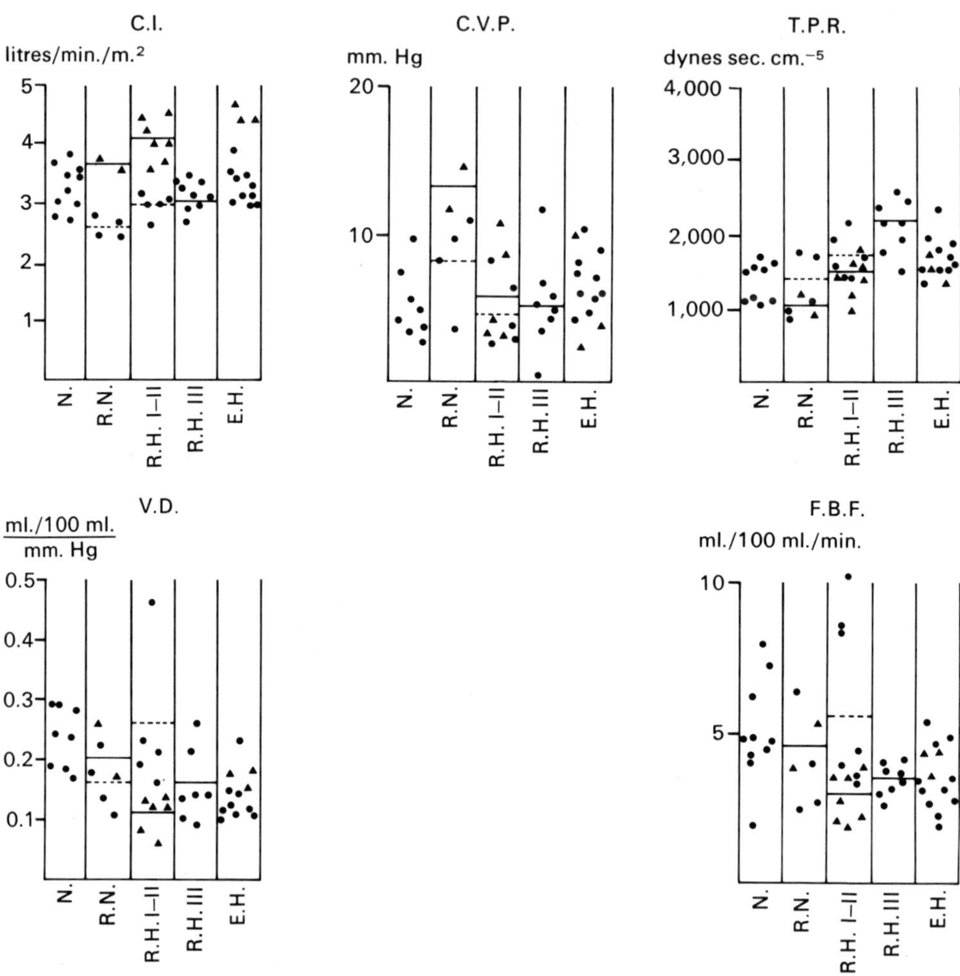

Fig. 6. Cardiac index (C.I.), central venous pressure (C.V.P.), total peripheral resistance (T.P.R.), venous distensibility (V.D.), and forearm blood flow (F.B.F.) in normotensive controls (N.), renal patients with normal blood pressure (R.N.), patients with renal hypertension (R.H.), and patients with essential hypertension (E.H.). The triangles represent subjects with a cardiac index of more than 3.5 litres/min./m.² (——— = mean values) and the circles those with a cardiac index of less than 3.5 litres/min./m.² (------ = mean values). (From: Brod, J., Cachovan, M., Bahlmann, J., Celsen, B., Sippel, R., Herbst-Falkenreck, I., Hundeshagen, H.: Peripheral blood flow and capacitance vessels in essential and renal hypertension. Paper presented at a symposium on hypertension, University of Nijmegen, 16.9.1974)

Table 1. Blood pressure and heart rate responses in conscious normotensive cats to intracerebroventricular (i.c.v.) administration of beta-blockers and local anaesthetics. "Initial rise" indicates the peak initial stimulant effect reached 5–15 minutes after the end of the intracerebroventricular infusion. "Prolonged fall" shows the maximum depressor effects recorded 1–1½ hours after the end of the infusion. Inhibition of isoprenaline-induced tachycardia was measured one hour after the administration of each blocking agent.

| | Dose (mg. i.c.v.) | Initial rise | | | Prolonged fall | | | % inhibition of tachycardia induced by isoprenaline i.c.v. | No. of cats |
| | | Blood pressure (mm. Hg) | | Heart rate (beats/min.) | Blood pressure (mm. Hg) | | Heart rate (beats/min.) | | |
		Systolic	Diastolic		Systolic	Diastolic			
DL-Propranolol	0.5	16.7 ± 3.2	16.2 ± 3.6	10.1 ± 2.2	24.7 ± 2.9	24.9 ± 2.8	30.7 ± 3.2	87	12
	1.0	25.4 ± 2.8	24.1 ± 3.2	22.3 ± 3.6	29.9 ± 3.1	30.3 ± 2.9	41.5 ± 3.6	100	10
D-Propranolol	0.5	18.4 ± 3.6	17.7 ± 3.8	12.4 ± 3.9	Nil	Nil	Nil	Nil	5
L-Propranolol	0.5	15.3 ± 3.3	15.1 ± 3.4	9.8 ± 3.5	25.8 ± 4.0	26.0 ± 3.8	38.6 ± 3.7	100	5
DL-Alprenolol	0.5	19.9 ± 2.4	20.1 ± 2.5	12.4 ± 2.1	20.3 ± 3.1	20.5 ± 3.2	24.5 ± 2.8	81	12
	1.0	28.4 ± 2.6	27.6 ± 2.3	25.7 ± 2.5	26.4 ± 2.6	26.5 ± 2.7	33.9 ± 3.0	100	10
D-Alprenolol	0.75	21.7 ± 3.9	20.2 ± 3.8	15.4 ± 4.2	Nil	Nil	Nil	Nil	5
Pindolol (LB 46)	0.75	27.8 ± 6.8	18.2 ± 3.0	35.0 ± 6.1	14.3 ± 1.2	16.0 ± 1.5	17.8 ± 2.8	92	5
	1.0	45.6 ± 9.4	27.0 ± 5.4	41.7 ± 8.0	22.1 ± 2.9	22.9 ± 3.2	26.9 ± 3.1	100	5
Practolol	2.0	10.3 ± 3.8	9.6 ± 3.7	7.3 ± 2.4	12.6 ± 2.2	12.6 ± 2.4	19.7 ± 3.7	66	5
	4.0	18.9 ± 4.1	15.9 ± 3.9	14.0 ± 4.4	20.3 ± 3.0	20.9 ± 3.2	29.4 ± 3.8	85	4
Oxprenolol	0.5	89.1 ± 13.6	62.0 ± 7.9	37.5 ± 6.5	14.7 ± 2.6	14.3 ± 2.1	25.0 ± 2.7	79	6
ICI 66,082	1.0	Nil	Nil	Nil	22.4 ± 3.1	22.6 ± 3.1	38.4 ± 4.6	71	6
	2.5	5.1 ± 2.1	4.8 ± 2.3	Nil	32.8 ± 3.7	34.2 ± 3.6	47.6 ± 5.0	88	5
Sotalol	2.0	16.8 ± 4.3	15.9 ± 4.1	8.4 ± 3.9	15.1 ± 3.8	15.9 ± 3.9	18.2 ± 3.5	67	5
Procaine	1.5	28.4 ± 4.6	26.1 ± 4.4	26.0 ± 5.1	Nil	Nil	Nil	Nil	4
Lidocaine	0.75	30.1 ± 2.3	29.3 ± 3.5	26.3 ± 3.6	Nil	Nil	Nil	Nil	5

M.D.Day: Dr. Brunner has mentioned the central nervous system as a possible site of action for the antihypertensive effects of beta-blockers, and, in the papers and discussions we have already heard, several other speakers referred to the central effects of these drugs. I would like to present a table summarising the effects of various beta-blockers and local anaesthetics when administered intracerebroventricularly in conscious, unanaesthetised, normotensive cats (Table 1). Now practically all the beta-blockers, when given into the lateral ventricles, produce initial sympathomimetic effects. This table is rather complicated, because it shows first of all the initial sympathomimetic effects, which usually reach their peak within 5–15 minutes from the start of the infusion, whereas on the right it indicates the much more prolonged effects in the shape of sustained bradycardia and a fall in pressure. All the beta-blockers induce an initial rise in pressure, as do also D-isomers – which don't have beta-blocking activity – and the local anaesthetics. However, it is only the substances with beta-blocking activity that produce a prolonged fall in blood pressure subsequent to their initial sympathomimetic effect. I believe that in man oxprenolol (®Trasicor) is the beta-blocker displaying the most potent intrinsic sympathomimetic activity. When given into the lateral ventricles, it is also the most potent stimulant; subsequently it causes a nice fall in pressure, as well as bradycardia. The only substance that we have used to date which has no intrinsic sympathomimetic activity is the new beta-blocker ICI 66,082; this compound, however, still produces a dose-dependent fall in blood pressure and heart rate after about 30 minutes or so.

Hence – though, as Dr. Brunner has pointed out, there are other explanations for the actions exerted by beta-blockers when given centrally – the fact remains that, when administered in this way, they do appear to be capable of selectively blocking the effect of beta-stimulants and they do produce as clinical effects a lowering of the blood pressure and of the heart rate.

P. Kielholz: I should also like to add a few remarks on this question of the central effect of beta-blockers with particular reference to the problem of anxiety. First of all, we have to decide precisely at what points an influence on anxiety can be produced at all. These possible points of attack are indicated in Figure 7: firstly, we can exert a central action on the psyche and, secondly, a peripheral action on the sympathetic nervous system; stimulation of the sympathetic system is liable to result in cardiac symptoms, which – if the patient becomes excessively preoccupied by them or is unable to verbalise the sensations to which they give rise – may well have adverse repercussions on the psyche, thereby setting up a vicious circle.

A central effect on the psyche can be achieved by administering minor or major tranquillisers and anxiolytic antidepressants; to combat sympathetic overstimulation, we can employ beta-blockers; and to overcome the feeling of isolation which prevents the patient from verbalising his problems, we can resort to psychotherapy. An interesting feature of the activity of the beta-blockers in this connection is the fact that they not only suppress the physical signs and symptoms, but also abolish or at least alleviate the central anxiety – and this is an effect which cannot be attributed solely to inhibition of sympathetic stimulation.

In a study in which we investigated 15 patients subject to acute anxiety reactions – e.g. students confronted with the prospect of an examination, persons worried at having to speak in public, and actors with stage fright on the opening night – we found that, when they were given 20–40 mg. oxprenolol two hours prior to the ordeal responsible for their apprehensiveness, not only did the cardiac symptoms fail to appear but the central component of the anxiety was also effectively counteracted. Moreover, the same type of response was observed in six patients suffering from states of anxious and agitated depression, in whom beta-blockade eliminated the anxious agitation but had no influence on the depression as such.

This, too, suggests that beta-blockers also have a central action. Finally, we have noted that certain types of tremor may respond to beta-blockers, a response which can likewise be mediated only via a central effect.

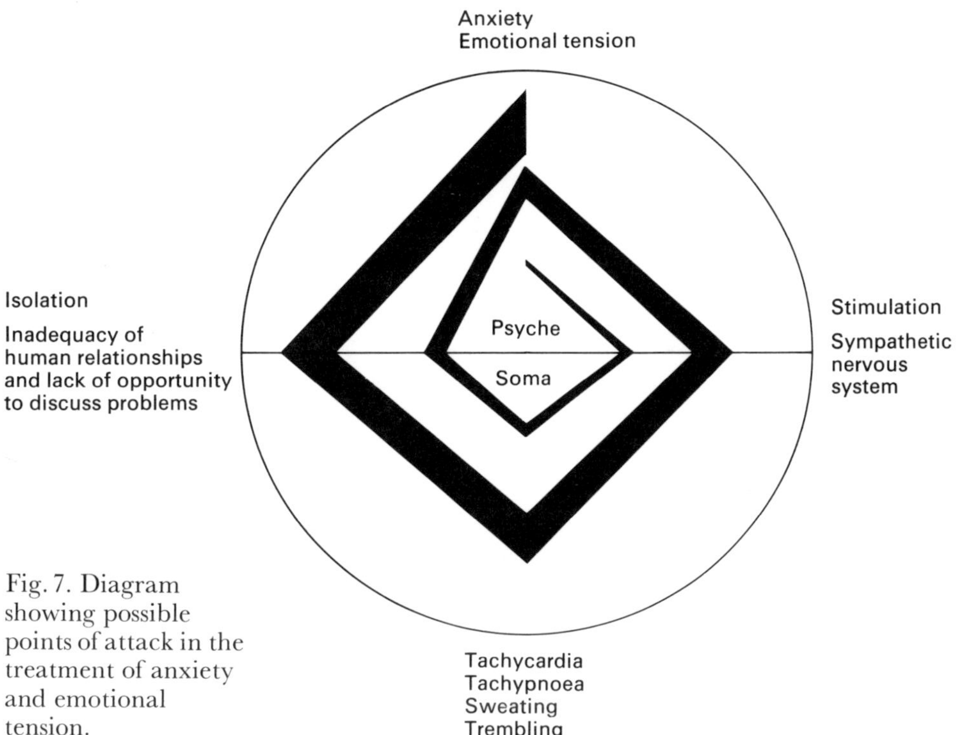

Fig. 7. Diagram showing possible points of attack in the treatment of anxiety and emotional tension.

These observations prompt me to pose three questions. Firstly, do the beta-blockers perhaps also exert a beneficial influence on hypertension by virtue of their central anxiolytic effect? Secondly, wouldn't it be correct to regard the beta-blockers as marking a big step forward in the treatment of hypertension, if for no other reason than that they are less liable to provoke depression than other antihypertensives, and that they actually combat anxiety, thus reducing the risk of depressive reactions and suicidal tendencies in hypertensive patients? Thirdly, how is this anxiolytic effect to be accounted for? I ask this last question because I have the impression that alleviation of the cardiac symptoms cannot be the sole explanation, but that a direct central action must also be involved. What is more, the anxiety-relieving effect appears to me to be particularly pronounced in the case of oxprenolol; this, too, seems to call for further investigation.

W. SCHWEIZER: Would you like at this point, Dr. GROSS, to give us your opinion on the anti-renin story?

F. GROSS: I think it is clear that renin – or the release of renin – is affected by beta-adrenergic blockers. On the other hand, I fancy it is an oversimplification to contend that they act mainly by inhibiting the release of renin and subsequently, what is more important, the production of angiotensin. All the other factors which Dr. BRUNNER has mentioned certainly also have to be taken into consideration. One thing that puzzles me is how one can argue that, in suppressed-renin responders, beta-blockers are useless and that in such cases one should only give diuretics, i.e. on the grounds that a diuretic will restore the response to renin. If this is so, I should have thought it would be far better in these cases to resort to diuretics plus beta-blockers from the very beginning, because then the response to renin would become normal, and – thanks to the concomitant diuretic treatment – you would thus succeed in inducing a state in which, according to the classification proposed by Dr. BÜHLER and his associates, the beta-blocker too would become active. I therefore feel that we should beware of trying to simplify things too much; instead, we should acknowledge that beta-blockers may act in various directions and on various mechanisms and that the renin-angiotensin system, though also one of them, is by no means the only one and probably not even one of the foremost. I believe that the action exerted by beta-blockers on the cardiovascular system is a more important factor in the treatment of hypertension than their effect on the renin-angiotensin system.

W. Schweizer: Thank you, Dr. Gross. Now let us move on to the second question: Are there any major differences in the antihypertensive effect of the various beta-blockers? Though I think we already know the answer to this question, may I ask Dr. Imhof just to sum up the evidence.

P. R. Imhof: At least five beta-blockers are now available on the market, and sooner or later there will be twice that number. How, then, is the general practitioner to decide which of them he should preferably prescribe as treatment for hypertension? From what has so far been said at this symposium it seems that – with the exception of practolol, which has a weaker action – all the beta-blockers exhibit antihypertensive properties of roughly the same potency and that neither cardioselectivity nor intrinsic sympathomimetic activity has any significant bearing on their antihypertensive effect. The decisive factor when considering which beta-blocker to select is that of long-term tolerability. Since the undesirable side effects of the beta-blockers vary in type, incidence, and severity from one drug to another, I suggest that this is a problem to which particular attention should be devoted in our present discussion.

B. N. C. Prichard: At the moment I don't believe there are any important differences other than the one Dr. Imhof mentioned, i.e. that practolol doesn't appear to be quite so effective. This is due to the amount of drug that can be tolerated, and is not indicative of any fundamental difference in activity. Since these drugs lower the blood pressure by means of beta-blockade, their effectiveness depends on how much beta-receptor blockade you can get out of a given drug. That in turn also depends on their side effects, and these side effects do indeed vary from one beta-blocker to another. For instance, for some unknown reason certain beta-blockers produce constipation rather readily; we have seen this more commonly with practolol and sotalol. While I agree that cardioselectivity would be an advantage in a beta-blocker, what one wants here is a drug that really is highly cardioselective and not just marginally so. Practolol is admittedly cardioselective – but, as the dosage is increased, its cardioselectivity becomes lost.
I suspect that apparent differences between beta-blockers as reported in the literature are probably due largely to the way in which the investigators have used the drugs. The dose-response relationship is more clear-cut, I think, than might have appeared from our proceedings during the session on hypertension. We ourselves, for example, performed a dose-response study with propranolol in six mild hypertensives and found a very clear dose-response relationship as the dosage was increased in logarithmic increments (in these particular patients, up to an average of 320 mg. per day). When they were also given half this dose and a quarter of it under double-blind conditions, a dose-dependent decline in blood pressure likewise occurred. Interestingly enough, when the daily dosage reached 320 mg. and the patients had become normotensive, doubling the dose produced no further increase in the response*.

L. Werkö: Again I would like to make a plea for more data before we decide, for example, that cardioselective drugs are no better than non-cardioselective ones. We have now had quite a lot of experience with propranolol, alprenolol, and oxprenolol,

---

* Prichard, B. N. C., Gillam, P. M. S.: Assessment of propranolol in angina pectoris. Clinical dose response curve and effect on electrocardiogram at rest and on exercise. Brit. Heart J. *33*, 473 (1971)

but haven't had as much experience with practolol and far less with any of the new cardioselective drugs. Before stating that there is no difference between these new compounds and the so-called first generation of beta-blockers, we shall have to study them more. Besides the problem of their main side effects, there is the question of bio-availability, for example, and there may well be other comparative aspects that would merit further investigation.

F. GROSS: I have one direct question for Dr. PRICHARD and for those who have had most experience with propranolol. At a meeting I recently attended, it was said that the dosage range for propranolol is much wider than, say, for pindolol or timolol. If I remember rightly, Dr. PRICHARD, you occasionally give propranolol in doses of up to 1 g. or more daily. Do you still do this, or do you consider it to be no longer necessary? Do you think you have already reached the maximum of effect at your average dosage level of 320 mg.?

B. N. C. PRICHARD: No, I did not want to give the impression that the average dose was 320 mg. It just happened to be the average dose in that small group of very mild hypertensive patients. Our average dose of propranolol for an optimum effect in angina pectoris is something like three-quarters of a gramme, and the same applies to sotalol. I don't think that the position is very much different in the case of hypertension. It's one of the myths that you need quite a difference in dose for hypertension as compared with angina; it's just not true. The maximum daily dosage that we have employed for propranolol is in fact not 1 g., but 4 g.; and the same goes for sotalol. I have not had a lot of experience with pindolol, but our intravenous human-pharmacology studies suggest that pindolol possibly does have a flatter dose-response curve.

R. C. TARAZI: I should like to raise a point on which the pharmacologists here may care to ponder – namely, whether there is any difference between the various beta-blockers as regards their ability to block extraneously administered catecholamines, as opposed to their ability to block the effects of sympathetic nerve stimulation. Our own clinical experience, like that of others, has been that practolol, for instance, will not block the tachycardiac effect of isoprenaline but that in equipotent doses it does very effectively inhibit the reflex tachycardia associated with head-up tilting or blunt that associated with static exercise, and in some animal experiments it can also block the effects of stellate ganglion stimulation.

J. SCHWARTZ: I, too, have a question for the pharmacologists: the majority of beta-blockers have a half-life of three to four hours. How is this to be reconciled with a twice-daily dosage schedule? In the case of minoxidil, the argument can be used that the half-life in the arteriolar wall differs radically from the half-life in the plasma. Is there, by analogy with this, any evidence that in the case of the beta-blockers a dissociation exists between the plasma half-life and the half-life at the beta-receptors?

P. R. IMHOF: Almost all beta-blockers have been shown to have a plasma half-life that is very much shorter than the duration of their beta-blocking effect. In the case of oxprenolol, for example, the plasma half-life works out at about 1½ hours, whereas the duration of effect ranges – depending on the dose administered and the parameter measured – from eight to 16 hours.

F. R. BÜHLER: I should like to revert for a moment to the question as to what position the beta-blockers can now be regarded as occupying in our antihypertensive armamentarium. At this symposium we have been told that treatment with beta-blockers alone proves effective in some 60% of hypertensive patients; and, in combination with diuretics and/or vasodilators, they have been reported to yield successful results in from 75% to 90% of cases. The main advantage offered by the beta-blockers is undoubtedly their good tolerability in comparison with other antihypertensives. We therefore have to ask ourselves whether the beta-blockers should be considered merely as a form of adjuvant medication, and their use confined chiefly to mild cases of hypertension, or whether, on the contrary, they do not deserve to become a basic element of modern antihypertensive therapy. Another point that also ought to be discussed is whether it is already possible at this stage to lay down guidelines for treatment with beta-blockers in hypertension. For example, should they always be prescribed together with a diuretic or a vasodilator from the very start? Alternatively, should the choice of treatment be dictated mainly by the biochemical or haemodynamic features peculiar to each of the various forms of hypertension? That is to say, should we aim at a sort of made-to-measure therapy? Finally, I think we should also discuss how treatment can best be initiated. Could one, for instance, commence with a beta-blocker alone? Or would it be preferable to play safe and combine the beta-blocker with a diuretic, in the hope that it might later become possible to manage with either the beta-blocker or the diuretic alone?

P. R. Imhof: As regards the question of monotherapy with beta-blockers in hypertension, I'd like to show you the results obtained by a group of Swiss general practitioners in a series of 84 patients suffering from mild to moderate hypertension (Figure 8). Of these 84 patients, 27 (32%) responded well to monotherapy with oxprenolol in a daily dosage of 240 mg., both their systolic and their diastolic blood pressures falling by more than 10%. The reduction was already clear-cut by the end of the first week of treatment and was only slightly more pronounced by the end of the second week. Consequently, in cases of mild to moderate hypertension neither the G.P. nor the patient need have any misgivings about starting monotherapy with beta-blockers, because the antihypertensive effect of these drugs sets in after only a few days. Should the patient fail to respond, recourse must then, of course, be had to combined treatment.

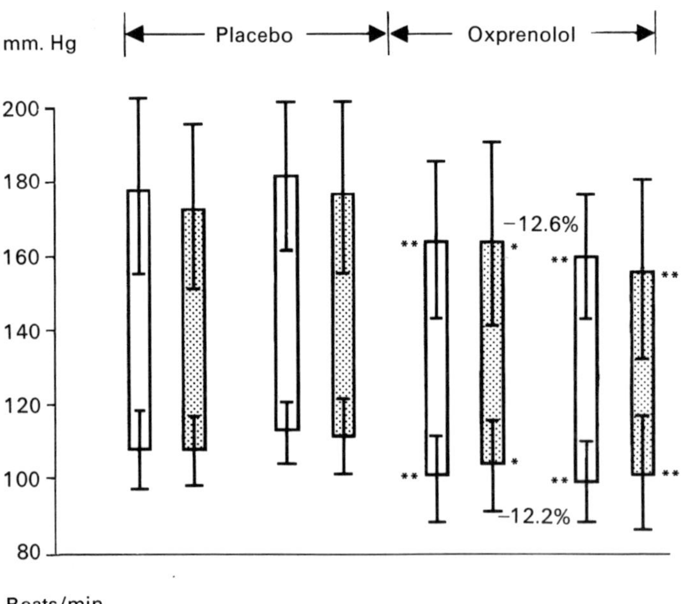

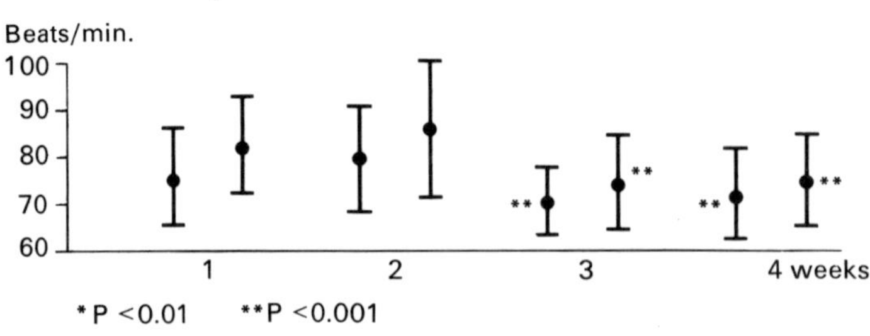

Fig. 8. Blood pressure and heart rate (mean values ± S.E.) in the lying (□) and standing (▨) positions in the 27 patients (out of a total of 84) suffering from mild to moderate hypertension who showed a good response to monotherapy with oxprenolol in a dosage of 80 mg. three times daily.

U. C. Dubach: At this point I'd like to say something about combined treatment with a beta-blocker and a vasodilator*. This type of combination strikes me as being particularly important for the management of hypertension, because vasodilators reduce peripheral resistance and thus serve to counteract the undesirable increase in peripheral resistance provoked by the beta-blocker. On the other hand, the beta-blocker also has the advantage of preventing reflex-induced tachycardia caused by the vasodilator. We tested oxprenolol (120–240 mg. daily) and dihydralazine (75–150 mg. daily), as well as a combination of both, in 18 cases of mild to moderate hypertension. Two of the patients had to be withdrawn from the trial because of side effects with the hydralazine. Three didn't respond satisfactorily to any of the treatments and have therefore likewise been excluded from the analysis of the results. There was a reduction in blood pressure of 14/8 mm. Hg with hydralazine, of 19/7 mm. Hg with oxprenolol, and of 21/8 mm. Hg when the two drugs were given in combination. The heart rate increased by 5 beats/min. with hydralazine, decreased by 7 beats/min. with oxprenolol, and was the same as the control values when the combination was given. Side effects due to hydralazine decreased in number and severity when oxprenolol was added. It seems to us that combination of a beta-blocker with a vasodilator offers a very good and promising approach to treatment in many cases of hypertension.

K. D. Bock: The question Dr. Bühler has raised as to precisely what role the beta-blockers can play in the treatment of hypertension has, I think, to be considered from four different angles:
Firstly, the potency of their antihypertensive effect. If we compare them with conventional antihypertensives, we have to admit that the beta-blockers display at best only the same potency as methyldopa, clonidine, or the diuretics. I have never yet met a patient with severe hypertension who was refractory to conventional drugs but who responded to a beta-blocker – whereas the converse is by no means rare. Thus, purely from the standpoint of the intensity of their effect, the beta-blockers don't constitute much of an advance.
Secondly, their subjective tolerability. Here, as various speakers have already stressed, the beta-blockers offer very decisive advantages over the conventional antihypertensives now at our disposal.
Thirdly, the extent to which they can be regarded as safe to use on a long-term basis. This is a question of cardinal importance in antihypertensive therapy. Among the conventional drugs there are at any rate a few – such as reserpine, as well as clonidine and methyldopa – which can be considered safe in this respect, at least insofar as they are unlikely to provoke any really dangerous, life-threatening, or irreversible side effects. Where the beta-blockers are concerned, some elements of uncertainty still exist on this score, and, bearing in mind experiences with practolol in particular, I feel that this problem of long-term safety should be thoroughly investigated in the case of each and every new preparation.
Fourthly, there is also the question, to which reference has already been made in the proceedings of this symposium, as to whether the beta-blockers – in contrast to the drugs available hitherto – are capable of lowering the incidence of deaths from coronary heart disease in hypertensive patients.
The advantages which, on balance, the beta-blockers certainly do offer, including especially their good subjective tolerability, have induced us to treat about one-third of our hypertensive cases with beta-blockers; this treatment, which is given in a special unit, takes the form partly of combined medication and partly of monotherapy.

J. Somerville: I'd like to support what Dr. Bock has said. While I think we can accept the fact that beta-blockers do lower the blood pressure, no one here has touched

---

* Änishänslin, W., Pestalozzi-Kerpel, J., Dubach, U. C., Imhof, P. R., Turri, M.: Antihypertensive therapy with adrenergic beta-receptor blockers and vasodilators. Europ. J. clin. Pharmacol. *4*, 177 (1972)

on the long-term problems which may arise with these drugs as regards their effects on the myocardium. Some of the actions of propranolol, for example, obviously affect the myocardium directly, and I therefore want to ask whether anything is known about their long-term effects on the muscle, determined electron-microscopically, or on actual myocardial performance, not after six weeks or three months, but after two or three years.

B. N. C. PRICHARD: There is not much relevant evidence, but we did do some studies on our patients in collaboration with Dr. HAMER* of St. Bartholomew's Hospital in London. Patients who had been on chronic treatment for hypertension for many years were exercised to quite high levels, and in these patients we found changes in cardiac output on exercise which were similar to those observed after acute administration of beta-blockers. I realise that this is not a full answer to your question, but it does suggest that the heart can function satisfactorily after long-term treatment with beta-adrenergic blocking drugs. I am not aware of any microscopic studies on the heart muscle of patients who had had chronic treatment.

May I just refer briefly to the C.N.S. activity of beta-blockers and the hypotensive effects which have been demonstrated in animals. What has been observed is an acute effect, irrespective of whether the drug is given into the cerebral ventricle** or by close-arterial injection***. This is quite different from what happens when a beta-blocker is administered to lower the blood pressure in a case of moderate or severe hypertension. Though in such cases you certainly do produce a full and immediate beta-receptor blockade, which causes a prompt and pronounced fall in the pulse rate, there is a definite time-lag – spreading over days or even weeks – before the blood pressure shows a marked fall.

W. GRÜTER: The findings obtained with beta-blockers in cases of acute psychosis likewise seem to suggest that these drugs exert an effect on the central nervous system. As ATSMON**** observed – and as we*****, among others, were able to confirm – the heart rate and blood pressure of patients suffering from acute psychoses of widely varying aetiology are not affected by conventional doses of beta-blockers (oxprenolol or propranolol). Only very high doses of, as a rule, between 3 and 5 g. daily lead to a reduction in heart rate and blood pressure, a reduction which is accompanied by the dramatic disappearance of the acute psychosis as well. When an attempt is made to withdraw the beta-blocker or to reduce its dose, however, the signs and symptoms of the psychosis recur and at the same time the effect of beta-blockade on heart rate and blood pressure also vanishes. This chronological correlation between psychic effects and effects on heart rate and blood pressure seems to suggest that the beta-blockers might also have a central mechanism of action in cases of hypertension.

---

* PRICHARD, B. N. C., SHINEBOURNE, E., FLEMING, J., HAMER, J.: Haemodynamic studies in hypertensive patients treated by oral propranolol. Brit. Heart J. *32*, 236 (1970)
** DOLLERY, C. T., LEWIS, P. J., MYERS, M. G., REID, J. L.: Central hypotensive effect of propranolol in the rabbit. Brit. J. Pharmacol. *48*, 343P (1973); abstract of paper
*** STERN, S., HOFFMAN, M., BRAUN, K.: Cardiovascular responses to carotid and vertebral artery infusions of propranolol. Cardiovasc. Res. *5*, 425 (1971)
**** ATSMON, A., BLUM, I., WIJSENBEEK, H., MAOZ, B., STEINER, M., ZIEGELMAN, G.: The short-term effects of adrenergic-blocking agents in a small group of psychotic patients. Preliminary clinical observations. Psychiat. Neurol. Neurochir. (Amst.) *74*, 251 (1971)
***** VOLK, W., BIER, W., BRAUN, J. P., GRÜTER, W., SPIEGELBERG, U.: Behandlung von erregten Psychosen mit einem Beta-Receptoren-Blocker (Oxprenolol) in hoher Dosierung. Nervenarzt *43*, 491 (1972)

The role of stress factors in ischaemic heart disease and their modulation by beta-receptor antagonists

by S. H. Taylor*

The basic problem in coronary heart disease is a disparity between myocardial energy demands and the supply of energy substrates and oxygen in the perfusing blood. The mechanisms controlling the balance between myocardial energy requirements and coronary blood flow are unknown, but it is certain that flow is chiefly regulated by local myocardial metabolic events.

In coronary heart disease it is often considered that the problem is entirely one of reduced blood supply, the assumption being that the myocardium and its energy demands are normal. This is not so, and the problem is in fact far more complex. A reduction in the blood supply to an area of myocardium rapidly induces a profound change in mechanical performance for two reasons. Firstly, actively contracting myocardium, like any other muscle, extracts a high percentage of oxygen from the blood perfusing it, so that it is extremely sensitive to any reduction in blood flow. Secondly, ischaemic myocardium undergoes a rapid and radical change in its metabolic cycle, switching from the utilisation of fat to that of carbohydrate. Whilst this switch reduces the amount of oxygen immediately required, it is achieved at a great sacrifice of energy production: on an equimolecular basis the oxidation of carbohydrate produces only a fraction of the energy available from lipid metabolism. These two factors – reduction in the oxygen supply and a change in the metabolic status of the myocardium – thus result in diminished energy production which leads to a decrease in contractile activity. The subsequent deterioration in the mechanical performance of the heart results in dilatation and in a reflex increase in sympathetic drive, both of which induce a further rapid escalation in the energy demand (Figure 1). Hence, although the primary problem is one of reduced blood supply, this in itself rapidly gives rise to profound changes in mechanical performance and in myocardial metabolism, which together summate to further impair contractile function of the heart and to increase its work load and energy requirements (Figure 2).

These events not only affect the mechanical activity of the heart but also upset its electrical stability. Ischaemia induces changes in myocardial cell-membrane integrity which rapidly lead to an increase in electrical excitability, a shortening of the refractory period, and a lowering of the fibrillation threshold (Figure 3). As a result, abnormalities of cardiac rhythm may arise, including ectopic ventricular beats superimposed on the T wave, which are liable to precipitate ventricular fibrillation and sudden death. Thus, obstructive disease of the coronary vessels is a far more complex process than is usually realised.

---

* University Department of Medicine, The General Infirmary, Leeds, England.

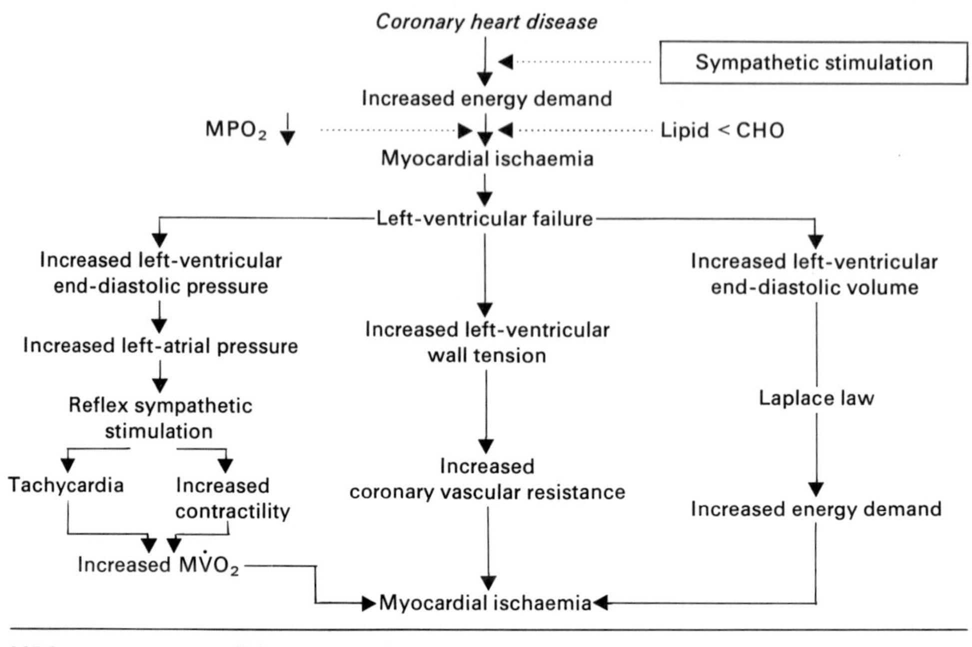

Fig. 1. Mechanical results of stress in coronary heart disease.

The work of the heart is significantly increased only when sympathetic drive is invoked; this is usual only during strenuous physical exercise or emotional stress. The increase in heart rate occurring during normal exercise (up to 110–120 beats/min.) is achieved largely by vagal withdrawal; little sympathetic stimulation is involved and there is relatively little extra energy cost to the myocardium. Consequently, the fact that it generally requires much greater exertion than this to precipitate anginal pain, and that such pain occurs only when considerable sympathetic drive is invoked, is not entirely coincidental.

The effects of catecholamines on the heart are quite different: they are entirely due to sympathetic stimulation, whereas vagal withdrawal is only minimally involved (Figure 4).

Sympathetic stimulation of the heart results in a number of interrelated changes in ventricular energy demand:

Factors increasing energy demand
1. Increased heart rate
2. Increased speed of ventricular contraction
3. Increased systolic pressure
4. Increased inotropic state.

Factors decreasing energy demand
1. Decreased ventricular volume.

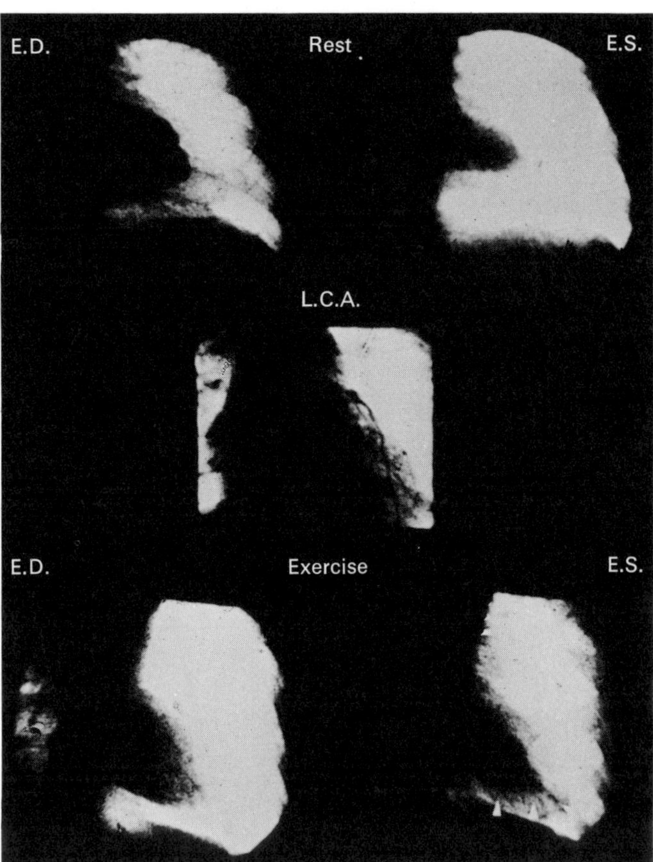

Fig. 2. The mechanical results of exercise stress in a heart afflicted with coronary artery disease. Angiograms taken in a patient with exercise-induced angina pectoris. The left coronary artery (L.C.A.) is severely obstructed by atheroma. At rest the end-diastolic (E.D.) and end-systolic (E.S.) configuration of the ventricle was normal, but during exercise systolic contraction was distorted by a large area of dyskinesia at the apex (△△).

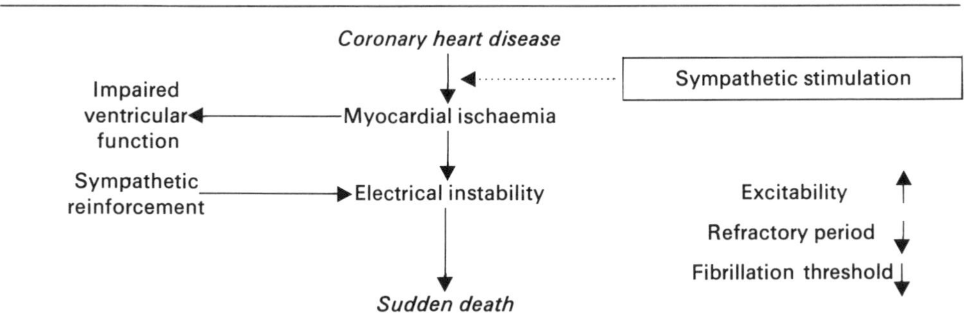

Fig. 3. Electrical results of stress in coronary heart disease.

The proportional importance of these various factors is unknown, but in the normal heart the factors increasing energy demand certainly outweigh the energy-sparing effect of the reduction in ventricular volume. However, since any increase in energy requirement can be met only by an increase in coronary blood flow, the normal efficacy of increased sympathetic drive is reversed in the ischaemic heart with a retarded blood supply. The detailed haemodynamic

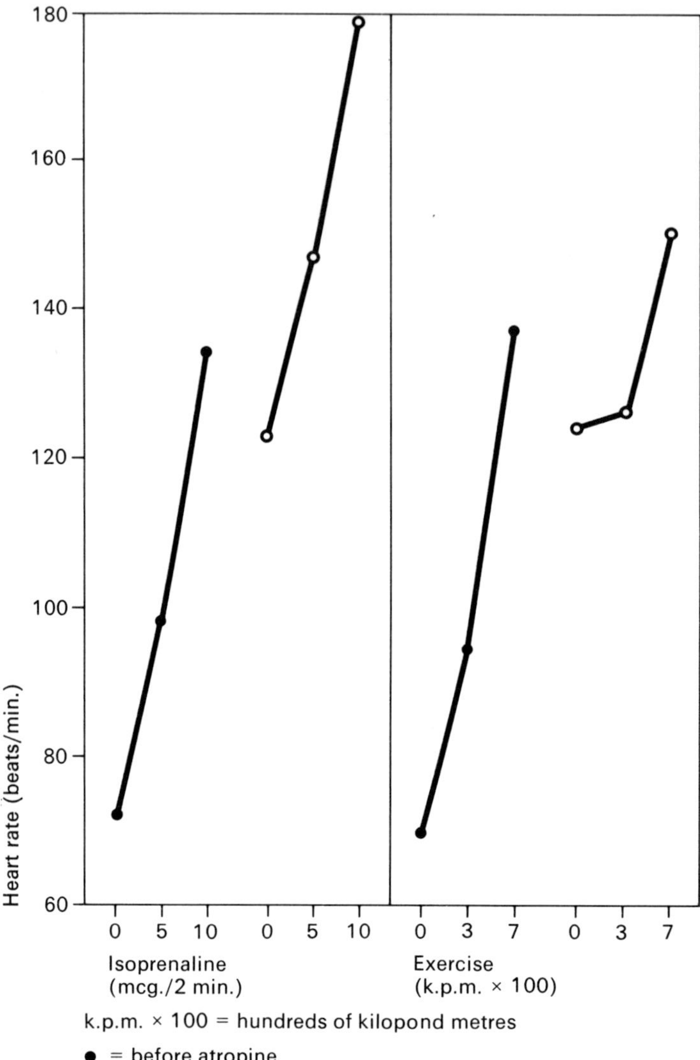

Fig. 4. Effect of vagal blockade on the heart rate response to isoprenaline (injected for two minutes at a rate of 5–10 mcg./min.) and to exercise in normal subjects.

and metabolic effects on the heart of repeated sympathetic stimulation associated with mental stress are unknown, but there is no doubt that they are associated both with a significant increase in myocardial energy demand and with the induction of electrical instability. The importance of this is particularly well illustrated by the effects which such sympathetic stimulation produces in patients with coronary heart disease. When exposed to mental or emotional stress these patients may complain of anginal pain, they often show evidence of ischaemic changes in the E.C.G., and the disturbed electrical activity of the heart in this situation is exemplified by the frequent occurrence of ventricular ectopic beats.

It is thus reasonable to conclude that events associated with sympathetic stimulation of the heart, such as heavy exercise or intense mental stimulation, may be hazardous in patients with coronary heart disease. But it is equally important to emphasise that there is no evidence that such stimuli are harmful in the normal heart or that they are causally related to the production of coronary atheroma. It may be that they are, but there is no categoric proof to corroborate such a speculation at present. Nor is there any evidence to support the contention that subjects with coronary heart disease are more sensitive or hyper-reactive to sympathetic stimuli than normal subjects of similar physical and mental characteristics.

It is against this background that the possible therapeutic usefulness of the beta-receptor antagonists in these patients must be judged. Blockade of the heart's sympathetic drive has various complexly interrelated effects which are the converse of those induced by sympathetic stimulation. It is the final equation between these complexly interrelated factors that determines both the overall effect of beta-receptor antagonists on the balance between myocardial blood supply and energy demand and presumably also their therapeutic effect. There is no longer any reasonable doubt that these drugs are effective in substantially reducing symptoms and electrocardiographic evidence of ischaemia in the majority of patients with exercise-induced angina pectoris and also in those with emotionally induced pain. But it is interesting to note that the dose of beta-receptor antagonist necessary to achieve relief of pain in angina due to exercise is considerably greater than the amount of drug required to relieve angina associated with emotion or mental stress. The explanation for this lies in the fact that, whereas in normal subjects relatively small amounts of these drugs are sufficient to counteract the effects of stress or of infused catecholamines, much larger amounts are needed to combat exercise-induced tachycardia (Figure 5). It is also known that beta-receptor antagonists are capable of reducing the rise in blood lipids and the increase in platelet adhesiveness that

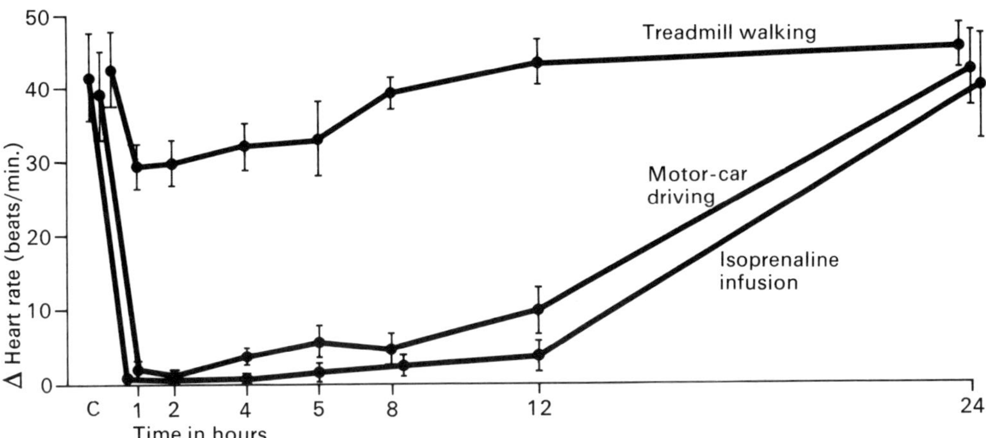

Fig. 5. Effect of 40 mg. oral oxprenolol (®Trasicor) on the heart rate increase during motor-car driving, isoprenaline infusion, and treadmill walking in normal subjects. The control values (C) are the average of the increases in heart rate measured throughout the day on placebo. Data expressed as mean of observations ± S.E. of mean.

are a feature of sympathetic stimulation. Until we know the relevance of these two factors in this situation, however, their importance will remain a matter of conjecture rather than fact.

But these observations must not be extrapolated out of context. Although it has been convincingly demonstrated that the beta-receptor antagonists can achieve a worthwhile reduction in symptoms of ischaemic heart disease due to exercise or to stressful mental stimuli, there is no conclusive evidence as yet that they are able in any way to alter the course of the underlying disease. However, bearing in mind the prevalence of ischaemic heart disease and the theoretical background to the possible beneficial effects of these drugs, it is most important to establish precisely what influence, if any, they do exert on the underlying condition. Some findings, albeit uncontrolled, suggest that beta-receptor antagonists *may* increase survival in patients with angina pectoris. In the light of the evidence presented here, it would perhaps be more reasonable to expect these drugs to reduce the incidence of sudden death in patients with coronary heart disease rather than to prevent recurrence of myocardial infarction. Sudden death, due presumably to electrical instability of the ischaemic myocardium, is an increasingly frequent cause of fatalities in patients with coronary heart disease. Beta-receptor antagonists *may* have a role to play in preventing the effects due to the mental stress of everyday living from affecting the myocardium. Fortunately, such an exciting conjecture lends itself to practical verification, and trials directly aimed at furnishing confirmation of this hope are now under way in many centres.

Cardiological aspects of beta-blockade in stress situations

by P. Taggart*, M. E. Carruthers**, and W. Somerville***

Evidence is accumulating to support the contention that emotion may exert important untoward effects on the heart both in normal subjects and in persons with coronary heart disease[1]. It is the purpose of this communication to examine the usefulness of beta-blockade in offsetting some of these effects and, in particular, to provide a rationale for extending the use of beta-blockers to the treatment of the normal person.

It is appropriate to consider what we mean by the term "normal heart". For example, if a 40-year-old man were to go to bed on a Sunday night perfectly well, symptom-free, and with a normal electrocardiogram, he would be classified as normal. If, while on his way to work the following morning, he were to experience angina for the first time, he would then be classified as abnormal. It is improbable that any massive atheromatous deposit could have developed overnight; a more likely explanation would be that this was the moment when his coronary disease became manifest. In other words, our classification depends on objective evidence, in the absence of which a person may be classed as normal and in the presence of which as abnormal.

We would suggest that this classification may be artificial, since coronary artery disease usually begins at, or even before, birth and progresses at a variable rate until death. The manifestations of the disease are largely dictated by the ratio of oxygen supply to oxygen demand. Emotional stress probably exerts its acute effects on the myocardium largely by increasing the oxygen demand through the intermediary of enhanced sympathetic activity. This being so, the person displaying severe atheromatous narrowing of the coronary arteries would be expected to compromise his myocardium with a lesser stress than the person displaying relatively little disease.

We would like to present evidence in support of this concept, using as illustrations three stress situations that we have studied and considering initially only subjects classified as normal. The first of these situations is driving in city traffic, chosen to represent a mild stress; the second is speaking to an audience, selected as a moderately severe stress; and the third is the anticipation of a parachute descent, chosen as an extreme stress. The first two investigations have been reported previously and are only included briefly here for purposes of comparison.

* Department of Cardiology, The Middlesex Hospital, London, and The Radcliffe Infirmary, Oxford, England.
** Department of Chemical Pathology (Research), St. Mary's Hospital, London, England.
*** Department of Cardiology, The Middlesex Hospital, London, England.

Table 1. Maximum heart rates (and range) observed in three groups of normal subjects exposed to stress situations.

		Beats/min.	
Road driving	(N = 32)	108	(75–155)
Public speaking	(N = 23)	151	(125–180)
Anticipation of a parachute descent	(N = 31)	175	(141–194)

Table 2. Plasma catecholamine concentrations (median values) determined in three groups of normal subjects exposed to stress situations.

	Noradrenaline + adrenaline		Noradrenaline		Adrenaline	
	Before	After	Before	After	Before	After
Road driving (N = 12)	0.81 P <0.01	1.03	0.79 P <0.01	0.91	0.04 P <0.01	0.12
Public speaking (N = 23)	0.90 P <0.01	1.20	0.81 P <0.01	0.90	0.09 n.s.	0.14
Parachute jumping (N = 8)	0.80 P <0.001	2.30	0.69 P <0.05	0.98	0.13 P <0.001	1.13

The mean maximum observed heart rates were 108 (range 75–155) for the road drivers, 151 (range 125–180) for the public speakers, and 175 (range 141–194) for the parachutists (Table 1). These figures serve as a crude but reasonable indication of the degree of arousal involved in each situation.

In all three circumstances total plasma catecholamine concentrations were increased as compared with samples taken beforehand. The increase was considerably greater in the severe stress situation than in response to the milder stresses (Table 2).

The parachutists included a group of beginners who showed much greater increases in plasma catecholamine levels than did those with experience.

The electrocardiograms recorded during these events were analysed for four parameters: J point widening, "J. E." pattern (see below), S-T segment depression, and ectopic activity. J point widening is illustrated in Figure 1, which shows the E.C.G. tracings of a subject speaking to an audience, and the J. E. pattern is exemplified in Figure 2 by the E.C.G. tracings of four subjects. In the J. E. pattern the S-T segment slopes upwards to the following P wave, and the T wave is largely, if not completely, absent. We coined the abbreviation "J. E." because the pattern seemed to us to be an "extended" J wave. This pattern is not simply rate-related, since the administration of atropine to produce a

Fig. 2. "J. E." pattern in the E.C.G. tracings of four normal subjects during exposure to stress situations.

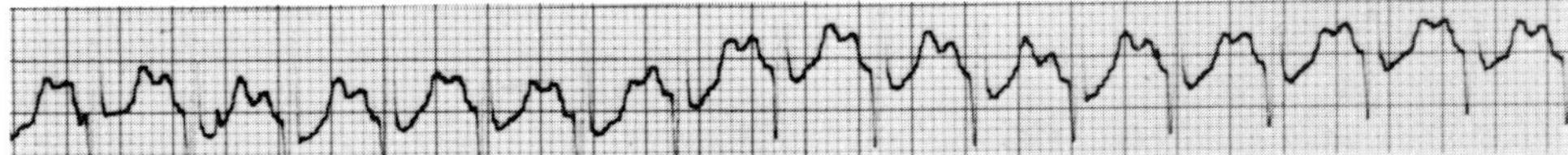

Begins to speak, heart rate 160

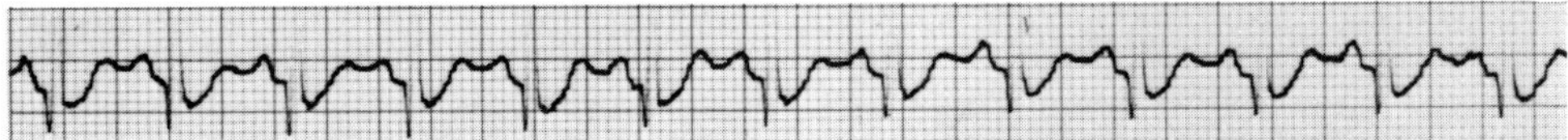

After 3 min., heart rate 130

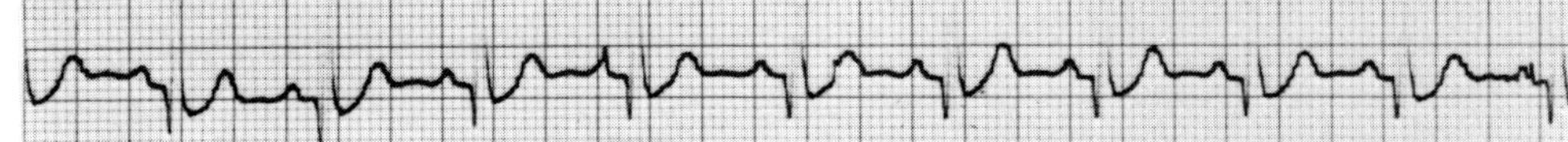

After 6 min., heart rate 100

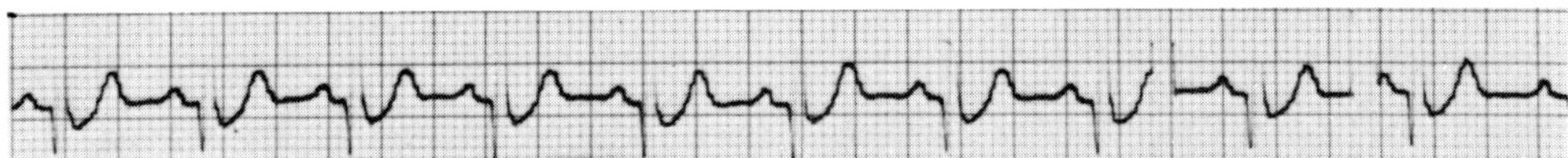

Questions, heart rate 110

Fig. 1. J point widening in the E.C.G. tracings of a normal subject during public speaking.

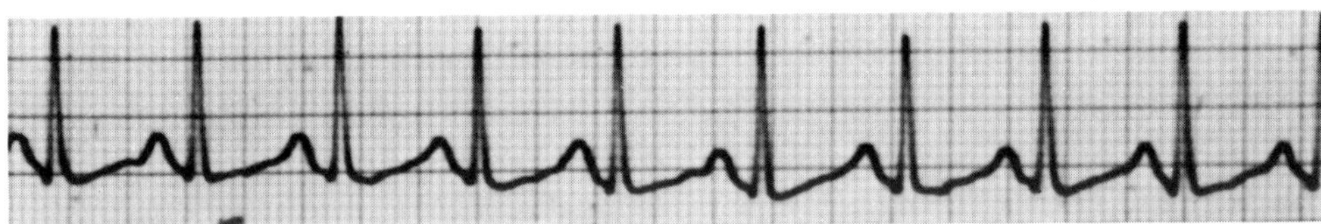

Male, 44 years, public speaking. Heart rate 125

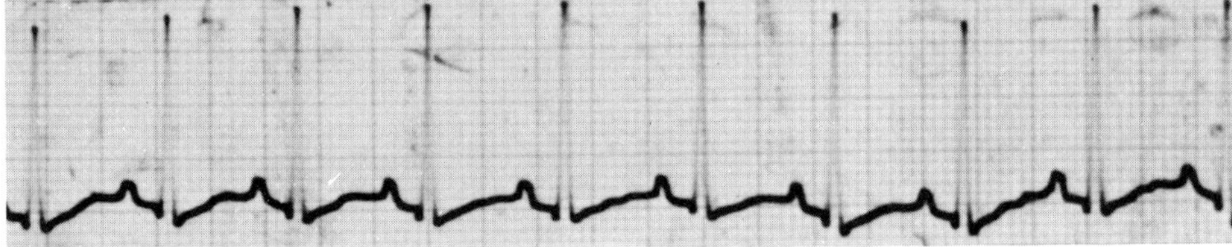

Female, 21 years, road driving. Heart rate 125

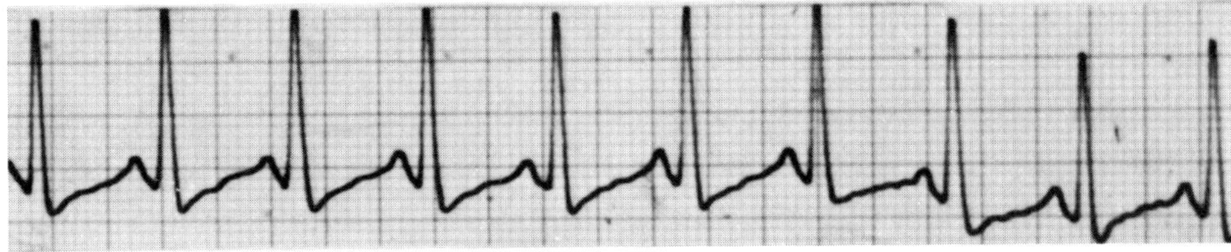

Male, 23 years, parachute jumping. Heart rate 125

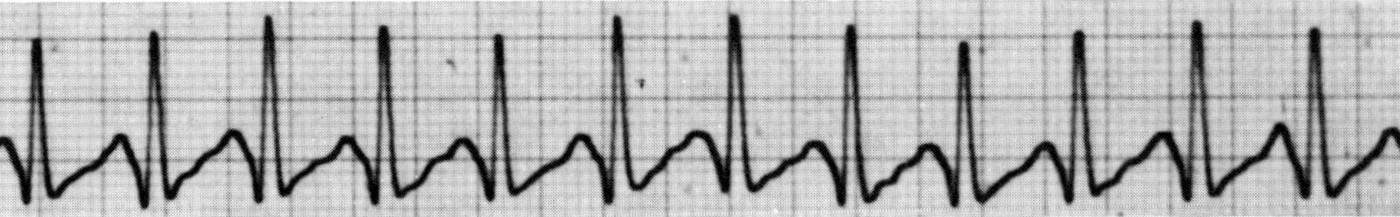

Female, 23 years, parachute jumping. Heart rate 160

Table 3. Electrocardiographic features in three groups of normal subjects exposed to stress situations.

	J widening	"J.E." pattern	S-T segment depression	Ectopic heart beats
Road driving (N = 32)	6 (19%)	3 (9%)	0	0
Public speaking (N = 23)	20 (85%)	17 (74%)	1 (4%)	6 (26%) (prolific in two)
Parachute jumping (N = 15)	13 (87%)	10 (66%)	7 (46%)	4 (27%)

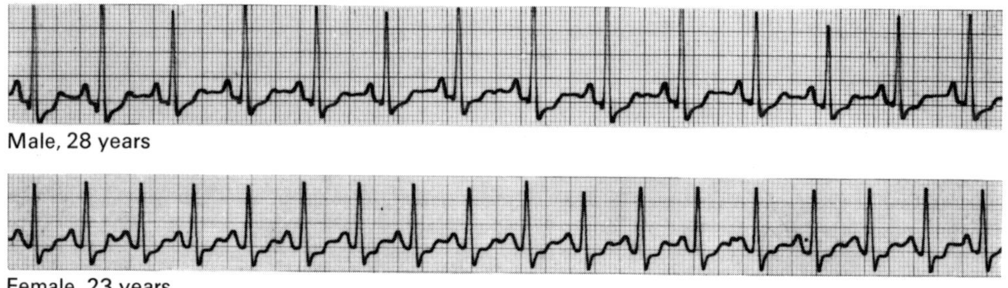

Fig. 3. S-T segment depression in the E.C.G. tracings of two normal subjects during the anticipation of a parachute descent.

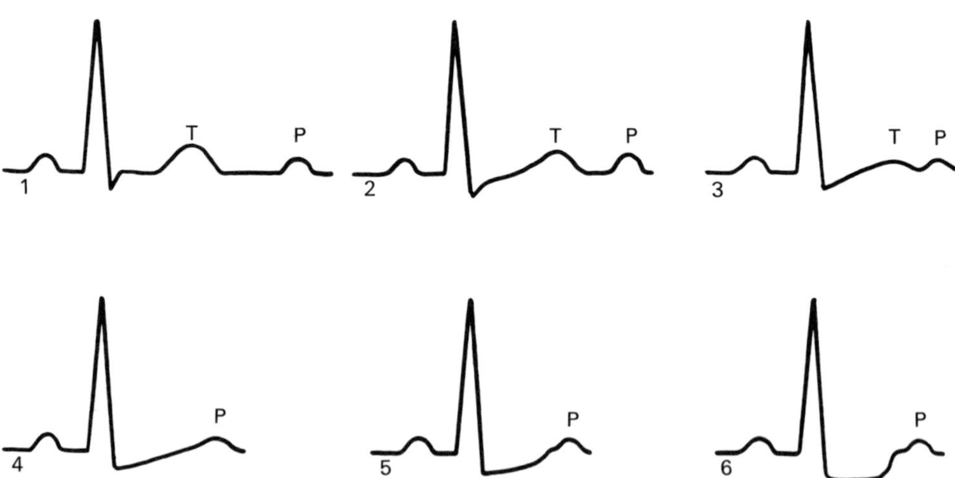

Fig. 4. Sequence of abnormalities commonly observed in the E.C.G. tracings of normal subjects exposed to severe stress situations. 1 = normal E.C.G.; 2 = J point widening; 3 = J point widening with loss of T amplitude; 4 = J.E. pattern; 5 = initially up-sloping S-T segment depression; 6 = horizontal or even downsloping S-T segment depression.

tachycardia comparable to the upper three tracings in Figure 2 did not affect the T wave, which remained normal.

Analysis of the electrocardiograms of the three groups of normal subjects is presented in Table 3. Of particular interest are, firstly, the high incidence of

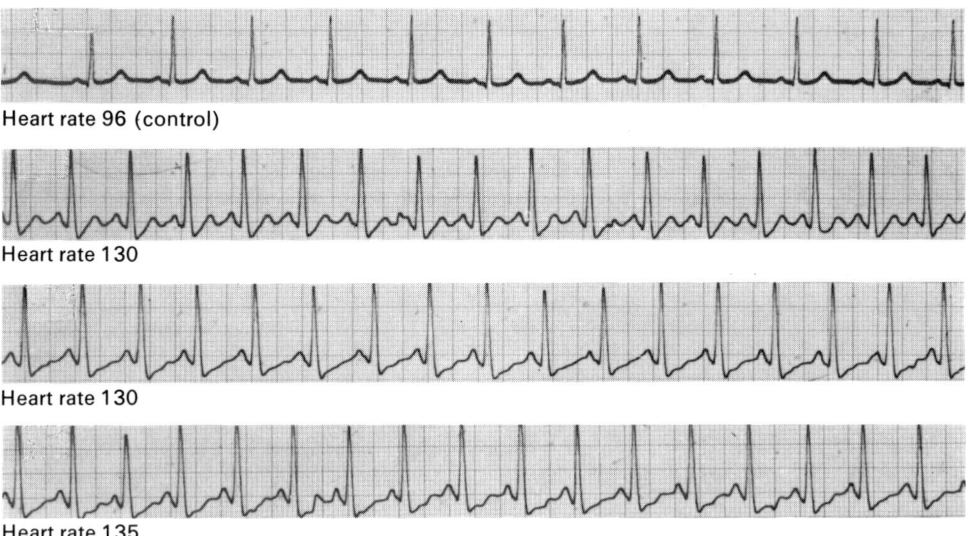

Fig. 5. Example of the sequence of E.C.G. abnormalities shown diagrammatically in Figure 4.

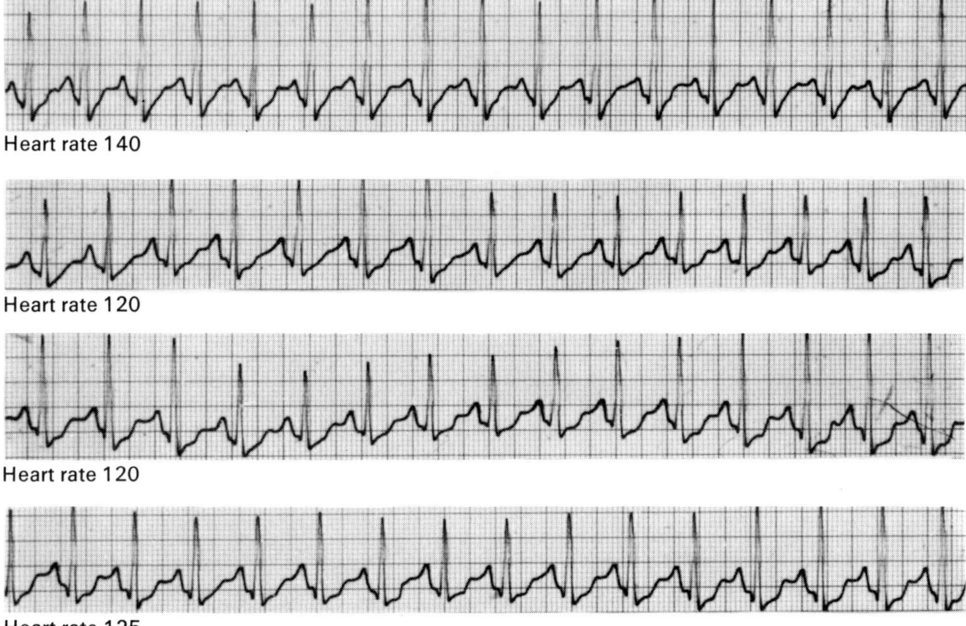

Fig. 6. Example of the sequence of abnormalities shown diagrammatically in Figure 4.

J widening and J.E. pattern in the two more severe stress situations, and, secondly, the fact that 46% of the parachutists showed S-T segment depression such that, if the circumstances had been an exercise test for the confirmation of coronary heart disease, the test would have been classed as positive. Ectopic beats occurred in roughly a quarter of the persons subjected to the two more severe stress situations and were at times prolific and multifocal.

E.C.G. tracings of two of the parachutists showing S-T segment depression are illustrated in Figure 3. The circumstances of this study were particularly fortunate inasmuch as they allowed us to monitor the transition from a normal E.C.G. to this square-wave S-T depression while the subjects remained stationary (i.e. sitting in the aircraft). Figure 4 illustrates the sequence commonly observed, i.e. normal E.C.G. (1), J point widening (2), J point widening with loss of T amplitude (3), J.E. pattern (4), and S-T depression which is initially upsloping (5) and then horizontal or even downsloping (6). Three examples of this sequence are presented in Figures 5, 6, and 7.

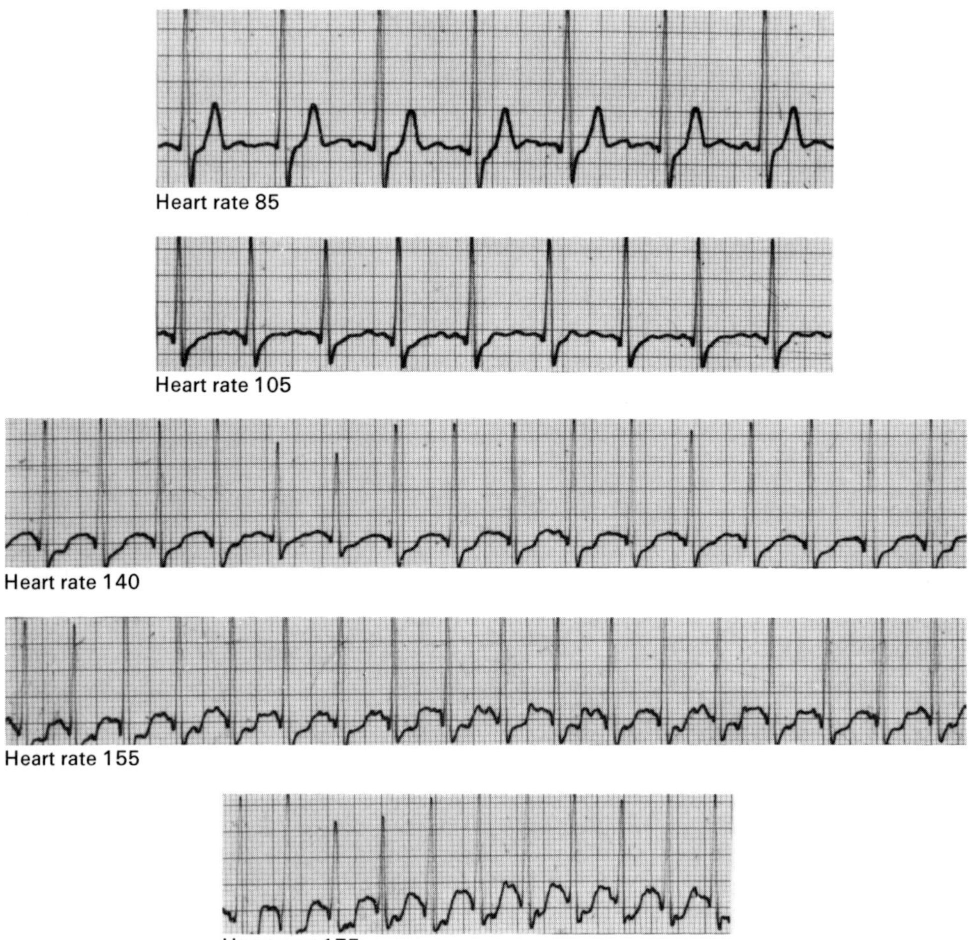

   Fig. 7. Example of the sequence of abnormalities shown diagrammatically in Figure 4.

Nearly half of the young, fit, healthy normal subjects exposed to an extreme emotional stress situation developed S-T segment depression resembling that encountered in ischaemic heart disease. This E.C.G. abnormality was invariably associated with greatly increased plasma catecholamine concentrations. It is tempting to assume a cause-and-effect relationship between enhanced sympathetic activity, increased myocardial oxygen demand, and the E.C.G. manifestations. If such a relationship were in fact to be demonstrated, it would support our suggestion that the concept of the normal and abnormal heart may in certain circumstances be inappropriate.

In a study in which people suffering from what is conventionally designated as coronary heart disease were investigated while driving in city traffic, we found a much greater incidence of S-T depression and ectopic activity amongst them than in a normal control group, despite comparable heart rates and plasma catecholamine concentrations[8].

Similar findings were observed in the two groups we studied speaking before an audience; again, each group had comparable heart rates and plasma catecholamine concentrations[6]. This is what one would expect, because in response to a given stress those with more advanced disease would be less able to increase their myocardial oxygen supply to meet the demand.

The tachycardia, S-T and T changes, and ectopic activity were conveniently and easily blocked with a single oral dose of 40 mg. oxprenolol ([R]Trasicor) in both normal subjects and persons with coronary heart disease speaking to an audience[6].

The foregoing comments have been largely concerned with the question of acute stress. Elevated plasma free fatty acids and triglycerides in response to emotion may well have an important bearing on the aetiology of atheroma[2]. A number of studies have shown that lipid levels rise in response to emotion and that this rise is associated with increased plasma catecholamine and urinary catecholamine concentrations[3]. Furthermore, a transient increase in circulating free fatty acids may well have an arrhythmogenic effect[4] and thus also be relevant to the acute situation.

To some extent, therefore, a rationale can be made out for the use of beta-blockade in normal subjects to relieve unpleasant symptoms of palpitation and anxiety and possibly to protect those whose coronary atheroma, though still subclinical, has advanced to a point where acute stress may be dangerous. The question of the long-term prophylactic use of beta-blockade to prevent atheroma and acute cardiac episodes remains controversial and is currently the subject of study.

In people with overt coronary heart disease beta-blockade is a logical and effective treatment for emotionally induced angina and palpitation, often allowing them as it does to take part, for example, in important meetings, marital relations, weddings, and other special events which they might otherwise avoid or find unpleasant and even dangerous. It would seem sensible to administer beta-blockade prophylactically even in the absence of angina, although statistical evidence of the value of such prophylactic treatment is still awaited. Beta-blockade is also a useful adjunct to digitalis in cardiac patients displaying atrial fibrillation, many of whom, though apparently well controlled at rest, may develop very rapid heart rates under emotional provocation[7].

The safety of beta-blockers, as well as the remarkably low dosages in which they need to be given in order to offset the effects of emotion, should encourage their wider use in this field.

Beta-blockers may well become unique in being the first group of drugs which, though developed in order to benefit the abnormal heart, can also be used to treat the normal person.

References

1 BURLEY, D. M., FRYER, J. H., RONDEL, R. K., TAYLOR, S. H. (Editors): New perspectives in beta-blockade, Int. Symp., Scanticon, Aarhus, Denmark, 1972 (Metropolis Press, London 1973)
2 CARRUTHERS, M. E.: Aggression and atheroma. Lancet *ii*, 1170 (1969)
3 LEVI, L. (Editor): Society, stress and disease, Proc. Int. interdiscipl. Symp., Stockholm 1970 (Oxford Univ. Press, London/New York/Toronto 1971)
4 OLIVER, M. F., KURIEN, V. A., GREENWOOD, T. W.: Relation between serum free-fatty-acids and arrhythmias and death after acute myocardial infarction. Lancet *i*, 710 (1968)
5 TAGGART, P., CARRUTHERS, M.: Endogenous hyperlipidaemia induced by emotional stress of racing driving. Lancet *i*, 363 (1971)
6 TAGGART, P., CARRUTHERS, M., SOMERVILLE, W.: Electrocardiogram, plasma catecholamines and lipids, and their modification by oxprenolol when speaking before an audience. Lancet *ii*, 341 (1973)
7 TAGGART, P., EMANUEL, R. W.: Beta-blockade as an adjunct to digitalis in the control of atrial fibrillation. In preparation
8 TAGGART, P., GIBBONS, D., SOMERVILLE, W.: Some effects of motor-car driving on the normal and abnormal heart. Brit. med. J. *iv*, 130 (1969)

Discussion

E. Sowton: One point I'd like to raise about Dr. Taggart's 40-year-old man who goes to sleep on Sunday night and has his first bout of angina on the Monday is this: in what way would beta-blockade over Sunday night have prevented the clinical appearance of angina on the Monday morning?

P. Taggart: Probably in the same way that it always does, or that it does in most cases, i.e. by raising the threshold at which angina occurs.

L. Werkö: Dr. Taylor speculated that the use of beta-blockers might also help to reduce the incidence of sudden death. We have just completed a controlled, randomised, double-blind study – lasting two years – on patients who had had a myocardial infarction and whom we treated with either alprenolol or a placebo. The code has only just been broken, so that what I can say now is only in the nature of a preliminary report. There seems to have been a significant difference in the incidence of sudden death in the two treatment groups, but there was no difference in the frequency with which myocardial infarction recurred.

B. N. C. Prichard: When describing his results, Dr. Taylor mentioned that, following atropine, he found the pulse rate on exercise to be at a different level. If I understood him correctly, he interpreted this finding as evidence that there was no sympathetic activity at low levels of exercise. The pulse rate – supine, standing, and at low levels of exercise – is in fact controlled both by the vagus and by the sympathetic system. Beta-blockers lower the supine pulse rate and the standing pulse rate, as well as the exercising pulse rate at very low levels of exercise.

S. H. Taylor: I entirely agree with Dr. Prichard that there is an interrelationship between the sympathetic and parasympathetic effects on the heart at all levels of exercise. The observations presented in my Figure 4 demonstrated that at low levels of exercise the change in pulse rate is preponderantly due to vagal withdrawal, whereas at high levels of exercise it is preponderantly due to sympathetic discharge. This is not a complete all-or-nothing situation.

A. Zanchetti: I must say I am not too happy about the use of the word "stress" – not for literary reasons, but because I think it conveys the wrong idea that the way in which the brain, or the mind, or behaviour can affect the body in general, and the cardiovascular system in particular, assumes only one sort of pattern. On the contrary, there seems to be good evidence to suggest that several different patterns can occur[*]. For instance, it is a well-known fact that stress can induce not only tachycardia but occasionally bradycardia instead. In studies on cats, for example, we found that when a cat is waiting in expectation of some form of aggression it generally shows bradycardia. Tachycardia supervenes when the cat actually starts to fight[**]. Of course, we couldn't study expectational anxiety in parachute-jumping cats! But it is known that, even in man, the expectation of a painful shock, for instance, is associated with bradycardia[***].

---

[*] Mancia, G., Baccelli, G., Zanchetti, A.: Hemodynamic responses to different emotional stimuli in the cat: patterns and mechanisms. Amer. J. Physiol. *223*, 925 (1972)

[**] Adams, D. B., Baccelli, G., Mancia, G., Zanchetti, A.: Cardiovascular changes during preparation for fighting behaviour in the cat. Nature (Lond.) *220*, 1239 (1968)

[***] Jenks, R. S., Deane, G. E.: Human heart rate responses during experimentally induced anxiety: a follow-up. J. exp. Psychol. *65*, 109 (1963)

Obrist, P. A., Wood, D. M., Perez-Reyes, M.: Heart rate during conditioning in humans: effects of UCS intensity, vagal blockade, and adrenergic block of vasomotor activity. J. exp. Psychol. *70*, 32 (1965)

HENRY et al.* recently published some extremely interesting findings obtained in colonies of mice. It appears that, when you put mice together in a colony, some of them develop a so-called dominant behaviour after a few weeks. These dominant mice carry out patrols and prevent the other mice from patrolling or from visiting the nests, with the result that the others become frustrated and display what is referred to as submissive behaviour. But, after a few weeks the dominant mice develop hypertension, whereas the submissive ones remain normotensive. In a further experiment, in which HENRY et al.** measured tyrosine hydroxylase, a sympathetically conditioned enzyme, they also demonstrated that the activity of this enzyme is strongly increased in the dominant hypertensive mice, but shows no increase in the submissive mice. Corticosterone levels and the weight of the adrenals, on the other hand, remain unchanged in the dominant mice, whereas they become increased in the submissive ones. Here, therefore, we have a very clear contrast between two different patterns of emotional behaviour. The one exhibited by the dominant mice is associated with cardiovascular changes and with an increase in the activity of enzymes connected with the adrenal medulla, but is not accompanied by any changes of the Selye stress type; in the submissive mice, by contrast, stress of the Selye type does develop, with endocrine changes involving the adrenal cortex, but no important alterations occur in the cardiovascular system. I believe that such findings may well also have a bearing on human pathology. If you take all the various manifestations of stress and lump them together in one single category, you lose sight of the differences and you get into great difficulties when trying, for example, to assess the relative importance of emotional factors in the causation – as opposed to the precipitation – of atherosclerosis and coronary disease. So I believe it's very important to try and distinguish between different behavioural patterns.

P. TAGGART: I think that's a very good point you have made, Dr. ZANCHETTI. There is a great deal of individual variation between stress and between people's reactions to it. The original Funkenstein hypothesis*** suggested that active aggressive emotional situations are associated predominantly with noradrenaline release, whereas the passive emotion of fear is associated with higher adrenaline levels. This has been challenged to some extent by FRANKENHÄUSER****, who suggests that it is the intensity rather than the quality of the emotion that determines the higher adrenaline secretion. Our findings – in an admittedly limited number of studies – seem to confirm the Funkenstein hypothesis. In our parachutists, for instance, who are confronted with a fear situation, it was only the adrenaline that showed an increase, whereas when we studied racing drivers – who are in an aggressive situation – we found that it was the plasma noradrenaline levels which rose. This is only one example of the differing response to different emotions that you mentioned, Dr. ZANCHETTI.
Now just a few words about the long-term implications of such differences as regards the aetiology of atheroma. Here, we would postulate roughly the following sequence of events: first you have emotion, followed by noradrenaline release and by the mobilisation of free fatty acids; in the absence of exercise, these free fatty acids are

---

* HENRY, J. P., ELY, D. L., STEPHENS, P. M.: Blood pressure, catecholamines, and social role in relation to the development of cardiovascular disease in mice. In Zanchetti, A. (Editor): Neural and psychological mechanisms in cardiovascular disease, Proc. Symp., Stresa, Italy 1971, p. 211 (Il Ponte, Milan 1972)
** HENRY, J. P., ELY, D. L., STEPHENS, P. M.: The role of psychosocial stimulation in the pathogenesis of hypertension. Verh. Dtsch. Ges. inn. Med. (printing)
*** FUNKENSTEIN, D. H.: Epinephrine-like substances in relation to human behavior. J. nerv. ment. Dis. *124*, 58 (1956)
**** FRANKENHÄUSER, M.: Experimental approaches to the study of human behaviour as related to neuroendocrine functions. In Levi, L. (Editor): Society, stress and disease, Proc. Int. interdiscipl. Symp., Stockholm 1970, p. 22 (Oxford Univ. Press, London/New York/Toronto 1971)

not metabolised as a source of fuel, but may be converted to triglycerides and so gradually lead to the formation of atheroma. If this assumption is valid, then it must be situations associated with the release of noradrenaline that trigger off the chain of events, because adrenaline is only a very weak mobiliser of free fatty acids.

Dr. ZANCHETTI also mentioned bradycardia as a stress-induced reaction. Some time ago, we did a study on people watching violent films, fully expecting to find that this would be a situation associated with marked tachycardia. Much to our amazement, however, the majority of these subjects had bradycardia. And this is perhaps not so strange after all, because on a number of occasions the persons watching this particular film actually fainted; they had to be taken out, and some of them vomited. In short, they had suffered a traditional vasovagal attack. Thus, as Dr. ZANCHETTI has said, there are some emotional situations which may indeed be associated with brady-cardia. This should also perhaps make one reflect a little on the random use of beta-blockers to treat "television angina"!

W. SOMERVILLE: May I comment on two points emerging from our paper presented by Dr. TAGGART. The first is the thesis that ischaemia may occur even in the normal heart, i.e. that you don't necessarily have to have coronary atherosclerosis or any ob-structive lesion of the coronary arteries in order to produce ischaemia. The second point is that we were not suggesting that all normal people should be "treated" with beta-blockers. A voice has just whispered in my ear: "Who wants to be without his or her catecholamines?" The answer is: certain people do! Consider a normal person addressing an audience or speaking at a board meeting who becomes aware of a tumult in his chest resulting from nervous tachycardia of 140 to 160 beats/min. It is our contention that he would benefit from medication which, besides enabling him to speak more effectively by shielding him from the distraction of his palpitation, might at the same time be protecting him from a deleterious influence on his coronary arteries if they were the seat of asymptomatic atherosclerosis.

Oxygen dissociation from haemoglobin

by K. H. WINTERHALTER* and P. TUCHSCHMID**

In higher animals haemoglobin assumes a cardinal role in the supply of oxygen to the tissues. Highly significant, even from the purely phenomenological standpoint, is the fact that the functional properties of the blood in every species – and, indeed, in every individual – are adapted to the requirements of the tissues. The molecular mechanisms of this adaptation have been largely elucidated in recent years[9,13]. From a medical point of view, pharmacological manipulation of oxygen transport would obviously be desirable, because it might be possible in this way to achieve a decisive improvement in the supply of oxygen to hypoxic tissues.

Since successful efforts have been made in the past ten years to gain an insight into the molecular mechanism of oxygen transport, a few introductory remarks on the structure and physiological functions of haemoglobin are perhaps indicated at this point.

The erythrocyte of adult man contains at least three different haemoglobins, each of which is composed of two pairs of polypeptide chains (Table 1). All haemoglobins have an α chain (comprising 141 amino acids)[4,7] in common, but they differ with respect to the second pair of chains (β, γ, or δ, each comprising 146 amino acids)[4,7]. In adult haemoglobin, at least three further modifications are found, one of which plays an important role in diabetes[14]. All three, however, are based on the amino acid sequence of the polypeptide chains.

Each polypeptide chain bears haem as its prosthetic group. Haem is a tetrapyrrole – i.e. a Type III protoporphyrin – with a divalent iron atom in the centre. The amino acid sequence of all chains is known and is termed the *primary structure*. On all four types of chain, eight distinct segments can be recognised, which are present in a so-called α helical arrangement[6]. They are designated by the letters A–H and each comprises 7–21 amino acids. They are interconnected by stretches of amino acids in a random coil arrangement. The random coil and the α helix are referred to as the *secondary structure*. The helical segments have an exactly defined spatial relationship to one another, which is known as the *tertiary structure*; the A helix, for example, is in the immediate vicinity of the G and H helix. On each chain the E and F helix form a deep hydrophobic cleft, into which the prosthetic group (or haem moiety) is inserted. The *divalent* iron atom (only in its divalent form can haemoglobin transport oxygen) has four of its coordination positions occupied by the nitro-

---

* Friedrich Miescher-Institut, Basle, Switzerland.
** Kinderspital, Zurich, Switzerland.

Table 1. Haemoglobins in the erythrocyte of adult man.

Type of haemoglobin	Amount present (%)	Composition	
Hb.A (adult)	~97	α_2	β_2
Hb.A$_2$	~2.5	α_2	δ_2
Hb.F (foetal)	<1	α_2	γ_2

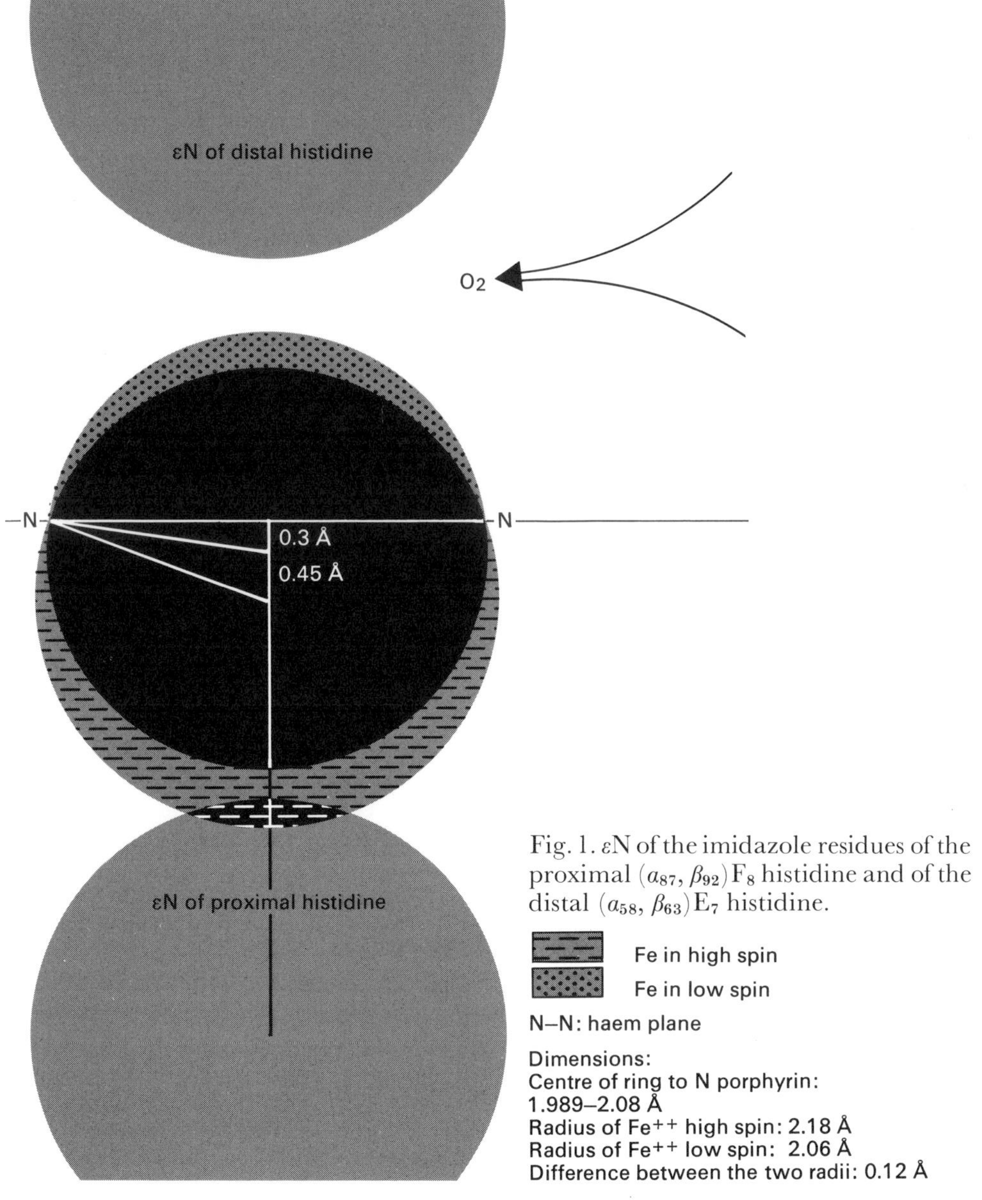

Fig. 1. εN of the imidazole residues of the proximal $(\alpha_{87}, \beta_{92})\,F_8$ histidine and of the distal $(\alpha_{58}, \beta_{63})\,E_7$ histidine.

gen atoms of the four pyrroles. The fifth coordination position is bound to the ε nitrogen of histidine, which in all chains is the eighth amino acid on the F helix [proximal histidine a_{87} or β $(\gamma, \delta)_{92}$]. The sixth coordination position is unoccupied in the deoxy state – with the result that the iron remains five-coordinated – and becomes occupied when ligands (O_2, CO) are taken up. This constitutes the principal difference between deoxyhaemoglobin and oxy-haemoglobin. The outer electrons of *five*-coordinated iron are in a state referred to as "high spin". This high spin state confers a radius of 2.18 Å on the iron atom (Figure 1). Hence, the iron atom is a little too thick to fit comfortably into the centre of the tetrapyrrole ring, and its centre is therefore displaced out of the haem plane towards His F_8 by about 0.75 Å. In response to the binding of oxygen – between the E helix and haem – the iron becomes *six*-coordinated and its state changes to one of "low spin". Its radius shrinks by 0.12 Å. Since it has now become smaller, its centre moves closer to the haem plane by about 0.45 Å. As a result, the bond between the iron and εN His F_8 becomes stretched. This stretching is transmitted to the entire polypeptide chain and is considered to be the *triggering mechanism* of the conformational changes[12]. As already mentioned, haemoglobin is made up of four chains. The relative positioning of these chains to one another (the *quaternary structure*) is distinctly different in the deoxy form from what it is in the oxy form. The binding forces between the chains consist of hydrophobic interactions, hydrogen bonds, and electrostatic interactions. There are no disulphide bridges. The deoxy form – also referred to as the T (short for "tense") structure – is stabilised by a number of salt bridges, which do not occur in the oxy or R (short for "relaxed") structure (Table 2).

The stability of these bridges depends of course on the pH of the medium and on the latter's ion concentration. It is also important to realise that the deoxy structure of haemoglobin has much less affinity for oxygen than the oxy structure.

A further mechanism operative in the stabilisation of the T structure – and one that plays a particularly important physiological role – is binding to organic phosphates. In the human erythrocyte 2,3-diphosphoglycerate is an especially important factor[2,5]. A gap opens in the T structure between the β chains, and in this gap the spatial arrangement of the two N-terminal ends of the β chains and the positively charged imidazole rings of the histidines in Position 143 on

Table 2. Salt bridges broken upon transition from the T to R structure.

	Deoxy structure (T)		Oxy structure (R)
α chain Arg HC$_3$ (141)α_1	COO$^-$ —$^+$H$_3$N	Val NA$_1$ (1) α_2	Broken
	Gua$^+$ —$^-$OOC	Asp H$_9$ (126) α_2	Broken
β chain His HC$_3$ (146) β_1	COO$^-$ —$^+$H$_3$N	Lys C$_6$ (40) α_2	Broken
	Imid.$^+$ —$^-$OOC	Asp FG$_1$ (94) β_1	Broken

both β chains is such that these groups can bind the negative charges of the organic phosphate (Figure 2). This of course enhances the stability of the T structure and consequently reduces the affinity of a given haemoglobin solution for oxygen[15].

What are the implications of all this for the organism? The oxygen-binding curve of haemoglobin is S-shaped (Figure 3). If this curve is displaced to the right, oxygen saturation in the lung may admittedly decrease somewhat, but oxygen unloading at the tissue level is decisively improved. In the human body this mechanism becomes operative within only a few hours at high altitudes, for example, but it also plays an important role in severe anaemia (e.g. pernicious anaemia). It is this which accounts for the fact that patients with a haemoglobin of only 5 g./100 ml. are still able to come to the doctor's surgery on foot.

The stability of the T structure is influenced in similar fashion by changes in pH (Figure 4) or also by changes in the salt concentration. The structure is most stable at a pH of 6.2. Both an increase and a decrease in pH causes the T structure to become unstable and thus shifts the balance in favour of the R structure, which has a greater affinity for oxygen. This of course is useful from the teleological standpoint as well, because in hypoxia acid metabolites are produced. As a result of these acid metabolites the intracellular pH decreases, and so does the affinity of the haemoglobin for oxygen. The release of oxygen

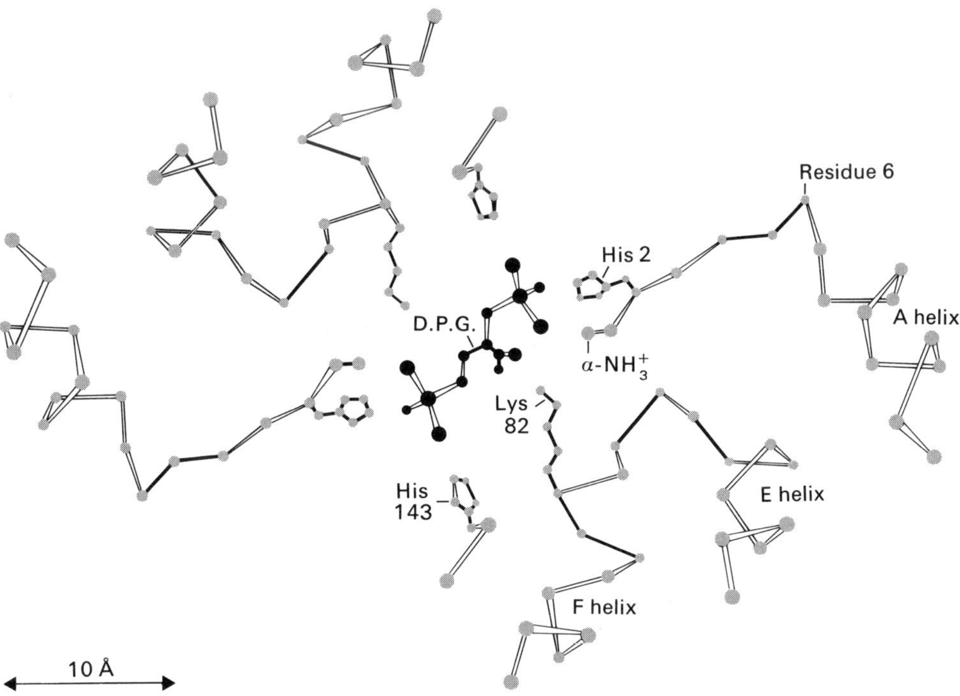

Fig. 2. 2,3-Diphosphoglycerate in the central cleft of the deoxy structure between the β chains.

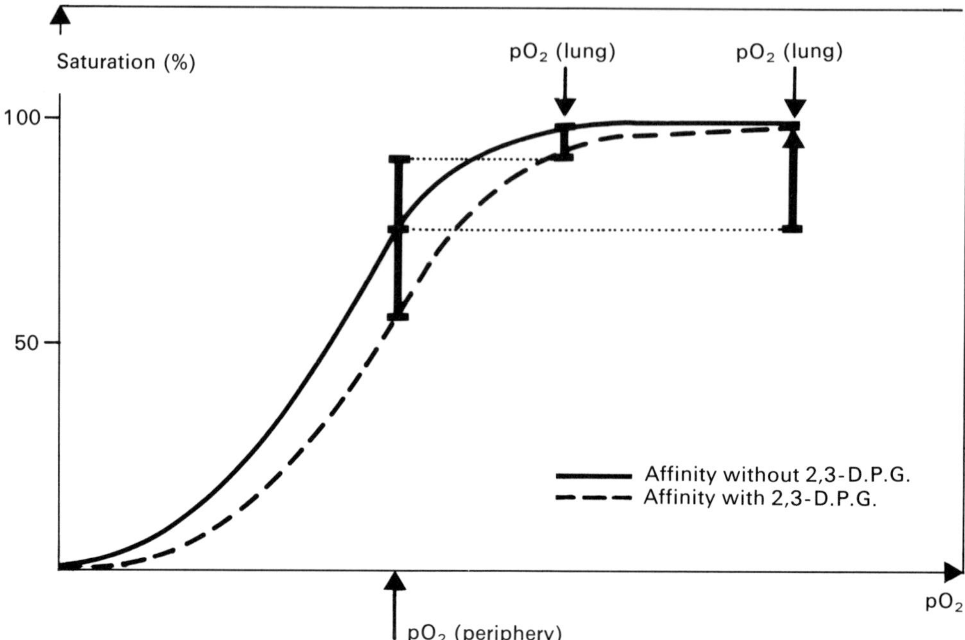

Fig. 3. Effect of 2,3-diphosphoglycerate (2,3-D.P.G.) on the affinity of haemoglobin for oxygen.

to the tissues is improved, and this in turn counteracts the production of acid metabolites. For any particular solution it is possible to establish the pO_2 value required to ensure that the haemoglobin is half saturated. This value is referred to as the pO_2 50 and serves as a measure of the system's affinity for oxygen. Consequently, there are at least three extramolecular, but intracellular, factors which diminish the affinity of haemoglobin for oxygen[15]:

1. Increase in 2,3-diphosphoglycerate concentration
2. Decrease in pH
3. Decrease in salt concentration.

In conclusion, it should also be pointed out that the binding of 2,3-diphosphoglycerate is likewise enhanced by a decrease in pH within the range from 6 to 9[8] or by a decrease in ion concentration.

In recent years it has been stated on a number of occasions in the literature that propranolol reduces the affinity of erythrocytes for oxygen both *in vivo* and *in vitro*[10, 11]. Experiments conducted in our own laboratory and by other authors have shown that this substance has no effect whatever on a haemoglobin solution. On the other hand, it does exert a considerable effect on erythrocytes (Table 3). The usual explanation advanced so far to account for the decrease in pO_2 50 in response to propranolol is that membrane-bound 2,3-diphosphoglycerate becomes displaced into the cytoplasm[10]. In point of fact, propranolol does not significantly influence the total cellular content of 2,3-diphospho-

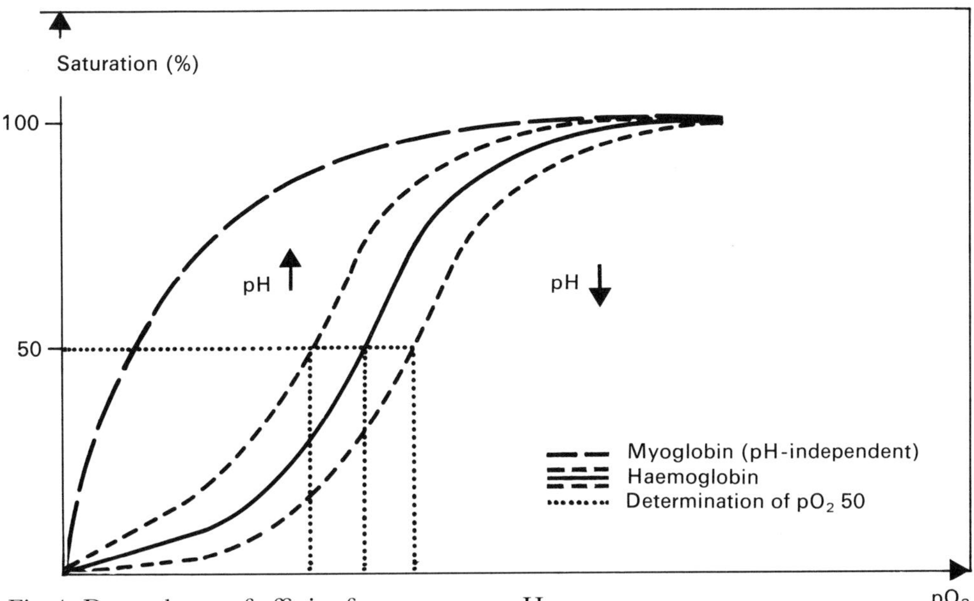

Fig. 4. Dependence of affinity for oxygen on pH.

Table 3. Effect of propranolol on pO_2 50 in washed erythrocytes*.

pO_2 50 (mm. Hg)		ΔpO_2 50	M 2,3-D.P.G.**/M Hb. tetramer	
Without propranolol	With propranolol		Without propranolol	With propranolol
25.3	32.0	6.7		
24.5	33.0	8.5		
20.1	24.2	4.1	0.75	0.74
20.0	24.3	4.3	0.77	0.74

*Conditions: Haematocrit of 25% prior to addition of propranolol. Cells washed three times in physiological buffer at pH 7.38. Addition of human serum albumin suspended in the same buffer.
**2,3-D.P.G. = 2,3-diphosphoglycerate

glycerate. But it does cause massive shrinking of erythrocytes; in our experiments the haematocrit of the erythrocyte suspension fell by more than 20%. It is therefore safe to say that propranolol causes a considerable loss of water and salt from erythrocytes.

Since neither we nor workers in other laboratories[3] were able to demonstrate the release of membrane-bound 2,3-diphosphoglycerate, the decrease produced by propranolol in the affinity of erythrocytes for oxygen is presumably due primarily to a change in their salt content. Studies are now being undertaken in an attempt to establish whether this salt depletion achieves its effect on the oxygen affinity of erythrocytes by lowering the intra-erythrocytic pH[1], by enhancing the binding of intracellular 2,3-diphosphoglycerate on the basis of reduced ionic strength, or by both these mechanisms.

References

1 AGOSTINI, A., BERFASCONI, C., GERLI, G. C., LUZZANA, M., ROSSI-BERNARDI, L.: Oxygen affinity and electrolyte distribution of human blood: changes induced by propranolol. Science *182*, 300 (1973)
2 BENESCH, R., BENESCH, R. E., ENOKI, Y.: The interaction of hemoglobin and its subunits with 2,3-diphosphoglycerate. Proc. nat. Acad. Sci. (Wash.) *61*, 1102 (1968)
3 BRANN, E. G., NEWMAN, D. J.: Oxygen affinity in red cells: inability to show membrane-bound 2,3-diphosphoglycerate. Science *179*, 593 (1973)
4 BRAUNITZER, G., GEHRING-MÜLLER, R., HILSCHMANN, N., HILSE, K., HOBOM, G., RUDLOFF, V., WITTMANN-LIEBOLD, B.: Die Konstitution des normalen adulten Humanhämoglobins. Hoppe-Seylers Z. physiol. Chem. *325*, 283 (1961)
5 CHANUTIN, A., CURNISH, R. R.: Effect of organic and inorganic phosphates on the oxygen equilibrium of human erythrocytes. Arch. Biochem. *121*, 96 (1967)
6 KENDREW, J. C.: Structure and function in myoglobin and other proteins. Fed. Proc. *18*, 740 (1959)
7 KONIGSBERG, W., GOLDSTEIN, J., HILL, R. J.: The structure of human hemoglobin. VII. The digestion of the β chain of human hemoglobin with pepsin. J. biol. Chem. *238*, 2028 (1963)
8 MANSOURI, A., WINTERHALTER, K. H.: Nonequivalence of chains in hemoglobin oxidation and oxygen binding. Effect of organic phosphates. Biochemistry *13*, 3311 (1974)
9 MUIRHEAD, H., GREER, J.: Three-dimensional Fourier synthesis of human deoxy-haemoglobin at 3·5 Å resolution. Nature (Lond.) *288*, 516 (1970)
10 OSKI, F. A., MILLER, L. D., DELIVORIA-PAPADOPOULOS, M., MANCHESTER, J. H., SHELBURNE, J. C.: Oxygen affinity in red cells: changes induced in vivo by propranolol. Science *175*, 1372 (1972)
11 PENDLETON, R. G., NEWMAN, D. J., SHERMAN, S. S., BRANN, E. G., MAYA, W. E.: Effect of propranolol upon the hemoglobin-oxygen dissociation curve. J. Pharmacol. exp. Ther. *180*, 647 (1972)
12 PERUTZ, M. F.: Stereochemistry of cooperative effects in haemoglobin. The Bohr effect and combination with organic phosphates. Nature (Lond.) *228*, 734 (1970)
13 PERUTZ, M. F., MUIRHEAD, H., COX, J. M., GOAMAN, L. C. G.: Three-dimensional Fourier synthesis of horse oxyhaemoglobin at 2·8 Å resolution: the atomic model. Nature (Lond.) *219*, 131 (1968)
14 RAHBAR, S., BLUMENFELD, O., RANNEY, H. M.: Studies of an unusual hemoglobin in patients with diabetes mellitus. Biochem. biophys. Res. Commun. *36*, 838 (1969)
15 WINTERHALTER, K. H.: Does hemoglobin breathe, and if yes, how? The T and R state of hemoglobin. New Engl. J. Med. *289*, 41 (1973)

Discussion

J. Hodler: In a double-blind trial on 12 healthy volunteers we ourselves, together with Dr. Imhof, studied the influence of various beta-blockers on the 2,3-diphosphoglycerate content of haemoglobin in an attempt to test the hypothesis postulated by Oski et al.*. According to this hypothesis, the binding of 2,3-D.P.G. to the membrane is dissolved by propranolol, with the result that more unbound 2,3-D.P.G. passes into a membrane-free haemolysate. To our great disappointment, however, we were unable to confirm this effect of beta-blockers; we prepared both membrane-free and non-membrane-free haemolysates, but we didn't find any difference between them as regards their content of 2,3-D.P.G. We then tested membrane-free haemolysates only, which of course should display Oski's effect. The double-blind trial revealed that propranolol, CGP 2175, and oxprenolol (®Trasicor) had no significant influence on the 2,3-D.P.G. in the supernatant membrane-free haemolysate after one, two, and 24 hours (Table 1). A slight, but not statistically significant, tendency for the 2,3-D.P.G. to increase was observed only with oxprenolol. I was, however, considerably reassured by the explanations Dr. Winterhalter gave in the last part of his paper, because I had almost feared that all my 2,3-D.P.G. determinations had been wrong.

Table 1. Mean values ± S.D. for 2,3-diphosphoglycerate in membrane-free haemolysates (expressed as mM/10 g.% Hb.) in 12 healthy volunteers following treatment with placebo and with three different beta-blockers.

	Control	1 hour	2 hours	24 hours
Placebo	0.125 (±0.008)	0.128 (±0.002)	0.125 (±0.003)	0.129 (±0.004)
Propranolol (3×80 mg.)	0.136 (±0.009)	0.137 (±0.008)	0.119 (±0.007)	0.130 (±0.008)
CGP 2175 (3×100 mg.)	0.124 (±0.008)	0.124 (±0.005)	0.135 (±0.010)	0.132 (±0.011)
Oxprenolol (3×80 mg.)	0.154 (±0.021)	0.164 (±0.007)	0.167 (±0.012)	0.168 (±0.011)

H. Brunner: Dr. Winterhalter, the release of potassium from the cells, to which you referred in the spoken version of the paper you presented, has been demonstrated both in animal experiments and in clinical studies. It is therefore not a phenomenon that occurs only *in vitro*.

K. H. Winterhalter: I fully agree. Several authors have also observed a change in the pO_2 50 *in vivo* following the administration of propranolol.

J. Hodler: As Dr. Bühler mentioned in his paper, there is a tendency for the blood potassium levels to rise in response to beta-blockade. In the light of this observation we ourselves conducted some tests on which I shall be reporting in my paper. The release of potassium from erythrocytes provides one possible explanation for this rise in the blood potassium levels. Can anyone here say how much potassium is released?

* Oski, F.A., Miller, L.D., Delivoria-Papadopoulos, M., Manchester, J.H., Shelburne, J.C.: Oxygen affinity in red cells: changes induced in vivo by propranolol. Science *175*, 1372 (1972)

K. H. Winterhalter: *In vitro* the release of potassium and the concomitant shrinkage of the haematocrit is linearly dose-dependent. At maximal stimulation approximately 30% of the intracellular potassium is released.

F. R. Bühler: We based our studies on the observation that, whereas propranolol suppresses renin secretion in every case, it doesn't reduce aldosterone secretion to the extent one would expect in the light of its excellent suppressant effect on renin. It was for this reason that we included the simultaneously determined plasma potassium concentration in our analyses. The increase in the plasma potassium concentration induced by beta-blockade appears to be large enough to stimulate aldosterone secretion and, at the same time, to suppress renin secretion still further. A similar dissociation is of course encountered in low-renin essential hypertension.

G. Hitzenberger: At what minimum concentrations did you see the changes in pO_2 50 that you reported, Dr. Winterhalter? In the spoken version of your paper you mentioned 10 mg./100 ml. solutions of propranolol. Did you observe such changes at concentrations that are attainable *in vivo* under physiological or pharmacological conditions?

K. H. Winterhalter: Determinations of the pO_2 50 are not all that easy from the technical point of view. Consequently, changes recorded at very low concentrations – i.e. with, say, 1 mg./100 ml. solutions – are not necessarily significant; these changes, however, can already be clearly demonstrated with 2–3 mg./100 ml. solutions.

P. R. Imhof: Dr. Winterhalter's findings should prompt us to investigate the problem of changes in viscosity. Reductions in the haematocrit of the type he mentioned could after all lead to a decrease in viscosity and thus to an improvement in rheological conditions in the periphery.

K. H. Winterhalter: The transition of the erythrocytes from a flat to a spherical shape would undoubtedly be bound to produce a decisive alteration in the internal viscosity of the blood. On the other hand, it must be borne in mind that a spherical erythrocyte will be much less liable to deformation (here, of course, the haemoglobin content, which normally already amounts to 33%, also plays a part), and this lack of deformability might under certain circumstances again lead to an increase in resistance in the capillary circulation.

J. Hodler: In the course of our experiments we studied the haematocrit readings before and after treatment with propranolol or oxprenolol in a group of volunteers. The readings remained exactly the same. This may be a question of dosage – that is to say, the dose of 80 mg. three times daily, which we administered, may not have been sufficient to produce an effect on the haematocrit.

K. H. Winterhalter: I think that haematocrit readings *in vivo* are only of significance if the total blood volume, or at least the plasma volume, can be determined separately at the same time.

Beta-blockers in the treatment of angina pectoris

by J.-L. RIVIER*

In recent years the use of beta-blockers for the treatment of angina pectoris has gained widespread acceptance[30, 45, 50, 72, 79]. Since the advent of propranolol, which was the first therapeutically effective beta-blocker, numerous other compounds belonging to the same group have given proof of their value in this indication. But the following paper, in which I shall discuss the position currently occupied by this form of medication, will be confined to five beta-blockers with which I have had personal experience, namely: propranolol[16], oxprenolol (®Trasicor)[17], alprenolol[62], pindolol[97], and practolol[27].

By way of introduction, a few essential notions concerning angina pectoris in general and angina of effort in particular should perhaps first of all be recalled. The problem that I shall be dealing with here is what happens as a consequence of coronary artery stenosis, the presence of which should always be borne in mind when discussing treatment. Though vascular obstructions of this type take quite a long time to develop, once they have reached significant proportions they have the effect of reducing the "coronary reserve" beyond the site of the obstruction[79]. Despite the existence of more or less efficiently functioning anastomoses, the arterial blood flow in the affected regions – and hence the supply of oxygen – cannot exceed a certain limit, even where the post-stenotic arterioles undergo maximal dilatation. If and when the oxygen consumption of the heart muscle oversteps this threshold, myocardial hypoxia sets in, resulting in an attack of angina pectoris characterised by painful sensations whose precise mode of origin is, incidentally, still unknown.

In recent years it has been demonstrated that the main factors determining myocardial oxygen consumption are the heart rate, the contractility of the myocardium, and the degree of intramyocardial tension, the latter being chiefly dependent on arterial blood pressure and ventricular volume[101, 110].

Treatment for angina pectoris should thus logically be aimed, firstly, at improving the coronary reserve in the affected area and, secondly, at reducing the myocardial oxygen consumption[15]. The first of these two objectives will not be discussed here: suffice it to say that surgical treatment specifically designed to achieve revascularisation of the myocardium by means of an aorto-coronary bypass seems to be proving increasingly successful[52], at least as a measure calculated to afford the patient effective relief for a certain number of years. With regard to the second objective, i.e. reduction of the myocardial oxygen consumption, it has long been known that the sympathetic nervous

---

* Département de cardiologie des Services de médecine et de chirurgie, Hôpital Cantonal Universitaire, Lausanne, Switzerland.

system exerts an important influence on the heart, both directly and via the catecholamines. By stimulating the beta-adrenergic receptors, it is capable of accelerating cardiac rhythm, increasing the speed of conduction in specific tissues, shortening the effective refractory period, and enhancing myocardial contractility[102]. Consequently, all sympathetic excitation – whether it be necessary, as in response to physical effort for example, or whether it be superfluous, as in the case of the reflex excitation induced by emotion – inevitably leads to an increase in myocardial oxygen consumption, an increase to which the sympathetic nervous system may also contribute indirectly by virtue of the role it often plays in raising the arterial blood pressure. Hence, any substance that is able to block the sympathetic inflow to the cardiac beta-receptors will, in principle, serve to diminish the oxygen requirement of the heart; but it must not be a substance which prevents the heart from performing sufficient work to meet the needs of the body in general. Experience acquired in recent years with beta-adrenergic blockers – including especially propranolol, the prototype of this class of compounds – has confirmed the correctness of this concept. Under treatment with a beta-blocker the heart is capable, within certain limits, of functioning more economically at a given level of physical effort thanks to the fact that the body adapts itself accordingly[64]. In other words, although blockade of the beta-adrenergic receptors results in a slowing of the heart rate and a decrease in cardiac output at all submaximal levels of effort, these two effects are at the same time offset by an adaptational response on the part of the peripheral circulation and by an increase in the systemic arterio-venous oxygen difference, the oxygen consumption of the body as a whole remaining unchanged. Since the heart thus has to work less hard and the myocardium therefore consumes less oxygen, the local coronary blood flow, in turn, is less severely taxed and consequently remains below its attainable maximum. In addition, beta-blockade serves to screen the heart against sudden bombardment with sympathetic nervous impulses, such as is liable to cause a sharp increase in myocardial oxygen consumption.

There are two drawbacks, however, which beta-blockade would *a priori* appear to involve. Firstly, by suppressing adrenergic stimuli it may diminish contractility of the heart muscle to such an extent that the ventricular volume increases[20], thereby causing a rise in myocardial oxygen consumption in accordance with the law of Laplace. Though this, of course, means that beta-blockers are contra-indicated in the presence of cardiac insufficiency, it is more of a theoretical than a practical disadvantage – judging at least from the positive results which beta-blockade has yielded in patients suffering from anginal pain. The second drawback is that inhibition of the coronary beta-receptors theoretically also entails some degree of vasoconstriction which could have the effect of reducing coronary blood flow[18, 20, 40]. In point of fact, this phenomenon too plays a negligible role, because such decrease in the blood flow as does occur appears to be quite commensurate with the diminished requirements of the heart muscle[81, 123]. Moreover, it has been demonstrated in animal experiments that beta-blockade is also capable of modifying the distribution of the coronary blood supply in such a way that irrigation of the ischaemic zones is actually improved[86]. Finally, it must be remembered that the bradycardia induced by all beta-blockers also serves to prolong diastole, an effect which likewise has favourable repercussions on coronary blood flow.

These few introductory remarks on some of the theoretical and experimental aspects of our topic should, I think, at this point be supplemented by a reminder that, from the clinical standpoint, any study on the therapeutic activity of a drug which is known or alleged to be effective in the treatment of angina pectoris is bound to pose difficult problems[53]. The reason for these problems is not only that the disease often runs a highly erratic and unpredictable course, but also that – against a background of anatomical lesions varying in number and severity – the manifestations to be combated by drug therapy for this condition consist either of pain, which is merely a variable of paroxysmal myocardial hypoxia[38], or of electrical changes whose development, as has long been known, does not necessarily parallel that of the angina pectoris itself. Clinical evaluation of the response to treatment therefore entails some element of uncertainty, even if – as should preferably be done – one employs the ergometric method for assessing effort tolerance as reflected in the occurrence of pain and in alterations in the E.C.G. pattern, and even if one also makes use of such indices as the tension-time index (T.T.I.)[101], the product of heart rate times systolic pressure[91, 92], or simply the heart rate alone, which also correlates very well with the myocardial oxygen consumption and coronary blood flow[64, 68]. In order to obtain homogeneous case material for study, it is necessary to select those patients whose angina pectoris is known or at least believed to be "stable"; and such cases account for only a limited proportion of all the various categories of patients suffering from the disease. Consequently, even under the strictest of experimental conditions and even if recourse is had to the assessment of effort tolerance, it is impossible to avoid certain dubieties, to which the placebo factor likewise adds. It is for this reason that, though repeated successes in open trials conducted in patients well known to the treating clinician may quite often already have convinced him of a drug's efficacy, double-blind studies involving a relatively small number of patients – all belonging to the same type – nevertheless remain indispensable as a means of affording proof of its value in angina pectoris[108].

Several 'open or double-blind clinical trials performed with propranolol[16] during the first few years after its appearance on the scene indicated that this compound was effective in the treatment of angina pectoris – a finding which further experience has done nothing to disprove[1, 45, 46]. Subsequently, other beta-blockers were also introduced, including oxprenolol[17], alprenolol[62], pindolol[97], and later practolol[27], all of which likewise proved to be effective anti-anginal drugs[50, 79]. An overall comparison of these five products, based on the relevant literature, shows (Table 1) that their coefficients of efficacy or, expressed in other terms, the percentages of successful responses elicited by them, are fairly similar, ranging from 74 to 90% in the open studies and from 69 to 75% in the double-blind trials[2, 5-10, 12-14, 19, 23-25, 29, 32, 35-37, 39, 42, 54, 57, 59-61, 63, 65, 67, 69, 71, 73-78, 80, 82, 84, 85, 88, 89, 93-96, 98-100, 103-107, 109, 111-122, 125-128, 131, 132]. The chief criterion employed in arriving at these results was relief from pain, as distinct

Table 1. Average percentages of successful responses (in terms of relief from pain) obtained with five different beta-blockers in the treatment of angina pectoris, as reported in the literature.

Beta-blocker	Open trials (response rate in %)	Double-blind trials (response rate in %)
Propranolol[1, 11, 28, 31, 43, 44, 46, 47, 55, 83, 130]	80	72
Oxprenolol[10, 19, 23, 35–37, 39, 71, 73, 76, 79, 82, 88, 89, 93–96, 99, 103, 105–107, 109, 113, 114, 117, 119, 125–127, 132]	88	75
Alprenolol[5, 8, 13, 14, 29, 32, 57, 60, 61, 63, 67, 74, 75, 77, 98, 104, 111, 116, 118, 121]	75	+ (statistically significant)
Pindolol[2, 6, 9, 24, 25, 54, 59, 65, 69, 78, 80, 100, 112, 115, 120]	74	69
Practolol[7, 12, 42, 84, 85, 122, 128, 131]	90	70

Beta-blocker	Daily dosage (in mg.)
Propranolol	10–1,280
Oxprenolol	120–480
Alprenolol	75–800
Pindolol	5–22.5
Practolol	100–2,400

Table 2. Daily dosages in which five different beta-blockers have been employed in the treatment of angina pectoris, as reported in the literature.

from such effect as the drug might possibly have had on the evolution of the disease. So far as the open studies are concerned, the differences in the response rates are not surprising in view of the fact that the investigators, the patients, the methodology, the duration of the observation periods, and the criteria of evaluation all varied – quite apart from the question of differences in the dosages prescribed, a point to which I shall revert later. The figures reported thus provide only a very approximative guide. What *is* striking about these results, on the other hand, is the consistently high proportion of favourable responses, which was almost invariably far in excess of the notorious 35% for responses to a placebo.

As for the controlled studies, the results of these are likewise most encouraging, with a few exceptions[4, 48, 50] that were probably due to the use of too low a dosage. The impression emerging from these studies, an impression also confirmed by my own experience, is that all these beta-blockers are therapeutically active, but that there will always be a certain number of patients, i.e. about 30%, who for some unknown reason derive no benefit from such medication. A closer look at the dosages employed (Table 2) reveals that in some instances they varied very considerably, the minimum and maximum dose being in a ratio of 1:100 or more in the case of propranolol, for example, and 1:24 and 1:10 in the case of practolol and alprenolol, respectively, whereas the ratios for oxprenolol and pindolol were less extreme. The reason for this progressive decrease in the ratio between the minimum and maximum dose may possibly have been the fact that oxprenolol and pindolol appeared on the market at a later stage, with the result that those employing them were able to profit from experience acquired with the use of earlier beta-blockers – although this argument obviously does not apply to practolol. In contrast to the relatively narrow therapeutic range of a drug such as digitalis, the situation with regard to dosage is quite different in the case of the beta-blockers and certainly more reassuring for the general practitioner: the physician resorting to a beta-blocker has plenty of "room to manoeuvre" and can rest assured that, not only in patients with hypertension but also in those suffering from angina pectoris, whatever dosage he prescribes will entail comparatively little risk of provoking dangerous side effects, including cardiac insufficiency in particular. On the other hand, the wide variations in the dosages in which beta-blockers are employed suggest, too, that the sympathetic nervous system, whose degree of activity is certainly not the same in all patients, varies considerably in its sensitivity to the effect of beta-blockers from one individual to another; it is also true to say that the severity and extent of the damage which the patient's heart muscle has already sustained may likewise have some influence in this connection[58, 60]. Beta-blockers should therefore be administered in progressively increasing doses in order to build up gradually to the dosage that is either therapeutically effective or no longer satisfactorily tolerated; not until this dosage level has been reached is it possible to decide whether the treatment is to be regarded as a success or a failure.

Although, in the light of the overall response rates achieved with each of the various beta-blockers in question, it can fairly safely be asserted that they do not differ greatly from one another as regards their efficacy, the results of numerous studies based on comparisons of their activity in animals do suggest that they differ with respect to certain of their properties[33]. By reference, for example, to the presence or absence of a cardioselective action on the one hand and an intrinsic sympathomimetic effect on the other, the beta-blockers can be classified as follows:

Non-cardioselective beta-blockers
– without an intrinsic sympathomimetic effect:
  propranolol
– with an intrinsic sympathomimetic effect:
  oxprenolol
  alprenolol
  pindolol.

Cardioselective beta-blocker
– with an intrinsic sympathomimetic effect:
  practolol.

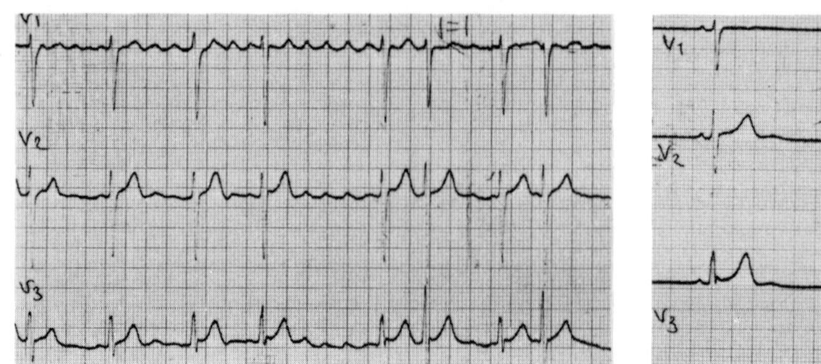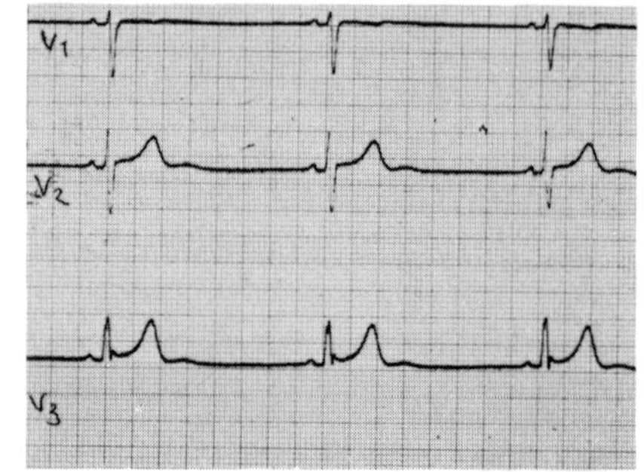

Fig. 1. Electrocardiographic recordings in a 75-year-old man suffering from angina pectoris with attacks of flutter fibrillation. *Left:* during a typical bout of arrhythmia. *Right:* under treatment with oxprenolol (the slow sinus rhythm of 37 beats/min. was well tolerated).

Experimental studies have shown that bradycardia at rest is most pronounced in response to propranolol, which can accordingly be regarded as displaying no intrinsic sympathomimetic action. This lack of intrinsic sympathomimetic activity would, on the face of it, appear to be an undesirable feature in cases where – as often happens in our experience – the patient already has a slow pulse rate of below, say, 50–55 beats/min. In practice, however, this is by no means always a factor of much importance: patients with coronary artery disease do in fact generally manage, with or without digitalisation, to tolerate a fairly low heart rate quite well (Figure 1), provided they do not also have a fall in blood pressure, which is usually less well tolerated. Moreover, for all practical purposes, the difference – dose for dose – between propranolol as compared with oxprenolol, alprenolol, pindolol, and practolol is insignificant[23, 49, 51, 66, 88, 89, 106, 114, 124, 132]. Where one beta-blocker proves to be less satisfactorily tolerated than another, this is not because it provokes more or less pronounced bradycardia at rest, but because of other side effects such as nausea (or possibly vomiting), diarrhoea, palpitation, hypotension, or even aggravation of anginal pain, the occurrence of which cannot be predicted in advance. It is thus as well to have several different beta-blockers at one's disposal, especially as their individual tolerability may vary from patient to patient.

On the other hand, complete suppression of adrenergic stimuli is not desirable in certain types of case, i. e. in patients in whom the heart is already functioning at the limit of its capacity and in whom even weak doses of a beta-blocker may – both at rest and particularly in response to effort – provoke an increase in left-ventricular pressure coupled with clinical manifestations of cardiac insufficiency. Consequently, where treatment has to be given to a patient bordering on cardiac decompensation – a situation which, incidentally, is seldom encountered in cases of pure angina of effort – it would appear preferable to resort to oxprenolol, alprenolol, or pindolol, because the negative inotropic effect of these preparations seems to be less marked[17, 21, 62, 90, 97]; the fact that their negative inotropic effect is weaker is probably related to their intrinsic sympathomimetic activity.

At this point it should be recalled that the depressant action exerted on the myocardium by a beta-blocker, in doses of the size employed in clinical practice, is due to suppression of adrenergic stimuli, and that very high doses, i.e. doses far higher than are required to achieve beta-blockade, have to be given in order to produce a direct, supplementary depressant effect of the quinidine-like type [26, 34]. In practice, cardiac insufficiency is less of a problem than it has been made out to be, because in suspect cases it is sufficient to prescribe diuretics or digitalis in addition. Digitalis has proved particularly effective in counteracting the depressant influence of beta-blockers on the myocardium, especially since its positive inotropic effect remains completely unaffected by beta-blockers [29, 41, 60]. Nevertheless, the treating physician should always bear in mind the possible risk of provoking cardiac insufficiency, this potential hazard being one of the reasons why treatment with a beta-blocker should be initiated with small, gradually increasing doses.

Except in the case of asthmatics and patients suffering from chronic respiratory insufficiency, in whom it is probably preferable to use practolol, comparisons between this cardioselective beta-blocker and the other non-cardioselective beta-blockers have not – at least from the clinical aspect – disclosed any major differences so far as treatment for angina pectoris is concerned. The property of "cardioselectivity", which experimental evidence has shown to result in the complete avoidance of coronary vasoconstriction, does not in fact seem to be of any importance in man from this point of view [22, 49, 51, 66, 88, 106, 124].

Another problem to which attention should be drawn is that of the patient who fails to respond to a beta-blocker. I am thinking here, not of failures which are due to the occurrence of side effects and which are bound to be encountered in a certain number of cases, but of genuine failures, i.e. of patients who tolerate the drug satisfactorily but who either derive no benefit from the medication or whose condition actually worsens when the dosage is increased [114, 130]. No clinical, haemodynamic, or haematological criterion has yet been found on the basis of which it would be possible to predict the response to treatment in this respect. Moreover, statistics relating to the number of genuine failures also vary. It would seem in fact that, with very few exceptions, such studies as have been devoted to this problem have been inadequate. Though the relevant literature does contain reports of failures, these generally concern isolated cases, the investigators' interest being chiefly concentrated on the patients who responded favourably. In view of this, it would appear necessary to pinpoint and study in greater detail – with particular reference to the behaviour of the ventricles and to the severity of arteriographically demonstrable coronary lesions – those cases which fail to respond to beta-blockers or which even deteriorate under treatment with them.

It must be acknowledged that, despite numerous studies, it has not yet proved possible to determine what is the most important factor accounting for the clinical improvement elicited by beta-blockers. Though the decrease in the heart rate certainly plays a role in this connection [88], it is impossible at present to assess the precise significance of this decrease, because no purely "chrono-selective" beta-blockers are available, i.e. no beta-blockers that simply reduce the heart rate without at the same time also modifying the impact of sympathetic nervous activity on myocardial contractility. Hence, it is in this direction that further research should, I feel, be undertaken. The possibility that the heart may contain two different types of beta-receptor – one for specific tissue and one for contractile myocardial tissue – is an hypothesis which, though merely speculative, cannot be rejected out of hand.

Finally, one last problem should also be touched upon, namely, the beneficial influence which prolonged treatment with beta-blockers might conceivably exert on the overall development of ischaemic heart disease[70, 129]. If therapy of this kind really can have such an influence, is this desirable effect bound up with the action which certain beta-blockers have on fibrinolysis and platelet aggregation[56, 87]? These are unsolved questions to which answers will not be found until a great deal more study has been undertaken.

In conclusion, viewing this whole problem from the practical standpoint, I would say that, although the chances of success with beta-blockers in cases of angina pectoris are of the order of some 70%, no criteria yet exist for selecting patients suitable for such treatment. It is, in fact, impossible to predict how well any one patient in particular is likely to respond; as regards dosage, however, the beta-blockers do at all events have a wide safety margin, and – provided the doses are built up gradually – their side effects are not of a serious nature.

References

1 AMSTERDAM, E.A., GORLIN, R., WOLFSON, S.: Evaluation of long-term use of propranolol in angina pectoris. J. Amer. med. Ass. *210*, 103 (1969)

2 AQUARO, G., PAVIA, M., BADUINI, G.: Il trattamento a lungo termine dell'angina pectoris con il pindolol. Minerva cardioangiol. *21*, 212 (1973)

3 ARONOW, W.S.: The medical treatment of angina pectoris. VI. Propranolol as an antianginal drug. Amer. Heart J. *84*, 706 (1972)

4 ARONOW, W.S., KAPLAN, M.A.: Propranolol continued with isosorbide dinitrate versus placebo in angina pectoris. New Engl. J. Med. *280*, 847 (1969)

5 ARSTILA, M., IISALO, E., KALLIO, V.: Alprenolol in angina pectoris. A clinical and ergometric study with a new beta-adrenergic blocking agent. Ann. clin. Res. *1*, 13 (1969)

6 ARSTILA, M., KALLIO, V., WENDELIN, H.: Propranolol and LB 46 (prindolol) in angina pectoris. A comparative long-term ergometric study. Ann. clin. Res. *5*, 91 (1973)

7 ATKINS, J.M., BLOMQVIST, C.G., COHEN, L.S.: Comparative study of beta-adrenergic blocking agents on exercise in ischemic heart disease. Circulation *42*, Suppl. III: 132, Abstr. 486 (1970)

8 AUBERT, A. NYBERG, G., SLAASTAD, R., TJELDFLAAT, L.: Prophylactic treatment of angina pectoris. A double-blind cross-over comparison of alprenolol and pentanitrol. Brit. med. J. *i*, 203 (1970)

9 BENYAMINE, R., JOUVE, A.: Traitement de l'insuffisance coronarienne chronique par le pindolol. Marseille méd. *109*, 91 (1972)

10 BIANCHI, C., LUCCHELLI, P.E., STARCICH, R.: Beta-blockade and angina pectoris: a controlled multi-centre clinical trial. Pharmacol. clin. *1*, 161 (1969)

11 BIRKETT, D.A., CHAMBERLAIN, D.A.: Beta-adrenergic blockade in angina pectoris: a method of treadmill assessment. Brit. med. J. *ii*, 500 (1966)

12 BJERREGAARD, P., HVIDT, S.: Praktolol (Eraldin) og iskaemisk hjertesygdom. En ergometerundersøgelse (The value of practolol [Eraldin] in ischaemic heart disease). Ugeskr. Laeg. *133*, 49 (1971)

13 BJÖRNTORP, P.: The treatment of angina pectoris with a new beta-receptor blocking agent (H 56/28). Acta med. scand. *182*, 285 (1967)

14 BJÖRNTORP, P.: Treatment of angina pectoris with beta-receptor blockade. Mode of action. Acta med. scand. *184*, 259 (1968)

15 BLACK, J.W.: The predictive value of animal tests to drugs affecting the cardiovascular system in man. In Wolstenholme, G., Porter, R. (Editors): Drug responses in man, Ciba Found. Symp., p. 111 [pp. 113, 121–122] (Churchill, London 1967)

16 BLACK, J.W., CROWTHER, A.F., SHANKS, R.G., SMITH, L.H., DORNHORST, A.C.: A new adrenergic beta-receptor antagonist. Lancet *i*, 1080 (1964)

17 BRUNNER, H., HEDWALL, P.R., MEIER, M.: Pharmakologische Untersuchungen mit 1-Isopropylamino-3-(o-allyloxyphen-oxy)-2-propanol-hydrochlorid, einem adrenergischen β-Receptorenblocker. Arzneimittel-Forsch. (Drug. Res.) *18*, 164 (1968)

18 BUSSMANN, W.D., RAUH, M., KRAYENBÜHL, H.P.: Coronary and hemodynamic effects of myocardio-selective beta-receptor blockade by ICI 50172 in closed-chest dog. Differentiation of coronary and myocardial beta receptors. Amer. Heart J. *79*, 347 (1970)

19 CANTÚ O., M., PIMENTEL M., J.D.: Uso de un nuevo bloqueador de los receptores adrenérgicos beta en la angina de pecho. Sem. méd. Méx. *67*, No. 861 (1971)

20 CHAMBERLAIN, D.A.: Effects of beta adrenergic blockade on heart size. Amer. J. Cardiol. *18*, 321 (1966)

21 CHOQUET, Y., CAPONE, R.J., MASON, D.T., AMSTERDAM, E.A., ZELIS, R.: Comparison of the beta adrenergic blocking properties and negative inotropic effects of oxprenolol and propranolol in patients. Amer. J. Cardiol. *29*, 257 (1972); abstract of paper

22 COLTART, D.J.: Comparison of effects of propranolol and practolol on exercise tolerance in angina pectoris. Brit. Heart J. *33*, 62 (1971)

23 COLTART, D.J.: Observations on exercise tolerance in anginal patients treated with beta-blockers. In Burley, D.M., et al. (Editors): New perspectives in beta-blockade, Int. Symp., Scanticon, Aarhus, Denmark, 1972, p. 167 (Metropolis Press, London 1973)

24 DEMIROGLU, C.: Les effets du Visken sur l'épreuve d'effort en cas d'insuffisance coronarienne. Ars Med. (Gand) *26*, 729 (1971)

25 DE VITA, C., CADEL, A., CARÙ, B., MAZZINI, C., PARENTI, G.F.: L'impiego di un beta-adrenolitico nel trattamento dell'angina pectoris. Minerva cardioangiol. *21*, 304 (1973)

26 DOLLERY, C.T., PATERSON, J.W., CONOLLY, M.E.: Clinical pharmacology of beta-receptor-blocking drugs. Clin. Pharmacol. Ther. *10*, 765 (1969)

27 DUNLOP, D., SHANKS, R.G.: Selective blockade of adrenoceptive beta receptors in the heart. Brit. J. Pharmacol. *32*, 201 (1968)

28 DWYER, E.M., Jr., WIENER, L., COX, J.W.: Effects of beta-adrenergic blockade (propranolol) on left ventricular hemodynamics and the electrocardiogram during exercise-induced angina pectoris. Circulation *38*, 250 (1968)

29 EKELUND, L.G., JOHNSSON, G., MELCHER, A., ORÖ, L.: Effects of Cedilanid-D and alprenolol in angina pectoris. Europ. J. clin. Pharmacol. *6*, 113 (1973)

30 ELLIOTT, W.C., STONE, J.M.: Beta-adrenergic blocking agents for the treatment of angina pectoris. Progr. cardiovasc. Dis. *12*, 83 (1969)

31 EPSTEIN, S.E., ROBINSON, B.F., KAHLER, R.L., BRAUNWALD, E.: Effect of beta-adrenergic blockade on the cardiac response to maximal and submaximal exercise in man. J. clin. Invest. *44*, 1745 (1965)

32 FINNSON, M., NYBERG, G.: Angina pectoris: klinisk värdering av profylaktika i

öppen vård (Angina pectoris: a clinical evaluation of prophylaxis in out-patient practice). Läkartidningen *66*, 5137 (1969)

33 FITZGERALD, J. D.: Perspectives in adrenergic beta-receptor blockade. Clin. Pharmacol. Ther. *10*, 292 (1969)

34 FITZGERALD, J. D., WALE, J. L., AUSTIN, M.: The haemodynamic effects of ($\pm$) propranolol, dexpropranolol, oxprenolol, practolol and sotalol in anaesthetized dogs. Europ. J. Pharmacol. *17*, 123 (1972)

35 FORREST, W. A.: A total of 254 cases of angina pectoris treated with oxprenolol in hospital practice – a monitored release study. Brit. J. clin. Pract. *26*, 217 (1972)

36 FORREST, W. A.: Trasicor (oxprenolol) in the treatment of 4,750 angina patients. Practitioner *208*, 412 (1972)

37 FORREST, W. A.: A subjective comparison between oxprenolol (Trasicor) and long-acting nitrates in the treatment of angina pectoris in general practice. J. int. med. Res. *1*, 253 (1973)

38 FROMENT, R.: Angine de poitrine. Arch. Mal. Cœur *65*, 291 (1972)

39 FÜLÖP, T., SZODORAY, P.: Beta-adrenerg receptor blokkolók a rhythmuszavarok és az angina pectoris kezelésében (Beta-adrenergic receptor blockers in the treatment of disturbances of cardiac rhythm and angina pectoris. Results obtained with CIBA 39,089-Ba). Orv. Hetil. *110*, 2387 (1969)

40 GANDER, M., VERAGUTH, U., KÖHLER, R., LÜTHY, E.: Cardiac haemodynamics under β-receptor blockade in dogs. Cardiologia (Basle) *49*, Suppl. II: 17 (1966)

41 GEBHARDT, W., BLÜMCHEN, G.: Haemodynamic effect of a combined therapy with propranolol and digoxin. Cardiologia (Basle) *52*, 190 (1968)

42 GEORGE, C. F., NAGLE, R. E., PENTECOST, B. L.: Practolol in treatment of angina pectoris. A double-blind trial. Brit. med. J. *ii*, 402 (1970)

43 GIANELLY, R. E., GOLDMAN, R. H., TREISTER, B., HARRISON, D. C.: Propranolol in patients with angina pectoris. Ann. intern. Med. *67*, 1216 (1967)

44 GIANELLY, R. E., TREISTER, B. L., HARRISON, D. C.: The effect of propranolol on exercise-induced ischemic S-T segment depression. Amer. J. Cardiol *24*, 161 (1969)

45 GILLAM, P. M. S., PRICHARD, B. N. C.: Use of propranolol in angina pectoris. Brit. med. J. *ii*, 337 (1965)

46 GILLAM, P. M. S., PRICHARD, B. N. C.: Propranolol in the therapy of angina pectoris. Amer. J. Cardiol. *18*, 366 (1966)

47 GINN, W. M., Jr., ORGAIN, E. S.: Propranolol hydrochloride in the treatment of angina pectoris. J. Amer. med. Ass. *198*, 1214 (1966)

48 GOLDBARG, A. N., MORAN, J. F., BUTTERFIELD, T. K., NEMICKAS, R., BERMUDEZ, G. A.: Therapy of angina pectoris with propranolol and long-acting nitrates. Circulation *40*, 847 (1969)

49 GOLDSTEIN, R. E.: A comparison of beta-blocking agents. Circulation *47*, 443 (1973)

50 GOLDSTEIN, R. E., EPSTEIN, S. E.: Medical management of patients with angina pectoris. Progr. cardiovasc. Dis. *14*, 360 (1972)

51 GOLDSTEIN, R. E., HALL, C. A., EPSTEIN, S. E.: Comparison of simultaneously determined inotropic and chronotropic effects of practolol and propranolol. Clin. Res. *19*, 316 (1971)

52 GOTT, V. L.: Outlook for patients after coronary artery revascularization. Amer. J. Cardiol. *33*, 431 (1974)

53 GREINER, T., GOLD, H., CATTELL, M., TRAVELL, M., BAKST, H., RINZLER, S. H., BENJAMIN, Z. H., WARSHAW, L. J., BOBB, A. L., KWIT, N. T., MODELL, W., ROTHENDLER, H. H., MESSELOFF, C. R., KRAMER, M. L.: A method for the evaluation of the effects of drugs on cardiac pain in patients with angina of effort. A study of khellin (visammin). Amer. J. Med. *9*, 143 (1950)

54 GUIRAN, J.-B., BAUDOUY, M., FELDMAN, G., SCHMITT, R.: Un nouveau β bloqueur en pratique cardiologique. Nice méd. *11*, 39 (1973)

55 HAMER, J., SOWTON, E.: Effects of propranolol on exercise tolerance in angina pectoris. Amer. J. Cardiol. *18,* 354 (1966)

56 HAMPTON, J. R., HARRISON, M. J. G., HONOUR, A. J., MITCHELL, J. R. A.: Platelet behaviour and drugs used in cardiovascular disease. Cardiovasc. Res. *1,* 101 (1967)

57 HANSEN, P. F., RASMUSSEN, P. A., NYBERG, G.: Alprenolol alone and in conjunction with pentanitrol in angina pectoris. A double-blind study with exercise tests. Acta med. scand. *193,* 419 (1973)

58 HARRISON, D. C.: Beta adrenergic blockade, 1972. Pharmacology and clinical uses. Amer. J. Cardiol. *29,* 432 (1972)

59 HECK, K.-J., HAAS, W., LANG, E.: Untersuchungen zur Arbeitstoleranz Koronarkranker unter dem Einfluss der Beta-Sympathikolyse. Med. Klin. *67,* 291 (1972)

60 HETHERINGTON, D. J., COMERFORD, M. B., NYBERG, G., BESTERMAN, E. M. M.: Comparison of two adrenergic beta-receptor blocking agents, alprenolol and propranolol, in treatment of angina pectoris. Brit. Heart J. *35,* 320 (1973)

61 HICKIE, J. B.: Alprenolol ("Aptin") in angina pectoris. A double-blind multicentre trial. Med. J. Aust. *57/ii,* 268 (1970)

62 JOHNSSON, G., NORRBY, A., SÖLVELL, L., ÅBLAD, B.: Studies on the potency and time-effect relation of beta-adrenergic antagonists in man. Acta physiol. scand. *68,* Suppl. 277: 97 (1966)

63 JONSSON, B., OLSSON, A. G., ORÖ, L.: Effects of alprenolol on central hemodynamics and exercise tolerance in patients with angina pectoris. Cardiology (Basle) *58,* 150 (1973)

64 JORGENSEN, C. R., WANG, K., WANG, Y., GOBEL, F. L., NELSON, R. R., TAYLOR, H.: Effects of propranolol on myocardial oxygen consumption and its hemodynamic correlates during upright exercise. Circulation *48,* 1173 (1973)

65 KALTENBACH, M., BECKER, H. J., GRAEF, V., HUNSCHA, H.: Zur Therapie der Angina pectoris mit Beta-Rezeptorenblockern. Med. Klin. *65,* 494 (1970)

66 KERBER, R. E., GOLDMAN, R. H., ALDERMAN, E. L., HARRISON, D. C.: Circulatory responses to beta-adrenergic blockade with alprenolol. Studies in patients with chronic heart disease. Amer. J. Cardiol. *29,* 26 (1972)

67 KEYRILÄINEN, O., NYBERG, G., UUSITALO, A. J.: Effects of alprenolol and isosorbide dinitrate in angina pectoris. A comparative study with methodological considerations. Acta med. scand. *193,* 281 (1973)

68 KITAMURA, K., JORGENSEN, C. R., GOBEL, F. L., TAYLOR, H. L., WANG, Y.: Hemodynamic correlates of myocardial oxygen consumption during upright exercise. J. appl. Physiol. *32,* 516 (1972)

69 KRÓL, W., KOLARZOWSKA, B., GABRÝS, B.: Porównanie działania propranololu i LB 46 w chorobie wieńcowej serca (Comparison of the action of propranolol and LB 46 in coronary disease). Przegl. lek. *26,* 726 (1970)

70 LAMBERT, D. M. D.: Beta-blockers and life expectancy in ischaemic heart-disease. Lancet *i,* 793 (1972)

71 LECEROF, H., MALMBORG, R. O.: Hemodynamic effects of oxprenolol alone and combined with nitroglycerin in patients with ischemic heart disease. Acta med. scand. *192,* 499 (1972)

72 LESCH, M., GORLIN, R.: Pharmacological therapy of angina pectoris. Mod. Conc. cardiovasc. Dis. *42:* 2: 5 (1973)

73 LEWIS, B. S., BAKST, A., GOTSMAN, M. S.: Oxprenolol in angina pectoris. S. Afr. med. J. *46,* 1209 (1972)

74 LIPPENHOLTZ, A., TALMASKY, S., BERCONSKY, I.: Acción del alprenolol en el tratamiento de la angina de pecho. Pren. méd. argent. *56,* 1357 (1969)

75 LORENTZEN, F., JØRGENSEN, P., NIELSEN, E., NYBERG, G., HVIDT, S.: En sammenlignende undersøgelse af alprenolol i tabletform (Aptin) og i "sustained release" form (Aptin Duretter) ved angina pectoris (Comparison between alprenolol in tablet form [Aptin] and in "sustained release" form [Aptin Durules] in angina pectoris). Ugeskr. Laeg. *135,* 1431 (1973)

76 Lorenzen, J.: Behandling af angina pectoris med Trasicor (CIBA 39'089-Ba). Nord. Med. *83*, 783 (1970)

77 Martin-Noël, P., Denis, P., Combalot, A., Courten, E. Lanney de: Etude clinique prolongée d'un nouveau bêta-bloquant. L'alprénolol dans l'angor et les troubles du rythme. Rev. méd. Alpes franç. *2*, 491 (1973)

78 Muzio, L.: Studio clinico-elettrocardiografico di un nuovo beta-adrenolitico. Riv. Pat. Clin. *28*, 11 (1973)

79 Nager, F.: Die medikamentöse Behandlung der Angina pectoris. Rationale Grundlagen und praktische Durchführung. Schweiz. med. Wschr. *102*, 1724 (1972)

80 Nager, F., Favre, H.: Double blind study with propranolol and LB 46 in angina pectoris. In Kaltenbach, M., Lichtlen, P. (Editors): Coronary heart disease, Int. Symp., Frankfurt 1970, p. 224 (Thieme, Stuttgart 1971)

81 Nayler, W., McInnes, I., Swann, J. B., Carson, V., Lowe, T. E.: Effects of propranolol, a beta-adrenergic antagonist, on blood flow in the coronary and other vascular fields. Amer. Heart J. *73*, 207 (1967)

82 Nellen, M.: Trial of two beta blockade drugs, CIBA 39,089-Ba and ICI 50,172, in angina pectoris. S. Afr. med. J. *43*, Suppl. to No. 48: 15 (1969)

83 Nestel, P. J.: Evaluation of propranolol ("Inderal") in the treatment of angina pectoris. Med. J. Aust. *53/ii*, 1274 (1966)

84 Nestel, P. J.: Practolol in the treatment of ischaemic heart pain. Results of a controlled trial. Med. J. Aust. *59/i*, 1033 (1972)

85 Phear, D. N.: Practolol in angina. Brit. med. J. *i*, 557 (1971); corresp.

86 Pitt, B., Craven, P.: Effect of propranolol on regional myocardial blood flow in acute ischaemia. Cardiovasc. Res. *4*, 176 (1970)

87 Ponari, O., Civardi, E., Potì, R.: Action of some β-blockers on plasma fibrinolysis in vitro and in vivo in man. Arzneimittel-Forsch. (Drug. Res.) *22*, 629 (1972)

88 Prichard, B. N. C., Aellig, W. H., Richardson, G. A.: The action of intravenous oxprenolol, practolol, propranolol and sotalol on acute exercise tolerance in angina pectoris: the effect on heart rate and the electrocardiogram. Postgrad. med. J. *46*, Suppl. (Nov.): 77 (1970)

89 Rapp, G.: Klinische Anwendung des Beta-Rezeptorenblockers Oxprenolol bei Angina pectoris. Aerztl. Prax. *23*, 2411 (1971)

90 Rivier, J.-L., Nissiotis, E., Jaeger, M.: Comparaison des effets hémodynamiques immédiats de trois médicaments béta-bloqueurs. Thérapie *25*, 245 (1970)

91 Robinson, B. F.: The relation of heart rate and systolic blood pressure to the onset of pain in angina pectoris. Thesis, London 1966

92 Robinson, B. F.: Relation of heart rate and systolic blood pressure to the onset of pain in angina pectoris. Circulation *35*, 1073 (1967)

93 Rondel, R. K.: Multi-centre trials of oxprenolol in angina pectoris, loc. cit. [23], p. 197

94 Rookmaker, W. A.: Enkele toepassingen van telemetrie in de cardiologie (Some aspects of telemetry in cardiology). Thesis, Groningen 1969

95 Rookmaker, W. A.: Etude comparative de cinq drogues antiangoreuses. Ars Med. (Gand) *26*, 1955 (1971)

96 Ruggiero, H. A., Krieg, R., Caprissi, L. F., Luzzi, R.: Empleo de un nuevo bloqueador beta-adrenérgico en la angina de pecho (Ba-39.089). Pren. méd. argent. *55*, 805 (1968)

97 Saameli, K.: Untersuchungen mit β-Rezeptorenblockern am isolierten Meerschweinchenvorhof. Helv. physiol. pharmacol. Acta *25*, 219 (1967)

98 Sakakibara, H., Matsutani, K., Mochizuki, S., Kitabatake, A., Takagishi, S., Matsuo, H., Aoki, K., Yoshioka, Y., Kubori, S., Kodama, K., Nimura, Y.: (Double-blind trial with propranolol and alprenolol on angina pectoris). Jap. Circulat. J. (En.) *34*, 303 (1970)

99 SANDLER, G., PISTEVOS, A.: Clinical evaluation of oxprenolol in angina pectoris. Brit. Heart J. *34*, 847 (1972)

100 SANER, R. P.: Die Behandlung der Angina pectoris mit einem neuen β-Rezeptoren-Blocker, dem Präparat LB-46 (4-[2-Hydroxy-3-iso-propylaminopropoxy]-indol). Schweiz. med. Wschr. *100*, 174 (1970)

101 SARNOFF, S. J., BRAUNWALD, E., WELCH, G. H., Jr., CASE, R. B., STAINSBY, W. N., MACRUZ, R.: Hemodynamic determinants of oxygen consumption of the heart with special reference to the tension-time index. Amer. J. Physiol. *192*, 148 (1958)

102 SARNOFF, S. J., MITCHELL, J. H.: The control of the function of the heart. In Hamilton, W. F., Dow, P. (Editors): Handbook of physiology, Sect. 2, Vol. I: Circulation, p. 489 (Amer. Physiol. Soc., Washington, D.C., 1962)

103 SCEBAT, L., BENSAID, J.: Long-term clinical results of oxprenolol in the treatment of cardiac arrhythmias, angina pectoris and hypertrophic subaortic stenosis. Postgrad. med. J. *46*, Suppl. (Nov.): 86 (1970)

104 SEALEY, B. J., LILJEDAL, J., NYBERG, G., ÅBLAD, B.: Acute effects of oral alprenolol on exercise tolerance in patients with angina pectoris. A dose-response study. Brit. Heart J. *33*, 481 (1971)

105 SHAH, S. J., BILLIMORIA, A. R., ANAND, M. P., HAVELIWALA, H. K., VIBHAKAR, B. B.: Objective and subjective parameters in angina pectoris (effect of beta-receptor blockade). Indian Heart J. *24*, 337 (1972)

106 SHARMA, B., MEERAN, M. K., GALVIN, M. C., TULPULE, A. T., WHITAKER, W., TAYLOR, S. H.: Comparison of adrenergic beta-blocking drugs in angina pectoris. Brit. med. J. *iii*, 152 (1971)

107 SHARMA, B., TAYLOR, S. H.: Comparison of the symptomatic, electrocardiographic, and haemodynamic effects of acute and long-term β-blockade in angina pectoris. Brit. J. Pharmacol. *43*, 470P (1971); abstract of paper

108 SHARMA, B., TAYLOR, S. H.: A critical review of the symptomatic, electrocardiographic and circulatory effects of adrenergic beta-receptor antagonists in angina pectoris, loc. cit. [23], p. 129

109 SLAFER, H., LEPERA, L., CONTI, L. A., MANTYKOW, E.: Tratamiento del angor pectoris con el CIBA 39.089-Ba. Sem. méd. (B. Aires) *135*, 635 (1969)

110 SONNENBLICK, E. H., ROSS, J., BRAUNWALD, E.: Oxygen consumption of the heart. Newer concepts of its multifactorial determination. Amer. J. Cardiol. *22*, 328 (1968)

111 SOWTON, E., SMITHEN, C.: Double-blind three-dose trial of oral alprenolol in angina pectoris. Brit. Heart J. *33*, 601 (1971)

112 STIEGLITZ, E.: Therapie der Koronarinsuffizienz mit Visken. Eine klinische Doppelblindstudie. Med. Welt *21*, 2081 (1970)

113 TAYLOR, S. H., THADANI, H., SHARMA, B.: Comparison of the symptomatic, electrocardiographic and circulatory effects of propranolol, oxprenolol and practolol in angina pectoris, loc. cit. [23], p. 177

114 THADANI, U., SHARMA, B., MEERAN, M. K., MAJID, P. A., WHITAKER, W., TAYLOR, S. H.: Comparison of adrenergic beta-receptor antagonists in angina pectoris. Brit. med. J. *i*, 138 (1973)

115 THORPE, P.: Pindolol (Visken) and angina pectoris: a double blind multicentre trial. N. Z. med. J. *76*, 171 (1972)

116 TRICOT, R., VALÈRE, P. E.: Utilisation prolongée de l'alprénolol dans l'insuffisance coronarienne. Ann. cardiol. Angéiol. *20*, 73 (1971)

117 TURNER, A. S., PEEL, J. S.: Clinical and haemodynamic studies with beta-blockade. Symp. Beta adrenergic blocking agents, Sydney, Australia, 1970, p. 51

118 VALÈRE, P. E.: Utilisation de l'alprénolol dans l'insuffisance coronarienne. Concours méd. *93*, 5122 (1971)

119 VÁRKONYI, G.: Trasicor az angina pectoris és a szív rhythmus-zavarainak kezelésében (Trasicor in the treatment of angina pectoris and in cardiac rhythm-disturbances). Gyógyszereink *23*, 12 (1973)

120 Vastesaeger, M. M., Rasson, G.: Etude de l'action immédiate de bêta-bloquants chez le coronarien en cours d'effort. Brux.-med. *51*, 429 (1971)

121 Waucampt, J.-J., Desruelles, J., Decalf, A., Delmon, A., Goethals, S., Pauchant, M.: Etude clinique d'un nouveau bêta-bloquant original: l'Aptine. A propos de 73 observations. Acquis. nouv. Path. cardiovasc. *15*, No. 2 (1973)

122 Westerlund, A.: Double-blind trial for evaluation of the therapeutic effect of practolol in patients with angina pectoris. Int. J. clin. Pharmacol. Ther. Toxicol. Suppl. 3: 44 (1971)

123 Whitsitt, L. S., Lucchesi, B. R.: Effects of propranolol and its stereoisomers upon coronary vascular resistance. Circulat. Res. *21*, 305 (1967)

124 Wilson, A. G., Brooke, O. G., Lloyd, H. J., Robinson, B. F.: Mechanism of action of β-adrenergic receptor blocking agents in angina pectoris: comparison of action of propranolol with dexpropranolol and practolol. Brit. med. J. *iv*, 399 (1969)

125 Wilson, D. F., Turner, A. S.: Clinical impressions of a new beta-adrenergic blocking drug, 39,089-Ba. N. Z. med. J. *66*, 682 (1967)

126 Wilson, D. F., Turner, A. S.: Trasicor (39,089-Ba) in angina pectoris. A single-blind trial. N. Z. med. J. *67*, 406 (1968)

127 Wilson, D. F., Watson, O. F., Peel, J. S., Turner, A. S.: Trasicor in angina pectoris: a double-blind trial. Brit. med. J. *ii*, 155 (1969)

128 Wiseman, R. A.: Practolol – an analysis of clinical trials in angina pectoris. Acta cardiol. (Brux.) Suppl. 16: 29 (1972)

129 Wolfson, S., Amsterdam, E. A., Gorlin, R.: Prognostic significance of angina therapy: preliminary report. Circulation *36*, Suppl. II: 274 (1967); abstract of paper

130 Wolfson, S., Heinle, R., Herman, M., Kemp, H. G., Sullivan, J. M., Gorlin, R.: Propranolol and angina pectoris. Amer. J. Cardiol. *18*, 345 (1966)

131 Wolfson, S., Phillips, S. L., Schecter, E.: Effects of specific myocardial beta blockade in angina pectoris. Amer. J. Cardiol. *26*, 666 (1970); abstract of paper

132 Zubillaga, R., Partidas S., A.: Empleo del betabloqueador (CIBA 39.089-Ba) en síndrome anginoso. Informe preliminar. Acta méd. venez. *17*, 141 (1970)

General discussion on the use of beta-blockers in myocardial ischaemia

W. Schweizer: Now we come to the general discussion on the use of beta-blockers in myocardial ischaemia. Once again, we have prepared a few questions that seem to be pertinent. Let us first of all consider the problem of patients with angina pectoris. Here, the first question is: Under what circumstances would you use beta-blockers and what considerations would lead you not to use them? This also involves the question of when to prescribe beta-blockers and when to resort to a bypass operation.

W. Somerville: When you are confronted with a patient complaining of angina pectoris, your first duty is to define what type of angina you are dealing with. It is my custom to divide angina pectoris into four grades: Grade 1 is angina in response to unusual effort; Grade 2 is angina coming on with usual and customary effort; Grade 3 is angina which occurs with less than ordinary effort; and Grade 4 is angina which sets in at rest and is repetitive. Angina triggered off by emotional stimuli may fall into Grades 4 or 3 and occasionally into Grade 2. On the basis of these grades, then, you attach a label to the patient, i.e. you add a qualifying adjective to the term angina.
Your next duty is to decide in which patients surgery is mandatory. In general, I should say that this calls for recognition of those patients with the intermediate coronary syndrome and those with accelerated angina. "Intermediate coronary syndrome" is the term I apply to a person whose history of anginal pain has lasted approximately four weeks and not much longer, and in whom the pain has occurred in a repetitive and crescendo fashion over that period. That is a particularly sinister type of angina and one in which coronary arteriography and aorto-coronary bypass surgery is often, if not invariably, indicated. "Accelerated angina", on the other hand, is a condition in which the history of anginal pain may go back for as many as ten years, but in which – at some time in the immediately recent past – the tempo of the attacks has speeded up and the anginal pains have been occurring more frequently. Intermediate coronary syndrome and accelerated angina appear to be two different conditions with only one common denominator, namely, the repetitive cardiac ischaemic pain. Each of them probably has a different pathophysiology. Carefully controlled observations on surgical treatment, carried out by Bertolasi et al.* in Buenos Aires, suggest that aorto-coronary bypass is strongly indicated in the intermediate coronary syndrome, but that it is not so strongly indicated – and certainly not invariably – in accelerated angina.
With regard to the placing of patients into Grade 1, 2, 3, or 4, I must mention at this point that the majority of patients do not fall into Grade 4 or Grade 3 at the time of presentation. Most of them are Grade 1 or Grade 2 cases. Now what part can be played by beta-adrenergic blockade in the treatment of these patients? In my priority list it ranks sixth. That is to say, there are five other things that must also be dealt with. Firstly, the patient's occupational environment may need to be corrected. Secondly, his home environment must be looked into. Thirdly, his weight may need reducing. Fourthly, his consumption of cigarettes and alcohol should be checked. Fifthly, one should explore the optimum use of glyceryl trinitrate.
It is profitable to advise patients in Grades 1 or 2 to take some form of exercise for a specific time every day, preceded by glyceryl trinitrate. Incidentally, there is an extra dividend to this recommendation: if the patient is told to take daily walks and finds that he can do so without experiencing pain, this gives him courage and may possibly also relieve some of the autonomic nervous influences that might have a deleterious effect on him.

---

* Bertolasi, C.A., Trongé, J.E., Carreño, C.A., Jalon, J., Ruda Vega, M.: Unstable angina – prospective and randomized study of its evolution, with and without surgery. Preliminary report. Amer. J. Cardiol. *33*, 201 (1974)

Now I come to the sixth point, i.e. to the beta-adrenergic blockers. The ones that I employ most frequently for an anti-anginal effect are propranolol and oxprenolol (®Trasicor). In my experience, the difference between them is negligible. On the other hand, I have found that the anti-anginal effect of practolol, dose for dose, is of a much lower order. Drug treatment might perhaps be referred to as Stage 2 in the management of angina pectoris, Stage 1 being recognition of those patients in whom surgery is mandatory. The length of time during which I keep patients under intensive medical treatment varies depending on their grade. For instance, in a person in Grade 1, i.e. with anginal pain that comes on only under unusual circumstances, I continue this treatment for two or three months. In Grade 2 the period of treatment is one month. In Grade 3, i.e. in which anginal pain already occurs in response to slight stimuli, the treatment lasts anything from a few days to a week. The patient's situation is kept constantly under review. If he does not respond – in other words, if his pain and/or dyspnoea do not completely disappear – then I put him up for coronary arteriography. Depending on the findings yielded by this arteriography, he is either given further medical treatment or undergoes surgery. One of the unanswered questions is whether these patients derive more benefit from medical or from surgical treatment. It has at all events been my experience that prolonged rest, plus beta-adrenergic blockade and other appropriate measures, may sometimes enable a patient to become symptom-free. A controlled multicentre trial involving a number of countries is at present being carried out in order to investigate the relative merits of medical and surgical treatment in such patients.

W. SCHWEIZER: In my priority list the beta-blockers rank much higher, i.e. in second position.

R. C. TARAZI: I'd like to ask Dr. SOMERVILLE a few questions. Is there any close correlation between the clinical picture of the patient and the extent of the coronary arterial disease as revealed by arteriography? Can a symptomatic response be taken as precluding the possible presence of advanced coronary lesions for which surgery would in fact be indicated? I am thinking here of three-vessel disease or a left main trunk lesion. Is it really sufficient just to rely on one's evaluation of the clinical picture, however thorough that evaluation may have been?

W. SOMERVILLE: These are very searching questions, Dr. TARAZI. I must admit that hardly a week goes by without my changing my ideas in this connection. The disclosure of severe and advanced obstructive lesions in an individual with relatively slight symptoms and with benign electrocardiographic changes must always be a disconcerting experience. In certain circumstances I myself resort to the exercise test as being a useful indicator of advanced three-vessel disease in patients with normal electrocardiograms. If slight exercise causes an S-T depression of 2, 3, or 4 mm. which is unaccompanied by tachycardia, this is usually a sign of severe obstructive disease.
Incidentally, the remark Dr. SCHWEIZER made about his own priority list prompts me to add that the six points I mentioned are, of course, all put into operation at the same time.

S. H. TAYLOR: Unlike Dr. SOMERVILLE, whose persuasive powers are well known, I think that I should find it quite impossible to persuade patients simultaneously to change their job, divorce their wives, give up smoking, go on a fat-free, sugar-free diet, walk every day in bleak weather, and avoid all excitement. Perhaps this explains why I use beta-blockers more often and possibly at an earlier stage than he does. I must also point out that there is no evidence that repeated exercise under controlled conditions, such as we impose during drug trials, produces any change in symptoms; nor is there any evidence that weight reduction does so either. Both may indeed do so, but we've no scientific evidence to substantiate it.

W. SCHWEIZER: The second question to be discussed is: Do the various beta-blockers differ with regard to their anti-anginal properties?

E. Sowton: Dr. Rivier said that his review of the literature had indicated that the five beta-blockers he mentioned were each very similar. We have compared these five drugs in the same 20 patients on a double-blind randomised cross-over basis. The results of total work carried out on the bicycle ergometer by all the patients on the different drugs showed no significant difference in performance, except in the case of the patients taking pindolol, who did less work. Practolol was just as effective as the other drugs. I should point out that these patients had all previously been stabilised on propranolol. The heart rate on pindolol was very similar to that on the other four drugs – a fact which suggests that the degree of beta-blockade was the same. It afterwards appeared to us that we had been using the wrong equivalent dose of pindolol. We accepted 5 mg. pindolol as being equivalent to 40 mg. propranolol, but 5 mg. is obviously far too high a dose. We believe this to be the reason for the poorer performance on pindolol; it is relatively easy to "go over the top" with certain beta-blockers, with the result that you reduce the exercise performance as you increase the dose. This has been demonstrated with alprenolol and by several centres with pindolol. As regards the therapeutic effect of beta-blockers in angina, we have found in various double-blind trials with these drugs that about 30% of the patients ceased to experience any pain at all on maximal exercise.

Finally, a brief comment on the fact that propranolol is almost invariably described as having no sympathomimetic activity. I don't think this is quite correct, because it is metabolised in the liver, and the 4-hydroxypropranolol produced is a potent beta-blocker, circulates in the plasma, and exerts quite a large degree of sympathomimetic activity*.

P.R. Imhof: The metabolite Dr. Sowton has mentioned, 4-hydroxypropranolol, undoubtedly possesses intrinsic sympathomimetic activity. It is, however, found only after oral, and not after intravenous administration of propranolol**. Following repeated oral treatment with propranolol, moreover, it can no longer be demonstrated***. According to Fitzgerald****, this metabolite is of no practical importance.

M. Ikeda: The results of a study published by Murakami et al.***** may be of interest in connection with the paper which Dr. Rivier has presented. This study took the form of a multicentre double-blind trial using the cross-over method, and the patients treated were all suffering from angina of effort. Cases in which beta-blockers were contra-indicated were, of course, excluded. The patients were divided into two groups composed of 52 cases in which oxprenolol was compared with placebo and 45 cases in which it was compared with propranolol. Each of the three treatments was given for two weeks, the oxprenolol in a daily dosage of 120 mg. (40 mg. t.i.d.) and the propranolol in a daily dosage of 60 mg. (20 mg. t.i.d.). The results were assessed on the basis of a statistical analysis of the following criteria: preference expressed by the treating physicians; physicians' global assessment; patients' symptomatology; frequency of anginal attacks; E.C.G. findings; patients' physical performance; and side effects. Evaluation of the results showed that oxprenolol was significantly more effective than

---

* Fitzgerald, J.D., O'Donnell, S.R.: Pharmacology of 4-hydroxypropranolol, a metabolite of propranolol. Brit. J. Pharmacol. *43*, 222 (1971)
** Paterson, J.W., Conolly, M.E., Dollery, C.T., Hays, A., Cooper, R.G.: The pharmacodynamics and metabolism of propranolol in man. Pharmacol. clin. *2*, 127 (1970)
*** Cleaveland, C.R., Shand, D.G.: Effect of route of administration on the relationship between β-adrenergic blockade and plasma propranolol level. Clin. Pharmacol. Ther. *13*, 181 (1972)
**** Fitzgerald, J.D.: Personal communication (1973)
***** Murakami, M., et al.: (Clinical effects of oxprenolol in angina pectoris. Comparison with placebo and propranolol by double blind trial). Nihon Rinsho (Jap. J. clin. Med.) *31*, 175 (1973)

placebo in affording relief from angina pectoris, the respective percentages in the case of the physicians' global assessment being 79% and 44.7%. There was no statistically significant difference, however, between the responses to oxprenolol and propranolol. Few side effects were encountered, but myocardial infarction occurred in three cases while the trial was in progress.

B. N. C. PRICHARD: I think one must distinguish between prolonged treatment with a beta-blocker and acute administration, whether oral or intravenous. Though you can probably give enough of a beta-blocker in an acute oral or intravenous dose to produce much the same anti-anginal effect as could be achieved with chronic medication, you also have to consider certain side effects that are time-related. Even the most complaining of Englishmen do not complain of being constipated after a single dose of, say, practolol – because, by the time they feel they must have their obligatory daily bowel motion, the drug's effect is beginning to wear off. But, if you give that particular drug on chronic oral administration, you see a very different picture. As I mentioned before, constipation is often a factor limiting the size of the doses in which you can give practolol. To get an idea of the comparative merit of these drugs in reducing the number of attacks of anginal pain, which is the purpose for which they would be employed clinically in angina pectoris, it is necessary to do a comparative oral study. In my view, this must be a variable-dose comparison. When we did this with practolol and propranolol a few years ago, we found that practolol was significantly less effective in maximum tolerated doses than propranolol*. We are now conducting a comparative trial along the same lines with oxprenolol, propranolol, and AH 5158, but this study is still in progress.

W. SCHWEIZER: Now we come to our third question: How do you judge the efficacy of beta-blockers in patients with angina? In other words, is pain alone sufficient as a criterion? Or do you also have to consider the patient's glyceryl trinitrate consumption or adopt more sophisticated methods of evaluation?

J. HODLER: What I want to tell you about now is more in the nature of an anecdote, rather than an example of genuine scientific work. It concerns a double-blind cross-over study with oxprenolol, pindolol, and placebo which we undertook in 18 patients with angina pectoris, who constituted what could be described as a run-of-the-mill selection of cases. After two control weeks, during which the patients received unchanged basic medication, they were allocated in randomised fashion to successive, two-week periods of oral treatment with oxprenolol (40 mg. t.i.d.), pindolol (5 mg. t.i.d.), or placebo (1 tablet t.i.d.). At weekly controls performed during treatment we recorded the attack rate, subjective symptoms, and nitroglycerin consumption, evaluated the patients' working capacity, carried out a physical examination, and did a 12-lead E.C.G. As is apparent from Table 1, the decrease in the frequency of attacks of angina pectoris was the same in response to both beta-blockers and did not differ significantly from that recorded following placebo. The same, moreover, applies to their effect on the patient's consumption of nitroglycerin pills (Table 2). However, when we examined the electrocardiograms of these patients, we found that even at rest there was a statistically very significant difference in the effect exerted by oxprenolol and pindolol on the S-T changes as compared with untreated controls and the placebo-treated patients. This difference was even more pronounced in cases where E.C.G. recordings were also made after a Master test. Finally, the blood pressure levels – even in the placebo-treated cases – also dropped, although the patients had no reason to suppose that the treatment they were receiving was likely to affect their blood pressure. What is the explanation for these results? Perhaps the patients were just gullible and the treatment made them feel better. This may account for the general effects (on the number of attacks of angina and on nitroglycerin consumption). As regards the

---

* PRICHARD, B. N. C., LIONEL, N. D. W., RICHARDSON, G. A.: Comparison of practolol and propranolol in angina pectoris. Postgrad. med. J. *47*, Suppl. (Jan.): 59 (1971)

Table 1. Comparison of the number of attacks of angina pectoris recorded in the four periods of the trial.

Period of trial	Number of patients	Number of attacks*	Length of period (days)	Average number of "attacks" per day
Control	18	378 (218)	252	1.52 (0.865)
Pindolol	18	225 (143)	255	0.894 (0.561)
Oxprenolol	18	215 (144)	250	0.9 (0.576)
Placebo	18	275 (151)	251	1.16 (0.602)

*The figures in this column are index figures which take account of both the severity and the number of attacks occurring during the various periods of the trial. They have been arrived at by scoring mild attacks as 1, moderate attacks as 2, and severe attacks as 3. The absolute figures for the number of attacks are indicated in brackets.
Significances: All treatments (incl. placebo) versus control: $P < 0.01$; treatment versus treatment: $P > 0.1$.

Table 2. Comparison of nitroglycerin consumption during the four periods of the trial. Four of the altogether 18 patients in the trial never required nitroglycerin tablets.

Period of trial	Number of patients	Number of nitroglycerin tablets	Length of period (days)	Average weekly consumption of nitroglycerin tablets
Control	14	280	252	1.11
Pindolol	14	161	255	0.63
Oxprenolol	14	163	250	0.65
Placebo	14	205	251	0.82

Significances: All treatments (incl. placebo) versus control: $P < 0.01$; treatment versus treatment: $P > 0.1$.

effects on the E.C.G. and on cardiac output, however, the fact that the patients were taking a pill which they believed to be effective may have induced them to overtax their working capacity, thus triggering off a mechanism which had the effect of causing both myocardial ischaemia and a fall in blood pressure during placebo medication. The point I want to make is that the administration of an ineffective drug – a placebo, in the present instance – may make the patient feel better, but at the same time it might actually be endangering him.

W. SCHWEIZER: The conclusion that seems to emerge from what Dr. HODLER has said is that the electrocardiogram is a much more reliable indicator than pain in cases of angina. Would anyone care to comment on this point?

A. ZANCHETTI: Dr. HODLER has very rightly drawn attention to the pitfalls of subjective methods of assessment. The electrocardiogram may prove very useful as a criterion, and, in particular, it offers an excellent objective method of evaluation in the type of angina known as Prinzmetal's variant. A few years ago some colleagues of mine* carried out a very elegant study in patients with this kind of angina: by moni-

* GUAZZI, M., POLESE, A., FIORENTINI, C., MAGRINI, F., BARTORELLI, C.: Left ventricular performance and related haemodynamic changes in Prinzmetal's variant angina pectoris. Brit. Heart J. *33*, 84 (1971)

toring the E.C.G. throughout the day, they were able to show that several episodes of S-T segment elevation may occur without any subjective symptoms. These authors subsequently demonstrated* that beta-blockers are extremely effective in relieving both the symptoms and the E.C.G. signs of this type of angina.

P. TAGGART: One difficulty encountered when basing evaluations on an E.C.G. recorded during exercise testing has been highlighted by the Seattle group**, who probably have the largest exercise series in the world, amounting to over 15,000. When they carried out coronary angiography and compared the angiographic findings with the electrocardiograms obtained on exercise, they discovered that it is very often the worst triple-vessel disease which shows no abnormalities in the S-T segment during performance of an exercise test. As a possible explanation for this, they point out that the surface electrocardiogram is the algebraic sum of the subendocardial minus sub-epicardial action potential groups; consequently, if both the inner and outer layers are equally compromised or sufficiently compromised, the disparity is abolished, i.e. the initial subendocardial ischaemia – which is usually the first thing to occur – is abolished and you thus get a normalisation of the S-T segment in the E.C.G.

B. N. C. PRICHARD: In the study to which you have just referred, Dr. HODLER, did you have any pre-trial stabilisation run-in period? I think that, when performing exercise tests, it's very important to make sure that the patient is properly trained; otherwise changes are liable to occur in the base-line performance and you may get improvements just as frequently on placebo as on active treatment. The same applies to the problem of getting patients to record the number of attacks of angina or their consumption of trinitrine. In many cases a spontaneous improvement may occur in response to the increased interest that the physician takes in the patients when he starts to perform a trial. You therefore need a stabilisation period lasting about two to three months, during the course of which the patient should be kept under observation under conditions corresponding as closely as possible to those of the full-blown trial; only at the end of this period should the double-blind phase be initiated. A one-month stabilisation period is, of course, better than none at all, but ideally this pre-trial phase should last for two to three months. Many of the less favourable results reported with beta-blockers in cases of angina are, I am sure, due to the investigators' failure to wait for the base-line to stabilise, i.e. to the fact that they have started on the trial without allowing for a proper run-in period. Another common error, of course, is the use of doses that are too small or of fixed dosage regimens.

J. HODLER: The patients on whom I reported, Dr. PRICHARD, were kept under observation for six months before the trial was started. They had already been participating in another study and were therefore quite used to the whole procedure.

S. H. TAYLOR: Regarding the E.C.G. as an aid to evaluation of treatment in angina patients given beta-blockers, I would also point out that many who completely fail to respond symptomatically nevertheless show considerable electrocardiographic improvement. If it is only the symptoms that we are trying to treat in these cases, then the E.C.G. can also be misleading in this connection.

W. SCHWEIZER: The next question you are now invited to discuss is: How do you explain the fact that in a minority of patients beta-blockers simply do not work, and what should be done about these patients?

---

* GUAZZI, M., MAGRINI, F., FIORENTINI, C., POLESE, A.: Clinical, electrocardiographic, and haemodynamic effects of long-term use of propranolol in Prinzmetal's variant angina pectoris. Brit. Heart J. *33*, 889 (1971)
** LAPIN, E. S., MURRAY, J. A., BRUCE, R. A., WINTERSCHEID, L.: Changes in maximal exercise performance in the evaluation of saphenous vein bypass surgery. Circulation *47*, 1164 (1973)
BRUCE, R. A.: Personal communication

B. N. C. PRICHARD: I can suggest two partial explanations. The first centres around the problem of diagnosis. AMSTERDAM et al.* published a report on quite a large series of cases treated with propranolol. They found that, when they excluded those patients who had no coronary artery disease and confined their attention to those with proven coronary artery disease, i. e. to those in whom the diagnosis of angina was more certain, then the response rate proved to be far higher. My second explanation relates to dosage. Dr. HODLER has described a trial in which oxprenolol was given in a fixed dose of 40 mg. t.i.d.; this is certainly a suboptimal dosage for oxprenolol. If you examine the reports on all the trials that have been published on propranolol, for example, you will find that, the higher the dosage used, the better the response rate. What is more, where the investigators used a flexible approach to dosage instead of sticking to a fixed dose, the response rate climbed to 80% or even higher. I am not saying that the problem of the non-responder doesn't exist at all; but I certainly do believe that you get far better results, firstly, if you do your best to eliminate misdiagnosis and, secondly, if you use the drug in its full and optimum dosage.

W. SOMERVILLE: Some of the patients to whom Dr. RIVIER referred had severe obstructive lesions – for instance, in the main left coronary artery, or perhaps two lesions in the left anterior descending and circumflex artery. It's obvious that in cases of this kind – regardless of what a beta-blocker does or does not do, and regardless of what pharmacological action glyceryl trinitrate may exert – ischaemia will inevitably occur, even in response to the challenge of mild exercise or slight emotional stress. Nothing but revascularisation can help these patients.

W. SCHWEIZER: Are you suggesting, then, Dr. SOMERVILLE, that the non-responders are the patients who have severe lesions?

W. SOMERVILLE: Not all of them, but many of them.

W. SCHWEIZER: Let us take a look now at a further question: Should painless myocardial ischaemia, proven by S-T segment depression or other E.C.G. changes, be treated with beta-blockers? We all know what Dr. TAGGART's answer to this question would be, but what other opinions are there?

E. K. CHUNG: When talking about S-T depression and S-T elevation, and particularly about S-T depression during exercise, we have to be careful to bear in mind that in many instances you can get so-called false positive findings. Perhaps the best example to quote in this connection is the effect of digitalis. In a patient taking digitalis, irrespective of the dose, you may well find significant S-T depression without coronary heart disease or any other cardiac disorder. Similarly, a left bundle-branch block, left-ventricular hypertrophy, or even some other cardiac disease may produce pronounced S-T depression in the absence of significant coronary heart disease. Another example is Wolff-Parkinson-White syndrome of Type B, in which you may, in rare instances, also find marked S-T depression without coronary heart disease.

J.-L. RIVIER: May I add that it is essential to carry out coronary arteriography in patients in whom the E.C.G. findings suggest the presence of myocardial ischaemia. One has to be particularly careful in the case of women, because their E.C.G. tracings are more often likely to be non-specific at rest, or even after exercise.

J. SCHWARTZ: In animals, the injection of a beta-blocker invariably leads to a decrease in coronary blood flow. While, in clinical use, it is true that these drugs yield beneficial results, I wonder whether in severe cases they may not sometimes give rise to untoward episodes which are simply not mentioned.

J. G. LEWIS: One problem that causes me a little concern is that of the occasional case in which increasing doses of beta-blockers have had to be given in order to cope

---

* AMSTERDAM, E.A., GORLIN, R., WOLFSON, S.: Evaluation of long-term use of propranolol in angina pectoris. J. Amer. med. Ass. *210*, 103 (1969)

with a situation of increasing angina. Some of these patients have subsequently had to be admitted to a coronary care unit in severe and sometimes fatal pulmonary oedema. This may, of course, be just coincidence, but other workers in this field have reported the same sort of thing. Perhaps only one person has actually made the observation, and others are simply repeating it. At any rate, I shouldn't treat such patients just with increasing doses of beta-blockers alone, but I should handle them as cases of impending cardiac infarction.

W. SCHWEIZER: I think I would agree with that, Dr. LEWIS. Now we come to our final question: Is there any evidence that beta-blockers alter the natural course of the disease by reducing the incidence of complications? If the answer were "yes", then this alone would also justify long-term treatment with beta-blockers even in patients complaining of little or no pain.

P. TAGGART: I think that basically the answer to this question is "no". Preliminary observations on the use of beta-adrenergic blockade as a prophylactic measure following myocardial infarction* are encouraging, though further work needs to be undertaken before one can say more than that. One of the problems at the moment is as follows:
Supposing you do an exercise test on a patient and you see – particularly in the post-exercise period – ectopic activity and/or an S-T segment shift which before the test were either absent or less pronounced. Is this a bad sign? Does it indicate a poor prognosis? Is this patient more liable to have a sudden "cardiac event"? I know that a joint study between a British hospital and a University in the U.S.A. is now being planned, in which a high-risk group of middle-aged subjects is to be investigated. By virtue of the risk factors, it is to be expected that, within the two and a half years which the study will cover, a proportion of these patients will have cardiac events. The purpose of the study will be to compare the prognosis in one group undertaking rehabilitative exercise, in another group receiving a beta-blocker, and in a control group. So I think we shall have to wait about three years for the answer to this final question.

J. BROD: May I ask Dr. SOMERVILLE whether the surgical treatment helps in any way to improve the prognosis in coronary heart disease. As far as I know, it makes life more tolerable for the patients and it also relieves the symptoms, but is there any evidence that it actually prolongs life?

W. SOMERVILLE: When the influence of surgery was studied over a three-year follow-up period, the quality of life was certainly found to have been improved, but there seems to be no reliable evidence that surgery mitigates the life-shortening effect of myocardial infarction.

E. SOWTON: I have some information which may be of interest to Dr. BROD. We have analysed groups of patients treated either medically or surgically, i.e. with saphenous vein bypass grafts; we matched the two groups according to the severity of the disease and followed the patients up over a three-year period. We found that the percentage mortality was related to the left-ventricular end-diastolic pressure; though the mortality was similar in both groups of patients, the important factor seemed to be this end-diastolic pressure. In patients with one-vessel disease there were no surgical deaths, either at or after operation, and – as was to be expected – there were also very few medical deaths. Among patients with two-vessel disease, on the other hand, some deaths did occur at or immediately following operation, but the surgical cases showed a better three-year survival rate than the medical ones. This phenomenon was even more marked in the case of patients with three-vessel disease. These findings would appear to suggest that surgery probably does improve the mortality – at least in the first three years.

---

* LAMBERT, D. M. D.: Beta-blockers and life expectancy in ischaemic heart-disease. Lancet *i*, 793 (1972); corresp.

J.-L. Rivier: I should just like to point out that in two homogeneous groups of patients followed up over a period of three years, one of which comprised 100 non-operated cases and the other 150 patients subjected to one or more coronary bypass operations, we made the same observations as Dr. Sowton.

H. Bricaud: I feel that in this discussion of ours too little emphasis has perhaps been placed on the importance of the beta-blockers in cases of coronary disease which are not amenable to surgery. I am referring here to severe cases in which coronary arteriography shows a bypass operation to be impracticable. We have not only given patients of this type beta-blockers, but we have also subjected them to physical rehabilitation measures, which seem to us to be absolutely indispensable in some severe cases. What is more, we found that these measures were greatly facilitated by the use of beta-blockers. In these cases, our aim was not primarily to reduce the mortality rate (in any event, not enough time has yet elapsed for us to know whether such a reduction can in fact be achieved), but simply to make the patients more comfortable. Even in completely incurable cases, we have invariably achieved good results with these physical rehabilitation measures.

Beta-blockers in the treatment of disturbances in cardiac rhythm

by E. K. CHUNG*

Introduction

During the past decade, thanks to the ready availability of direct-current cardioverters and artificial pacemakers, there has been a marked improvement in the results of treatment for various cardiac arrhythmias. The use of large amounts of anti-arrhythmic drugs can thus often be avoided – an advantage which assumes growing significance in the light of current knowledge. Certain pharmacological agents, used in conjunction with direct-current cardioverters and artificial pacemakers, are nevertheless still indispensable in order to prevent, as well as to treat, many types of arrhythmia.

It is only when a precise diagnosis of the arrhythmia has been established, however, that the best therapeutic results can be obtained, because some drugs are more effective, or even virtually specific, in certain forms of arrhythmia[4]. In addition to a precise diagnosis of the arrhythmia, there are also underlying aetiological factors which likewise significantly influence the outcome of treatment. For example, catecholamine-induced arrhythmias are best treated with beta-adrenergic blocking agents, whereas digitalis-induced arrhythmias respond best to diphenylhydantoin or potassium[3]. On the other hand, any cardiac arrhythmia provoked by heavy smoking, excessive ingestion of coffee or tea, etc. should preferably be treated by eliminating the direct causative agent rather than by giving the patient some particular anti-arrhythmic drug[4].

Prevention of the recurrence of tachyarrhythmia is another important aspect of management. For this reason, maintenance therapy with anti-arrhythmic drugs is often necessary for long periods of time, or even indefinitely. Similarly, various predisposing factors, of cardiac as well as non-cardiac origin, must be avoided or eliminated as quickly as possible. These include emotional and physical stresses, excessive consumption of tobacco, coffee, and alcohol, as well as indigestion and constipation.

Since the introduction of pronetalol[1] in 1962 and of propranolol[2] in 1964, experimental and clinical investigations have shown that beta-blockers are capable of abolishing a variety of cardiac arrhythmias, including those induced by digitalis and those resistant to digitalis[9-12, 15, 22]. Though pronetalol was initially used for this purpose, propranolol was subsequently found to be more potent, i.e. ten times more active, and less toxic; propranolol has proved effective in the treatment of supraventricular tachycardia, ventricular tachycardia

---

* Heart Station, Division of Cardiology, Department of Medicine, Thomas Jefferson University, Philadelphia, Pennsylvania, U.S.A.

216

or fibrillation, and premature ventricular contractions. Pronetalol, on the other hand, is now no longer recommended for clinical use, since it was discovered to be carcinogenic in mice[1, 2].

At the moment, propranolol is the only beta-blocker commercially available in the United States of America, and it is therefore with propranolol that the present paper will primarily be concerned. Other beta-blockers under investigation include practolol, alprenolol, oxprenolol (®Trasicor), sotalol, metalol, bunolol, and triprenolol. The pharmacological aspects of the beta-blockers have been described by a number of authors[8, 14, 18], and an excellent in-depth review article on propranolol appeared only recently[14].

Indications

Propranolol has been shown to be effective in terminating and preventing various tachyarrhythmias, including those of the digitalis-induced and digitalis-resistant type, but it has seldom been employed as a drug of primary choice[3, 4, 9, 10, 12, 14, 15, 22].

The anti-arrhythmic activity of propranolol is attributable to two effects: firstly, inhibition of adrenergic stimulation of the heart and, secondly, a direct action on the electrophysiological properties of cardiac tissue. Thus, the overall effects of the drug usually entail a reduction in automaticity, including a reduction in the sinus rate, and prolongation of the atrial and atrioventricular conduction time. In the treatment of digitalis-induced tachyarrhythmias, it is the direct action of propranolol on the myocardial cell-membrane that is primarily responsible for its therapeutic effect. In addition, propranolol is very effective in the treatment of tachyarrhythmias precipitated by exercise, emotional distress, or excessive sympathetic stimulation, as well as in the treatment of supraventricular tachyarrhythmias associated with Wolff-Parkinson-White syndrome. It is also of value in slowing the ventricular rate in cases of atrial fibrillation or flutter in which it proves difficult to bring the rapid ventricular rate under control with adequate digitalisation and quinidinisation.

1. Wolff-Parkinson-White syndrome
It is a well-known fact that Wolff-Parkinson-White syndrome is frequently associated with various forms of supraventricular tachyarrhythmia[4], the most common of which is a rapid and regular supraventricular tachycardia (Figure 1). According to present knowledge, tachyarrhythmias would appear to be due to a re-entry phenomenon operating through an accessory conduction pathway in the presence of a normal atrioventricular conduction system. The QRS configuration during tachyarrhythmias will be normal, i.e. narrow, when cardiac impulses are conducted via a normal atrioventricular pathway, whereas it will be wide and bizarre when conduction occurs via an anomalous pathway[4].
Propranolol is very effective in the treatment of supraventricular tachycardia with a normal QRS complex in Wolff-Parkinson-White syndrome. This applies

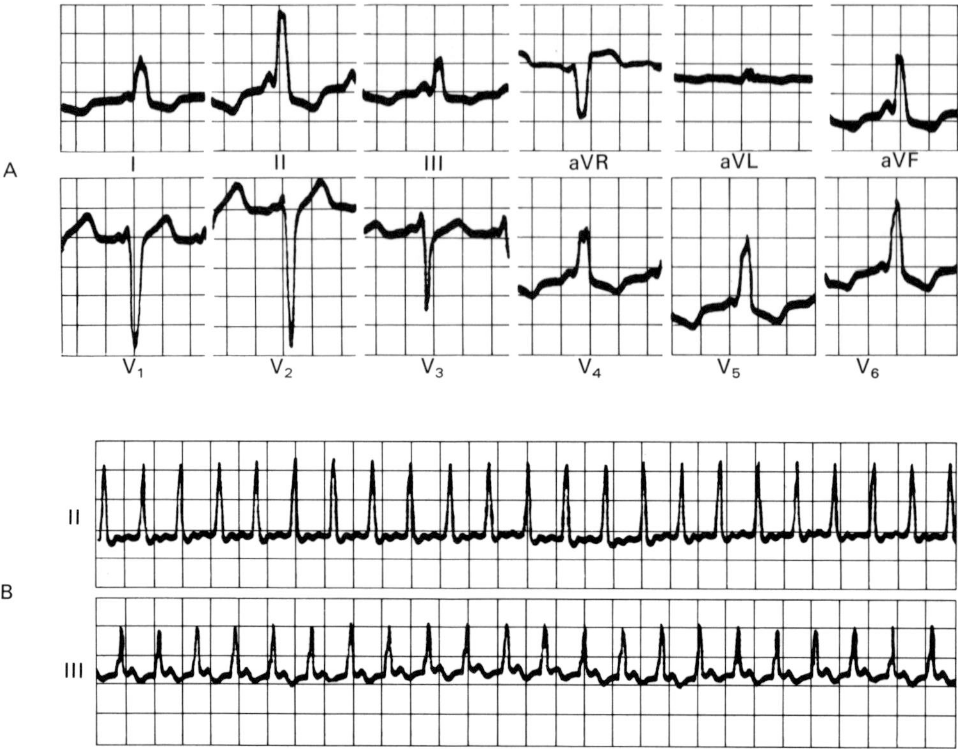

Fig. 1. Tracings A and B were obtained from the same patient on different occasions. Tracing A shows Wolff-Parkinson-White syndrome (Type B). Tracing B reveals paroxysmal supraventricular tachycardia with a heart rate of 214 beats/min.

particularly to young individuals with no demonstrable heart disease[5]. In Wolff-Parkinson-White syndrome, propranolol is therefore normally the drug of choice for the treatment of supraventricular tachycardia with a narrow QRS complex. Given orally in a low dosage, it is also highly effective in preventing supraventricular tachycardia in such patients. On the other hand, in cases of Wolff-Parkinson-White syndrome involving elderly individuals with a diseased heart, digitalis is found to be more effective in terminating this type of tachycardia with a narrow QRS complex[3].

Treatment with propranolol alone is usually not sufficient to terminate or prevent atrial flutter or fibrillation – with or without anomalous atrioventricular conduction – in Wolff-Parkinson-White syndrome. Similarly, the effect of propranolol is less reliable in supraventricular tachycardia with anomalous atrioventricular conduction[4]. Propranolol nevertheless does prove of value when used in conjunction either with digitalis or, at times, with quinidine[5].

2. Catecholamine-induced tachyarrhythmias

Propranolol exerts a very good effect when employed in the treatment of various tachyarrhythmias precipitated by emotional distress, excessive sympathetic stimulation, or physical exercise. This, once again, applies particularly

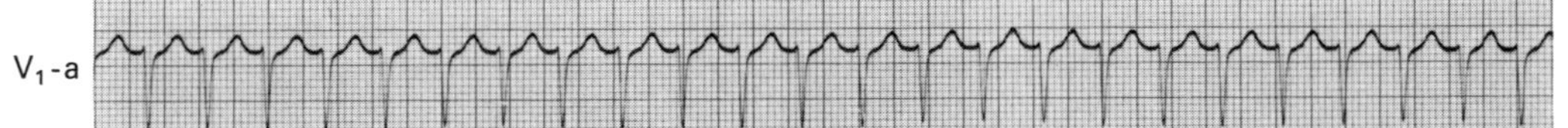

Fig. 2. Leads V₁-a and V₁-b are not continuous. Paroxysmal supraventricular tachy-cardia (heart rate: 185 beats/min.) was terminated by an intravenous injection of 1 mg. propranolol. Note the three premature ventricular contractions (marked V) before restoration of sinus rhythm.

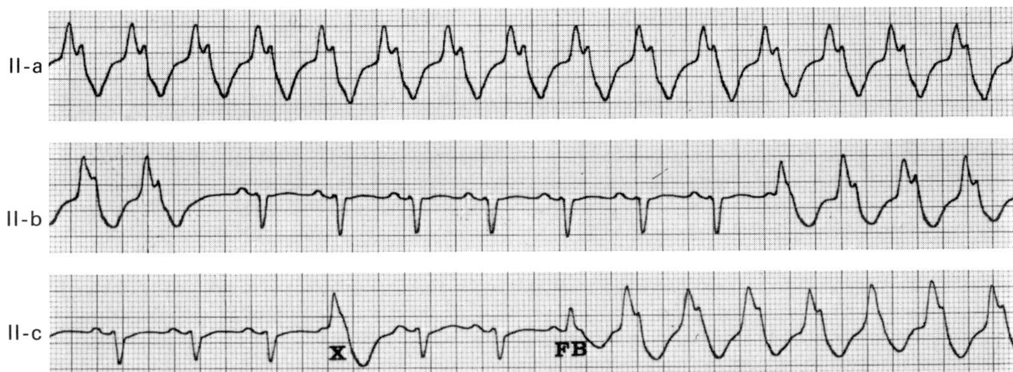

Fig. 3. Leads II-a, II-b, and II-c are not continuous. The tracing shows sinus tachy-cardia (heart rate: 102 beats/min.) with intermittent ventricular tachycardia (heart rate: 125 beats/min.). Note the two ventricular fusion beats (marked X and FB) which are of differing degree.

to young patients with no demonstrable heart disease, in whom the drug is especially effective in combating extrasystoles of varying origin and supraven-tricular tachycardia[5] (Figure 2). Ventricular tachycardia provoked by stress can also sometimes be successfully controlled with propranolol (Figure 3). Beta-blockers are therefore the drugs of choice both for the treatment and for the prevention of catecholamine-induced tachyarrhythmias.

Propranolol has also been employed as treatment for tachyarrhythmias in-duced by increased sympathetic activity during anaesthesia. Here, however, its use can be hazardous, particularly if the anaesthetic administered is one that produces myocardial depression[14]. In addition, propranolol may be prescribed in conjunction with an alpha-blocking agent in the management of tachyarrhythmias associated with phaeochromocytoma.

Although, in some cases of hyperthyroidism, tachyarrhythmias may be ter-minated by recourse to propranolol, a very large dosage is often required for this purpose unless the underlying disease is treated simultaneously. Proprano-lol is extremely useful as a means of slowing persistent and marked sinus tachycardia in hyperthyroid patients and in young individuals suffering from anxiety neuroses or from hyperkinetic heart syndrome[5]. It can often also be employed to good effect in order to slow the ventricular rate in atrial fibrillation associated with hyperthyroidism; but atrial fibrillation itself cannot be converted to a sinus rhythm by administering propranolol.

3. Digitalis-induced cardiac arrhythmias

As mentioned previously, it is the direct action of propranolol on the myocardial cell-membrane that is primarily responsible for the drug's therapeutic effect in digitalis-induced supraventricular and ventricular tachyarrhythmias[9, 10, 12, 15, 22]. In my experience[3], however, potassium and diphenylhydantoin offer the most effective means of terminating various digitalis-induced tachyarrhythmias. Potassium should be used first if there is marked hypopotassaemia, whereas diphenylhydantoin is preferable in cases where the digitalis has given rise to significant atrioventricular block. Except in the presence of pronounced hypopotassaemia, diphenylhydantoin is probably the drug of choice in the treatment of digitalis-induced ventricular tachyarrhythmias[3]. But propranolol can be used in digitalis-induced arrhythmias in situations where potassium or diphenylhydantoin are either ineffective or contraindicated. Needless to say, immediate discontinuation of digitalis is the first and foremost measure to be taken when treating digitalis intoxication.

4. Atrial fibrillation and flutter

One of the most important purposes served by propranolol is to reduce the ventricular rate in atrial fibrillation or flutter in cases where a rapid ventricular response (120–200 beats/min.) persists despite treatment with normal therapeutic doses of digitalis[4]. The effect exerted by propranolol in such conditions is due to the fact that the drug increases the refractory period in the atrioventricular junction. This effect of propranolol can be exploited therapeutically regardless of any underlying heart disease and regardless of the aetiology of the disturbance. It should be pointed out, however, that a synergistic action of digitalis and propranolol on the atrioventricular conduction time is liable to result in an extremely slow ventricular rate.

5. Miscellaneous indications

Propranolol often proves useful if and when a patient with idiopathic hypertrophic subaortic stenosis develops supraventricular tachyarrhythmia, particularly if he is also suffering from Wolff-Parkinson-White syndrome. Idiopathic hypertrophic subaortic stenosis is one of the commonest congenital heart diseases associated with Wolff-Parkinson-White syndrome, and both conditions frequently produce supraventricular tachyarrhythmias[6]. Under these circumstances, it is hazardous to administer digitalis, because this may well aggravate the clinical picture of the hypertrophic subaortic stenosis[3].

Finally, propranolol may also be of value for the prevention and treatment of tachyarrhythmias associated with angina pectoris, since it serves to combat both the tachyarrhythmia as well as the angina pectoris. Recently, there have been several reports (unpublished) indicating that beta-blockers are very effective in terminating multifocal atrial tachycardia in some cases.

Administration and dosage

Propranolol may be administered either intravenously or orally, depending on the clinical situation. In urgent cases where a rapid result is required, the drug can be given by slow intravenous injection under electrocardiographic monitoring (Figure 2); the rate of administration should not exceed 1–2 mg. per minute. The dose may if necessary be repeated after five to ten minutes, but the total dosage should not exceed 10 mg. If pronounced bradycardia (Figure 4) develops, 0.4–1 mg. atropine should be administered intravenously. If marked hypotension is observed following propranolol, isoprenaline should be given by slow intravenous infusion. In children, the dosage of propranolol can be adjusted according to the patient's age and body weight; but not enough clinical data are yet available upon which to base exact dosage recommendations for children.
When employed by the oral route, propranolol can be administered in a dosage of 10–40 mg. three to four times daily. Oral medication is recommended either in non-urgent situations or as maintenance therapy for prophylactic purposes.
Where cardiac arrhythmias are associated with hyperthyroidism, angina pectoris, or hypertension, considerably higher dosages – ranging from 160–240 mg. up to as much as 480 mg. daily – are often required[14].
It should be noted that a poor correlation exists between the oral dosage and the degree of beta-adrenergic blockade[9, 10, 12, 15, 22]. Consequently, the dosage should be individualised in the light of the therapeutic response.

Side effects and contra-indications

Propranolol is liable to produce various untoward reactions, particularly in the cardiovascular and respiratory systems.

1. Cardiovascular system
Two major cardiovascular side effects of propranolol include slowing of the heart rate and either the development of congestive heart failure or aggravation of pre-existing congestive heart failure[4].
Slowing of the heart rate in response to propranolol almost always takes the form of sinus bradycardia (Figure 4). The drug may also produce very unstable sinus activity with periods of sinus arrest, sino-atrial block, atrioventricular junctional or ventricular escape rhythm, and finally ventricular tachycardia and fibrillation followed by ventricular standstill (Figure 5). In cases where marked bradycardia occurs, atropine should be administered intravenously, supplemented by such cardiopulmonary resuscitation measures as may be

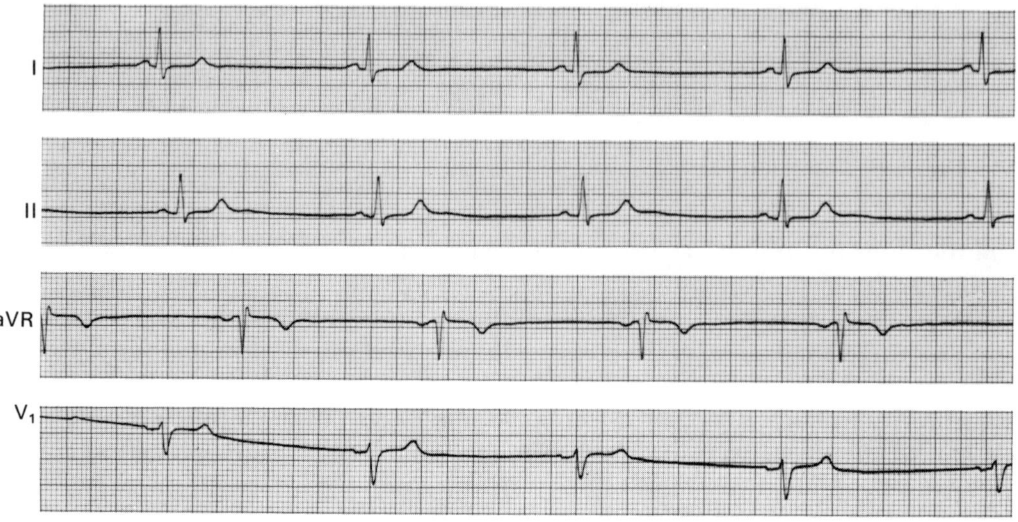

Fig. 4. Pronounced sinus bradycardia (heart rate: 38 beats/min.) induced by propranolol.

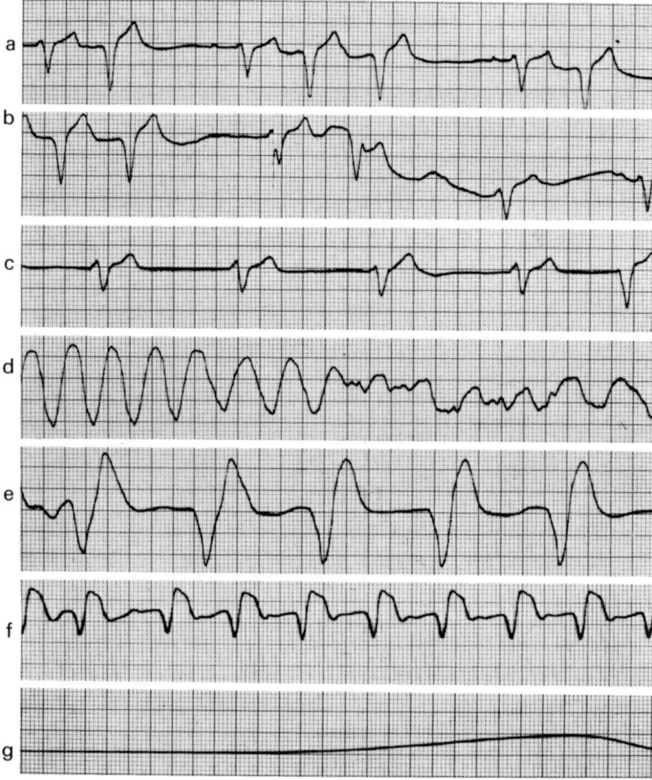

Fig. 5. Markedly unstable sinus activity with atrioventricular junctional escape rhythm followed by ventricular tachycardia and fibrillation, and ventricular escape rhythm followed by ventricular standstill, in a patient treated with propranolol. The strips a, b, c, d, e, f, and g were obtained from the same patient, but are not continuous.

necessary. In patients with atrioventricular block, propranolol may induce cardiac arrest; complete ventriculo-atrial block, without impairment of forward atrioventricular conduction, has also been reported in response to propranolol[17]. Propranolol is therefore contra-indicated in patients with significant sinus bradycardia, unstable sinus activity, sinus arrest, sino-atrial block, or second or third degree atrioventricular block.

Because of its negative inotropic action, propranolol may aggravate frank or borderline congestive heart failure – in the presence of which its use should thus, if possible, be avoided. In instances where propranolol is considered to be definitely indicated despite the existence of congestive heart failure, adequate digitalisation plus diuretic therapy is essential, and the propranolol should not be given in large doses. It must always be borne in mind that congestive heart failure is the most serious untoward effect of propranolol.

Propranolol may also give rise to pronounced hypotension, in which case intravenous infusion of isoprenaline should be carried out immediately. Owing to the competitive nature of beta-blockade, large doses of isoprenaline may sometimes be required for this purpose.

There have been several reports suggesting that acute myocardial infarction is liable to occur following rapid withdrawal of propranolol in patients with angina pectoris[14]. The dosage of propranolol should therefore be reduced gradually in cases where the drug has to be withdrawn.

2. Respiratory system

Propranolol should not be administered to patients with asthma or bronchitis or with a strong family history of bronchial asthma[5], the reason for this precaution being that the drug increases airway resistance and may thus induce asthmatic attacks in such individuals.

3. Miscellaneous

Propranolol may produce hypoglycaemia in patients undergoing insulin therapy, in patients recovering from anaesthesia, and in paediatric age groups during periods of restricted food intake[14]. Untoward neurological reactions which have been observed in response to propranolol include light-headedness, fatigue, lethargy, hallucinations, mental depression, insomnia, and peripheral neuropathy[5, 6, 8, 14, 17, 18, 23]. In some patients, propranolol may cause nausea, vomiting, diarrhoea, constipation, fever, erythematous rashes, thrombocytopenia, and agranulocytosis[5, 14, 23]. The occurrence of alopecia following administration of propranolol has also been reported in several cases[16].

References

1 BLACK, J.W., CROWTHER, A.F., SHANKS, R.G., SMITH, L.H., DORNHORST, A.C.: A new adrenergic beta-receptor antagonist. Lancet i, 1080 (1964)
2 BLACK, J.W., STEPHENSON, J.S.: Pharmacology of a new adrenergic beta-receptor-blocking compound (nethalide). Lancet ii, 311 (1962)
3 CHUNG, E.K.: Digitalis intoxication (Excerpta Medica Found., Amsterdam 1969)

4 CHUNG, E. K.: Principles of cardiac arrhythmias (Williams & Wilkins, Baltimore 1971)

5 CHUNG, E. K.: Cardiac arrhythmia: management (Tape Series). (Williams & Wilkins, Baltimore 1973)

6 CHUNG, E. K.: Electrocardiography: practical applications with vectorial principles (Hagerstown, Harper and Row 1974)

7 FINEGAN, R. E., MARLON, A. M., HARRISON, D. C.: Circulatory effects of practolol. Amer. J. Cardiol. *29*, 315 (1972)

8 FITZGERALD, J. D.: Perspectives in adrenergic beta-receptor blockade. Clin. Pharmacol. Ther. *10*, 292 (1969)

9 GIANELLY, R., GRIFFIN, J. R., HARRISON, D. C.: Propranolol in the treatment and prevention of cardiac arrhythmias. Ann. intern. Med. *66*, 667 (1967)

10 GIBSON, D., SOWTON, E.: The use of beta-adrenergic receptor blocking drugs in dysrhythmias. Progr. cardiovasc. Dis. *12*, 16 (1969)

11 HILLESTAD, L., ANDERSEN, A.: Beta-receptor blockade in maintenance of sinus rhythm obtained by electroversion. Acta med. scand. *185*, 535 (1969)

12 IRONS, G. V., GINN, W. N., ORGAIN, E. S.: Use of beta adrenergic receptor blocking agent (propranolol) in the treatment of cardiac arrhythmias. Amer. J. Med. *43*, 161 (1967)

13 JOHNSTONE, M.: Cardiac dysrhythmias during anaesthesia with special reference to practolol (Eraldin). Acta cardiol. (Brux.) Suppl. XV: 293 (1972)

14 KOSMAN, M. E.: Current status of propranolol hydrochloride (Inderal). J. Amer. med. Ass. *225*, 1380 (1973)

15 LUCCHESI, B. R., WHITSITT, L. S.: The pharmacology of beta-adrenergic blocking agents. Progr. cardiovasc. Dis. *11*, 410 (1969)

16 MARTIN, C. M., SOUTHWICK, E. G., MAIBACH, H. I.: Propranolol induced alopecia. Amer. Heart J. *86*, 236 (1973)

17 MAYTIN, O., CASTILLO, C. A., CASTELLANOS, A., Jr.: Complete V-A block (without impairment of forward conduction) produced by propranolol. Amer. J. Cardiol. *27*, 687 (1971)

18 MORAN, N. C.: Adrenergic receptors, drugs, and the cardiovascular system. Mod. Conc. cardiovasc. Dis. *35*, 93 and 99 (1966)

19 PONTINEN, P. J., EGGERT, H., KEMILA, S.: Alprenolol in resuscitation. Brit. med. J. *iv*, 723 (1971)

20 RYDÉN, L., NYBERG, G., SAETRE, H., KARLSSON, L., LINDÉN, L., SJÖHOLM, J., SUNDQUIST, O.: Alprenolol as maintenance therapy for the protection against arrhythmia relapse in patients with DC-converted chronic atrial fibrillation. Acta med. scand. *191*, 43 (1972)

21 RYMASZEWSKI, Z., POPŁAWSKA, W., PREIBISZ, J., JANUSZEWICZ, W.: Clinical evaluation of practolol in acute arrhythmias. Brit. Heart J. *34*, 260 (1972)

22 SLOMAN, G., STANNARD, M.: Beta-adrenergic blockade and cardiac arrhythmias. Brit. med. J. *iv*, 508 (1967)

23 STEPHEN, S. A.: Unwanted effects of propranolol. Amer. J. Cardiol. *18*, 463 (1966)

General discussion on the use of beta-blockers in disturbances of cardiac rhythm

W. Schweizer: Before we discuss the use of beta-blockers in disturbances of cardiac rhythm, there are three remarks I wish to make. Firstly, in this discussion, we should state clearly whether we are referring to curative or to prophylactic treatment. Take atrial fibrillation, for instance: everyone will agree that the best method for terminating an attack of atrial fibrillation is cardioversion; but the best method of reducing the risk of further attacks is probably to prescribe quinidine. Secondly, we should not spend time discussing the indications for drug treatment in general in the different disorders of cardiac rhythm. In a number of these disorders, of course, there is no need for any drug treatment. I hope you will agree, then, that, when discussing tachycardia for example, we should confine our attention to the patient who for one reason or another does in fact require a drug. Thirdly, when debating the value of beta-blockers, we should try to compare them with conventional drugs with a view to answering the question: How reliable are the beta-blockers and do they offer any real advantages over the drugs we have hitherto had at our disposal?

E. K. Chung: One of the indications for the use of beta-blockers is as a replacement for quinidine in patients who can't tolerate this drug. As you know, quinidine produces a lot of side effects, including even sudden death due to quinidine poisoning. In these cases of sudden death, you get a prolonged Q-T interval which leads to an R-on-T phenomenon. Another major indication for beta-blockers in this field, as I explained in my paper, is catecholamine-induced tachyarrhythmias. Finally, in Wolff-Parkinson-White syndrome, beta-blockers are definitely the drugs of choice, especially in the treatment of regular tachycardia with narrow QRS complexes.

E. Sowton: I'd like first of all to ask Dr. Chung about Wolff-Parkinson-White syndrome. His-bundle recordings and stimulation studies have shown that beta-blockers exert their effect largely at the atrioventricular node and, to almost an equal degree, at the bypass. Drugs such as verapamil, on the other hand, act primarily at the atrioventricular node and not at the bypass*, whereas drugs like ajmaline have a blocking effect at the bypass but not at the atrioventricular node. Our experience has been that drugs of the verapamil and ajmaline type are far more effective in terminating Wolff-Parkinson-White tachycardias. I think the concept of the selective action of anti-arrhythmic drugs is now well established and that blunderbuss therapy belongs to the past. During your presentation, Dr. Chung, you made an off-the-cuff remark to the effect that many people continue to regard digitalis as contra-indicated in atrial fibrillation; in my opinion, there are still valid grounds for believing this. Patients who have Wolff-Parkinson-White syndrome with one or two bypasses, and perhaps even James or Maheim fibres, and who go into atrial fibrillation do frequently conduct very fast through one or other of the bypasses. If we use digoxin, which – in another aside – you mentioned as being often effective alone and which blocks primarily at the atrioventricular node, a situation can arise in which the atrial fibrillation leads very rapidly to ventricular fibrillation. Thus, although I would agree that digoxin is sometimes effective by itself, I think it is sometimes fatal too – also by itself.

E. K. Chung: Digitalis, Dr. Sowton, is often extremely effective when given alone to treat atrial fibrillation or flutter, especially if the QRS complex is normal, i. e. narrow. What is more, in the light of my experience – I am thinking here of 100 documented cases of Wolff-Parkinson-White tachyarrhythmia, most of which were of Type A with atrial fibrillation or atrial flutter – I still don't understand why some people

---

* Spurrell, R.A.J., Krikler, D.M., Sowton, E.: Effects of verapamil on electrophysiological properties of anomalous atrioventricular connexion in Wolff-Parkinson-White syndrome. Brit. Heart J. *36*, 256 (1974)

believe that digitalis is contra-indicated. When I was following all the cases of Wolff-Parkinson-White syndrome, I found that none of them really developed ventricular fibrillation either directly or indirectly as a result of having received digitalis. As you also know, Dr. Sowton, beta-blockers prove very effective in cases where the QRS complex is normal, i.e. if the impulse is primarily conducted through the normal A-V system; here, they exert a better effect than in the presence of anomalous A-V conduction.

E. Sowton: When the QRS is normal, we presume that conduction is occurring down the ordinary conducting system; since digoxin and digitalis are known to act at the A-V node, I agree that they would be effective in this situation. But in the course of your presentation you showed one E.C.G. which was not included in the written version of your paper and which revealed conduction down the bypass; in this case digoxin may not be effective, because it does not actually work at the bypass.

E. K. Chung: Let me put it as follows, Dr. Sowton. I think we both agree that fibrillation or flutter occurring in a case of Wolff-Parkinson-White syndrome with a normal QRS can be, and often is, treated with digitalis alone. Even if, as in the E.C.G. you referred to (Figure 1), you have atrial fibrillation with anomalous A-V conduction, digitalis alone may still be effective; but if, as often happens, it does not have an effect, then you have to add quinidine and/or a beta-blocker – and this usually produces the desired response. One more comment: when a patient has a very rapid ventricular

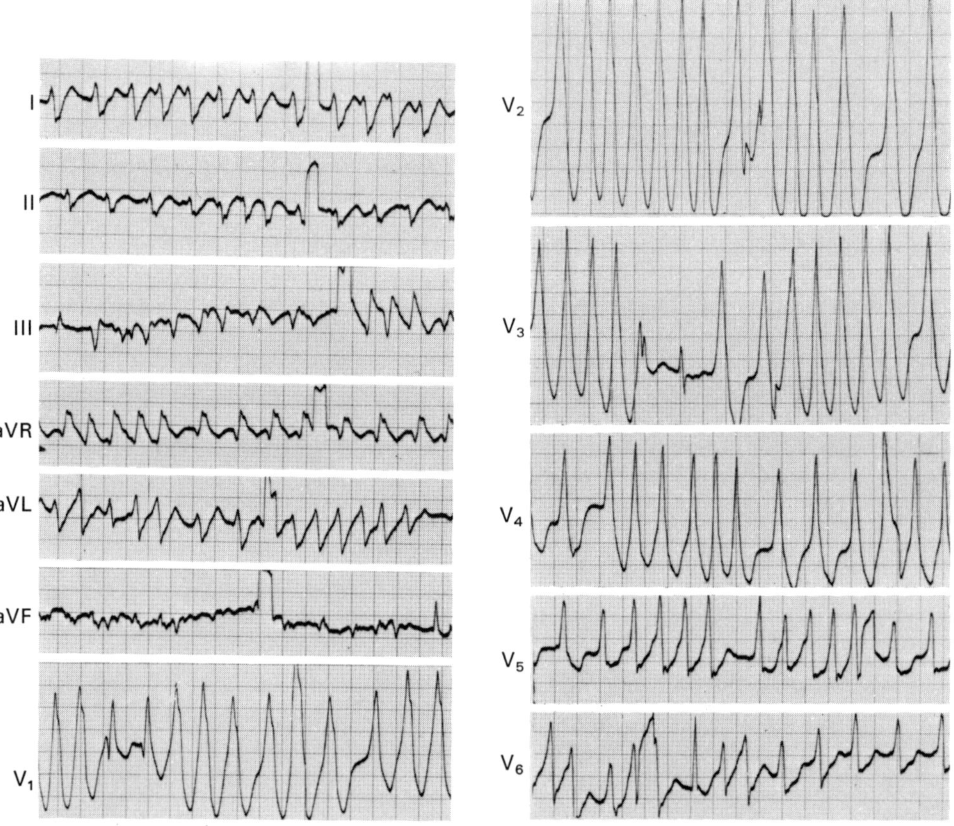

Fig. 1. Atrial fibrillation with very rapid ventricular response and anomalous A-V conduction in Wolff-Parkinson-White syndrome of Type A.

rate, we try D.C. shock first, provided a D.C. shock machine is available. Then we routinely give digitalis and quinidine, or digitalis and a beta-blocker. Using this approach, I have never really had a failure up to this point.

J. SOMERVILLE: I think that here some confusion may have arisen between, on the one hand, treatment for an acute attack of dysrhythmia in an otherwise apparently healthy patient and, on the other hand, more chronic treatment designed to suppress a dysrhythmia. What I want to say now is based on my experience with young patients under the age of 30 who present with paroxysmal atrial tachycardia, in some 40% of whom this is subsequently shown to be associated with Wolff-Parkinson-White syndrome. I think that digitalis certainly has a place to occupy, and it is always the paediatrician's first line of treatment; though it helps in the patient who presents with failure, it does not usually revert those with an associated Wolff-Parkinson-White syndrome. The ones who revert are those who have a narrow QRS and usually not in association with a Wolff-Parkinson-White syndrome. I completely agree with Dr. SOWTON that when you have a wide complex, or when there's conducting down the bypass, digitalis should not be used.

Regarding the suppression of recurrent attacks in the Wolff-Parkinson-White syndrome, I think that at present you can decide the choice of drug on the toss of a coin. Dr. CHUNG is probably at a disadvantage in not having had access to other beta-blockers besides propranolol. Generally speaking, for the suppression of attacks associated with a Wolff-Parkinson-White syndrome in young patients, I myself have found practolol more effective than propranolol.

E. SOWTON: I would agree that practolol is very effective in such cases – and not only in young patients. With reference to practolol, I have another question for Dr. CHUNG. In his paper he has attributed much of the anti-arrhythmic activity of propranolol to its direct action on the myocardial cell-membrane. Practolol has virtually no such action, but is nevertheless a very effective anti-arrhythmic drug. I'd appreciate Dr. CHUNG's comments on this point.

E. K. CHUNG: Let me reply first to Mrs. SOMERVILLE. All I can say is that, the more patients you have to treat, the more experience you acquire. There was a time when I used to think that digitalis was the drug of choice for any regular form of tachycardia – which usually means supraventricular tachycardia, with a normal QRS – regardless of the patient's age and regardless of what underlying heart disease there might be. But a few years ago I had cause to change my opinion slightly, because I had used propranolol for younger people without heart disease and had found it almost 100% effective in these cases. If an older patient has the same type of regular, rapid, supraventricular tachycardia with or without a Wolff-Parkinson-White syndrome, I still often try digitalis first. But in younger people without any heart disease, I try propranolol – and, as I have just said, the success rate is almost 100%. Incidentally, one of the main reasons why the tachycardia in young people with Wolff-Parkinson-White syndrome is often best treated by using beta-blockers may be that it is largely due to a rise in catecholamine rather than to other factors.

P. KARHUNEN: One type of indication for beta-blockers which may perhaps be of some interest comprises the syndromes characterised by prolongation of the Q-T interval and syncopal attacks due to ventricular fibrillation following violent emotional or physical stress. When associated with congenital deafness, this is known as Jervell-Lange-Nielsen syndrome. Where the patient's hearing is normal, it is referred to as Romano-Ward syndrome. Both syndromes have a very high mortality rate, and the steady increase in the number of reports on them suggests that they are more frequent than was previously suspected. Uncertainty about the pathogenetic mechanisms involved in these illnesses led to attempts at treatment with a wide variety of drugs, such as digitalis, diphenylhydantoin, quinidine, procainamide, and so on; but the results obtained were very poor. Now, however, beta-blockers – administered in ordinary doses – have been shown to be very effective in these conditions, and most of the patients become quite free of symptoms. In some cases, but not in all, the Q-T interval itself can be shortened, especially if there is TU fusion with a bizarre T wave. A very similar situation is liable to arise in association with tricyclic antidepressant medication, especially if the tricyclic has been taken in an overdosage or in an attempt at suicide. Here, use of beta-blockers can very often prevent potentially fatal dysrhythmias.

E. K. CHUNG: Besides this particular syndrome that Dr. KARHUNEN has mentioned, there are broad categories of the same kind in which a long Q-T interval may be associated with sudden death. This is always due to R-on-T phenomena, and, since sudden death is often the first "symptom", you may already be finished before you can start, as it were. Here, prevention is therefore extremely important. But I haven't had much personal experience of the syndrome that Dr. KARHUNEN described.

E. SOWTON: May I just point out that my colleagues and I are publishing a report on patients who have a Type A Wolff-Parkinson-White syndrome but who use a Type B bypass during their tachycardia*. This means that any analysis based on ordinary surface E.C.G.s is of no use in further discussion of the relative merits of different treatments.

E. K. CHUNG: I agree with you, Dr. SOWTON, but not for quite the same reason. There are many patients, including many of my own, who have more than one bypass. Moreover, there are some cases in which, when you examine the surface E.C.G., you can't really decide what type the patient belongs to. At times they fall inbetween the two types; perhaps one could even put them into a separate category and call them Type C! Obviously, certain patients have more than one bypass; at one time they may have a Type B and at some other time a Type A, or possibly a mixture of the two. There are a whole variety of fusion beats, which in turn may produce a variety of QRS complexes.

E. SOWTON: I'd like to hear from Dr. CHUNG what evidence there is that propranolol does in fact maintain sinus rhythm after D.C. counter-shock.

E. K. CHUNG: I have been able to confirm this in a very large number of cases. As a matter of interest, we gave some patients quinidine after their dysrhythmia had been terminated by D.C. shock, whereas others were given propranolol – in addition to digitalis. In some patients, of course, we cannot be sure that digitalis alone will do a good job. We never know 100%. But I did get the impression that a combination of propranolol and digitalis is superior to digitalis alone after atrial fibrillation or atrial flutter has been terminated in more severe cases of rheumatic heart disease.

W. SCHWEIZER: Speaking for myself, I must say that we have been disappointed by the results we have obtained when using a combination of beta-blockers and quinidine for the prevention of atrial fibrillation. We have not had the impression that beta-blockers add anything to this prevention of atrial fibrillation. Has anybody any further comments to make on possible differences in the anti-arrhythmic properties of the various beta-blockers?

E. SOWTON: The only work we have done in this respect showed that there was no significant difference between them, apart from certain differences in the size of the dose required to produce a given effect.

B. N. C. PRICHARD: I think one should draw attention to the obvious exception of thyrotoxicosis. Here, drugs with intrinsic sympathomimetic activity are best avoided: those without this activity are better at controlling any dysrhythmia – sinus tachycardia or atrial fibrillation – that may be associated with thyrotoxicosis.

---

* SPURRELL, R.A.J., KRIKLER, D.M., SOWTON, E.: Problems concerning assessment of anatomical site of accessory pathway in Wolff-Parkinson-White syndrome. Brit. Heart J. *37*, 127 (1975)

W. SCHWEIZER: If there are no further comments on this question, I'd now like to ask Dr. WERKÖ to tell us how, and in what indications, he uses beta-blockers as anti-arrhythmic agents.

L. WERKÖ, My indications would be largely the same as those that have already been discussed here. I quite agree that there is no proof that beta-blockers exert any preventive action in patients liable to go into atrial fibrillation. Like you, Dr. SCHWEIZER, we too have been greatly disappointed at the results we got when we used both propranolol and alprenolol, either alone or in combination with quinidine, in an attempt to prevent such fibrillation.

J. BROD: Is there any point in using beta-blockers in the hope of preventing arrhythmias following myocardial infarction?

L. WERKÖ: I think controlled studies would have to be carried out first before one could satisfactorily answer that question, Dr. BROD. We have in fact done a controlled study of this kind, but the results are at present still under statistical evaluation. For a period of two years following their myocardial infarction, the patients were given alprenolol or placebo in random fashion. Though, as I say, evaluation of the findings has not yet been completed, it rather looks as if this preventive treatment with alprenolol did have an influence on the incidence of sudden death during the two-year period, but not on recurrences of myocardial infarction. This problem will have to be studied in larger series before one can arrive at any final judgment.

S. H. TAYLOR: I'd like to add a brief comment on the intravenous use of beta-blockers in the treatment of dysrhythmias. There seems to be a preference for the cardioselective drug practolol, but there is no theoretical support whatsoever for this preference; in fact, there is evidence to the contrary, in that practolol is far less active in antagonising catecholamines than propranolol or oxprenolol (®Trasicor). In the clinical situation, moreover, intravenous practolol may prove hazardous, since it can induce a very marked fall in blood pressure. This has to be expected in immobile sick patients, in whom there may be large amounts of circulating catecholamines. The abrupt decrease in the cardiac stimulating effects of these catecholamines results in a relative preponderance of their peripheral vasodilator effects; hence the very pronounced fall in blood pressure that intravenous treatment with practolol is liable to elicit. I therefore suggest that it might well be preferable to use the indiscriminate blockers, such as propranolol or oxprenolol, for intravenous administration – not only for the reason I have just stated, but also because, in addition, they block the rise in free fatty acids, which may be a factor contributing towards the production of dysrhythmias.

E. K. CHUNG: Since we have been asked in this discussion to comment on possible differences in the anti-arrhythmic properties of the beta-blockers, I'd like to attempt a brief summary. As you all know, only one beta-blocker, propranolol, is available in the U.S.A.; my experience is therefore largely confined to this one agent. Now we also have practolol under investigation, and the results obtained with it in the small number of patients we've so far studied seem to us to suggest that there's not much difference between the anti-arrhythmic properties of this drug and propranolol. I'd say that the two main indications for propranolol as an anti-arrhythmic agent are, firstly, catecholamine-induced tachyarrhythmias and, secondly, Wolff-Parkinson-White syndrome, especially in young people with a rapid regular tachycardia and a normal QRS complex. Another possible indication is the prevention of atrial fibrillation, or, in the presence of persistent atrial fibrillation, the maintenance of an ideal ventricular rate; here, propranolol is usually given together with digitalis. I should also like to emphasise once again that, whenever the ventricular rate is slower than 60 beats/min., and particularly when it is slower than 50 beats/min., you must bear in mind the possibility that you are dealing, not simply with atrial fibrillation, but with high-degree A-V block.

Beta-adrenergic blocking drugs in the treatment of hypertrophic cardiomyopathy

by E. Sowton*

Introduction

This paper deals with the influence exerted by beta-adrenergic blocking drugs on haemodynamic variables and on symptoms in 20 patients with hypertrophic cardiomyopathy. At the time of cardiac catheterisation the patients (16 males and four females) were aged from 16 to 60 years (mean: 27.4 years). The diagnosis was initially made clinically either on the basis of a history of chest pain, palpitation, or syncope, or else on the basis of a symptomless systolic ejection murmur together with left-ventricular hypertrophy and a jerky or normal carotid pulse. Chest X-ray showed bulky hearts in all cases without appreciable increase in the cardio-thoracic ratio, and E.C.G. tracings confirmed the presence of left-ventricular hypertrophy.

The diagnosis was further corroborated by cardiac catheterisation, left-ventricular angiography, and, where indicated, selective coronary arteriography performed in order to exclude coronary artery disease. An outflow subvalve gra-

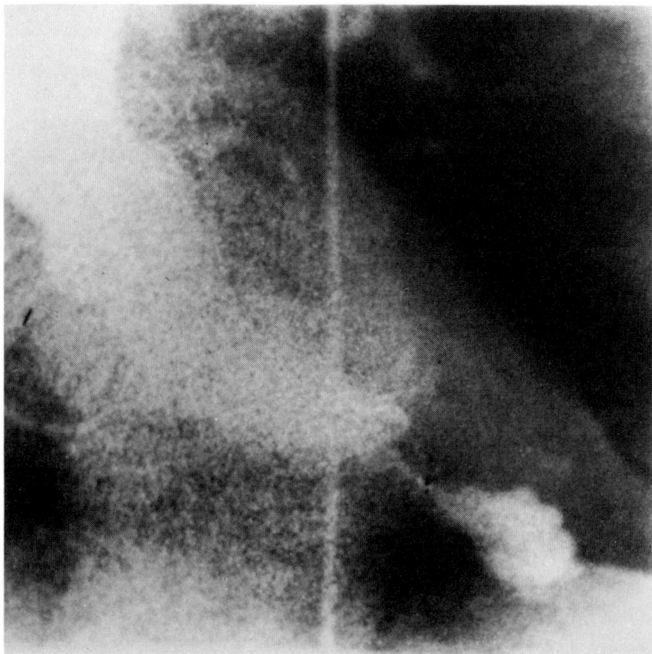

Fig. 1. Severe left-ventricular shutdown in systole with apical "pocketing". Gradient 100 mm. Hg.

---

230   * Cardiac Department, Guy's Hospital, London, England.

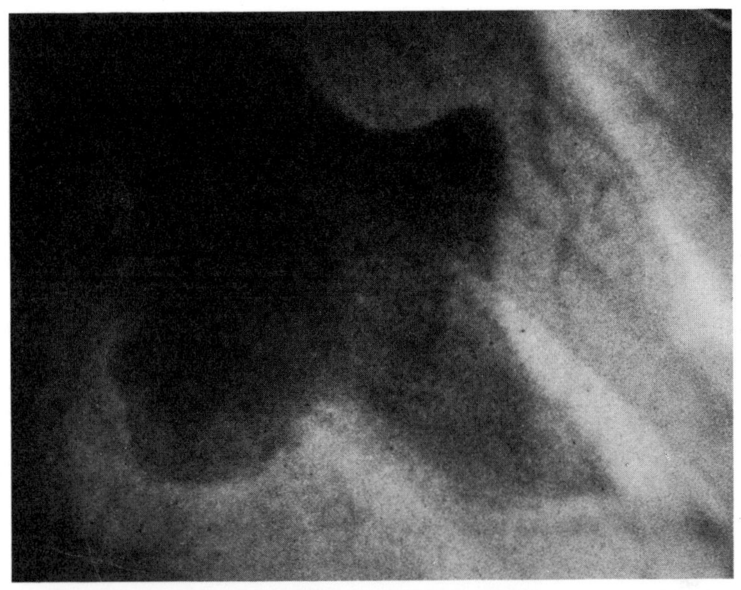
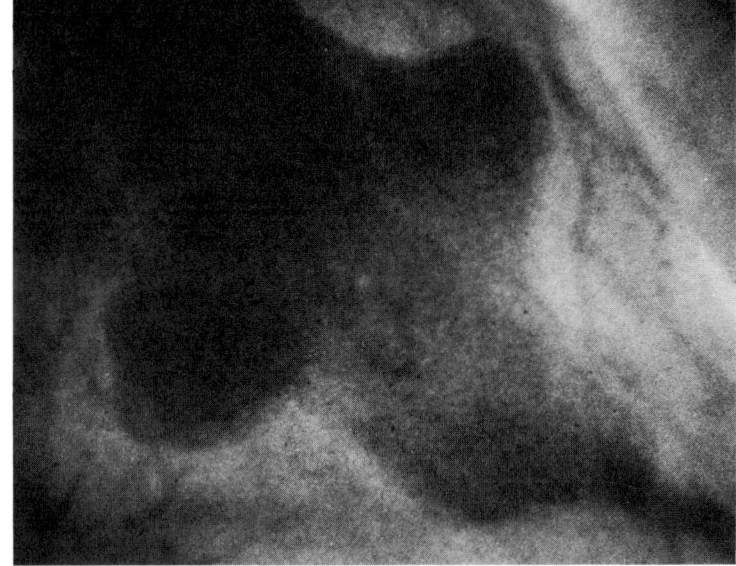

Fig. 2. Left-ventricular angiograms during systole *(above)* and diastole *(below)* in a patient with severe hypertrophic cardiomyopathy but no outflow gradient.

Table 1. Haemodynamic findings recorded at cardiac catheterisation in three groups of patients with hypertrophic cardiomyopathy.

	No. of patients	Age at catheterisation (mean values and ranges)	Cardiac output (litres/min.)	Systolic gradient from left-ventricular cavity to aorta (mm. Hg)
Mild (early) group	9	19 (16–21)	7.7	27
Intermediate group	8	31 (16–48)	5.3	68
Severe (late) group	3	46 (28–60)	4.5	90

dient was usually present but was not considered essential for the diagnosis in the presence of typical angiographic features of abnormally hypertrophied muscle masses and thick left-ventricular wall, together with normal coronary arteries and normal or high cardiac output. The left-ventricular pressure traces showed an early dip followed by a plateau in diastole (the "square root" sign), and the angiographic features are illustrated in Figures 1 and 2, which relate to cases with and without muscular obstruction; both show thick left-ventricular wall muscle with abnormal intrusion into the left-ventricular cavity.

On the basis of the physical signs, severity of symptoms, and, in particular, the catheter and angiographic findings the cases were classified into three groups – mild (early), intermediate, and severe (late). Details of these patients are presented in Table 1. Since an attempt was made to obtain a group of patients with mild clinical disease, and since "mild" implies that the disease was presumably at an early stage in its natural history, the diagnosis cannot be so secure in these cases as in those classified as intermediate or severe. However, as shown in Table 1, this mild group as a whole had an unusually high cardiac output and a distinct left-ventricular subvalve outflow gradient.

Symptoms and follow-up

The mild (early) group were largely symptom-free, although two patients gave a history of palpitation occurring in particular during emotional stress. Four of the nine patients had been treated with propranolol and one with practolol. None of these patients had syncope or angina by the time of cardiac catheterisation, although one developed atypical chest pain and had one syncopal episode subsequently. Two complained of effort dyspnoea. The time of follow-up since catheterisation in this group varied from one to 12 years (mean: 5.3 years).

Of the eight patients in the intermediate group, five experienced angina, four undue dyspnoea on effort, and six palpitation. Only one gave a history of syncope, but two others developed syncope when placebo was substituted for propranolol. Five of the eight were already being treated with beta-blockers.

All three patients in the severe group had angina, dyspnoea, palpitation, and S-T segment depression on the E.C.G. at rest; upon exercise, this depression increased beyond 3 mm. in all cases. Two of the three had syncopal episodes, which became much more frequent in one during placebo medication. All three were already being treated with propranolol. The time of follow-up was from six months to seven years (mean: 3.1 years).

Beta-blockers at catheterisation

The acute changes produced by beta-blocking drugs administered intravenously during catheterisation are well recognised, and a characteristic find-

232

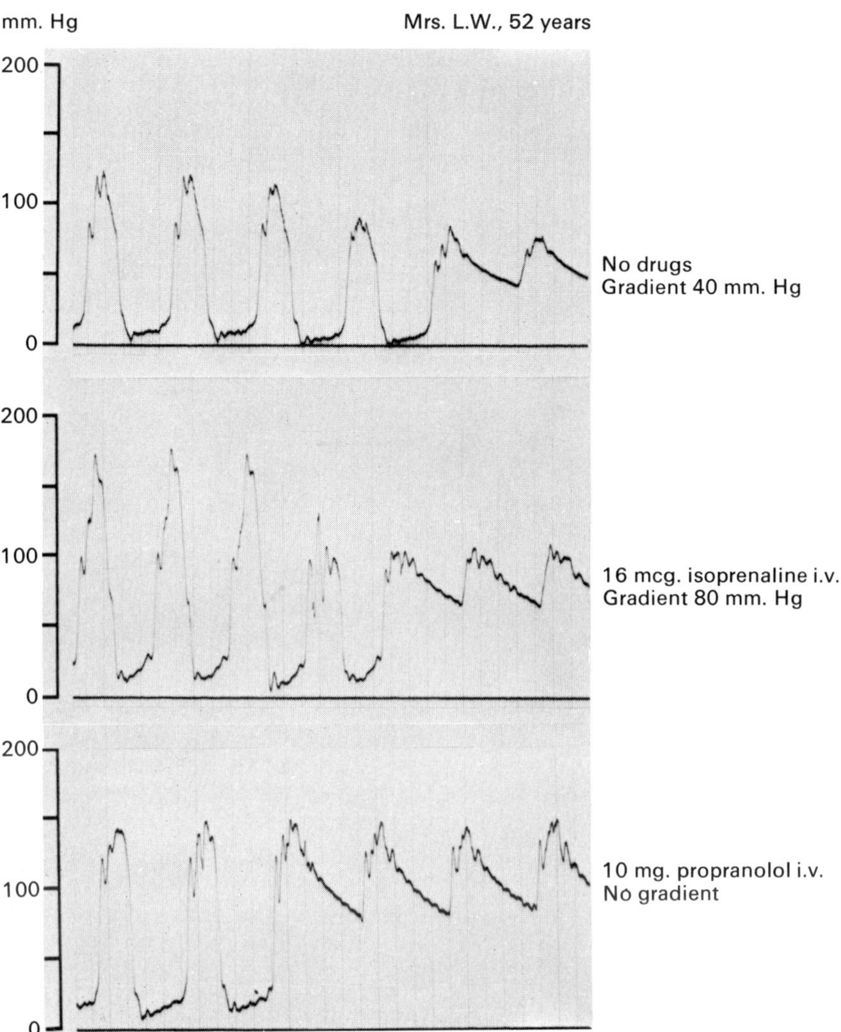

Fig. 3. Withdrawal traces from the left ventricle showing augmentation of the outflow gradient by isoprenaline, and reduction after propranolol (10 mg. i.v.), in a case of hypertrophic cardiomyopathy.

Table 2. Left-ventricular end-diastolic pressure (mm. Hg) in two studies performed six years apart in a patient belonging to the severe group.

Year of catheterisation	Pre-A wave		Post-A wave	
	Before	After propranolol (10 mg. i.v.)	Before	After propranolol (10 mg. i.v.)
1967	12	10	25	14
1973	24	13	40	25

ing is shown in Figure 3. This illustrates the increase in left-ventricular cavity gradient produced by isoprenaline and the total abolition of the gradient following intravenous administration of propranolol.

The acute effect of intravenous beta-blockers on left-ventricular end-diastolic pressure has also been described previously. Data from two studies performed six years apart on a patient in the severe group are presented in Table 2. Both "pre-A wave" and "post-A wave" values were reduced following the intravenous injection of 10 mg. propranolol.

Exercise studies

In the present investigation use was made of two protocols involving exercise testing. On every occasion the exercise test followed a standard pattern in which the patients first of all performed a warm-up [two minutes' exercise at 210 kilopond metres (k.p.m.) per minute] in the erect position on a bicycle ergometer (Monark), after which they rested until the heart rate had returned to normal. In typical cases this period of rest amounted to about half an hour. The definitive exercise test was then carried out with the work load increased at three-minute intervals. Blood pressure by the cuff method and an E.C.G. were recorded at the outset, after one minute at each work load, and also after three minutes at each work load – i.e. immediately before the work load was increased.

In an acute trial the patients first performed the warm-up and then, following recovery, carried out a full exercise test without being given any drugs. They then received a placebo tablet and one hour later repeated the exercise test. Following this, they were given 80 mg. oxprenolol (®Trasicor) by mouth, and the test was repeated for a third time after a further hour. The total work performed on each occasion was used as an index of effort tolerance.

A prolonged trial was then carried out in which, after effort tolerance had first been established, the patients were given medication three times daily for two weeks, whereupon the exercise test was repeated; they then received a second medication for a further two weeks, after which the exercise test was again repeated. In the case of the patients who were not already being treated therapeutically with beta-blockers the two medications consisted of placebo and oxprenolol. In some instances in which propranolol was already being administered for therapeutic reasons, attempts to treat the patients with placebo led to a severe exacerbation of symptoms; in these cases, therefore, a straightforward comparison was made between the results of the exercise test conducted during propranolol treatment and the results obtained after two weeks of medication with oxprenolol. Although the trial was planned as a double-blind one, it was realised that this method of study cannot be properly adhered to when beta-blocking drugs are compared with placebo, because the resting and exercising heart rates immediately identify the active drug. However, this effect was not obvious to the technicians and doctors until the end of the trial, and it was therefore felt to be of some value to continue with this protocol.

During the exercise tests there were several features which became apparent early on. As expected, the blood pressure – particularly the systolic levels – increased with the work load, but there was a distinct tendency for it to be reduced at the final work load, and this is illustrated in Figure 4 for each of the three procedures (i.e. no drugs, placebo, and oxprenolol) employed in the acute study. The effect was not confined to patients with severe disease and was seen irrespective of the procedure used.

In some patients, exercise also produced a noticeable effect on conduction, as illustrated in Figure 5.

A different type of E.C.G. abnormality – namely, S-T segment depression – was noted in some cases, an example of which is given in Figure 6. In this patient, who belonged to the intermediate group and remained completely

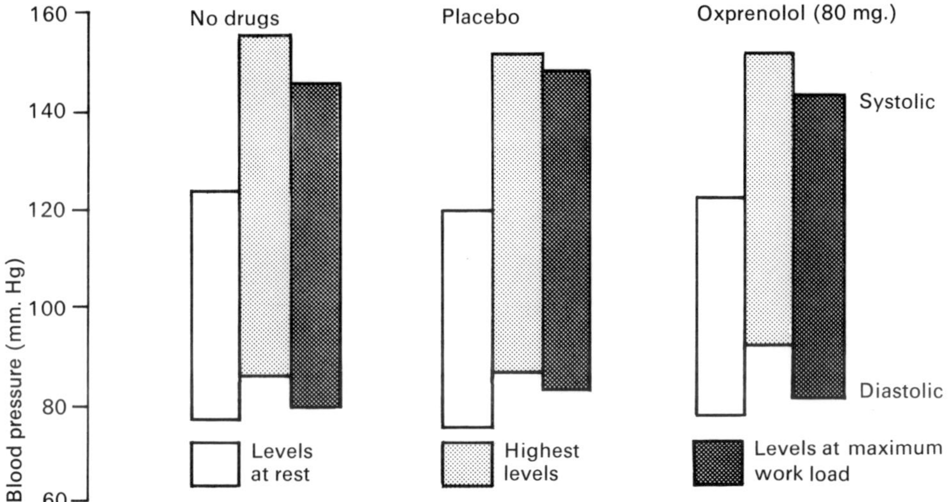

Fig. 4. Blood pressures in all patients following each of the three procedures. Note that in each case the levels at maximum work load were lower than the highest values measured.

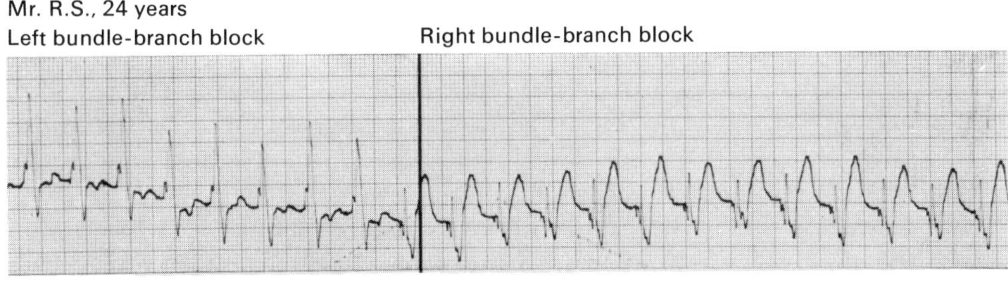

Fig. 5. Conduction disturbance produced by exercise following placebo in a case of symptomless hypertrophic cardiomyopathy.

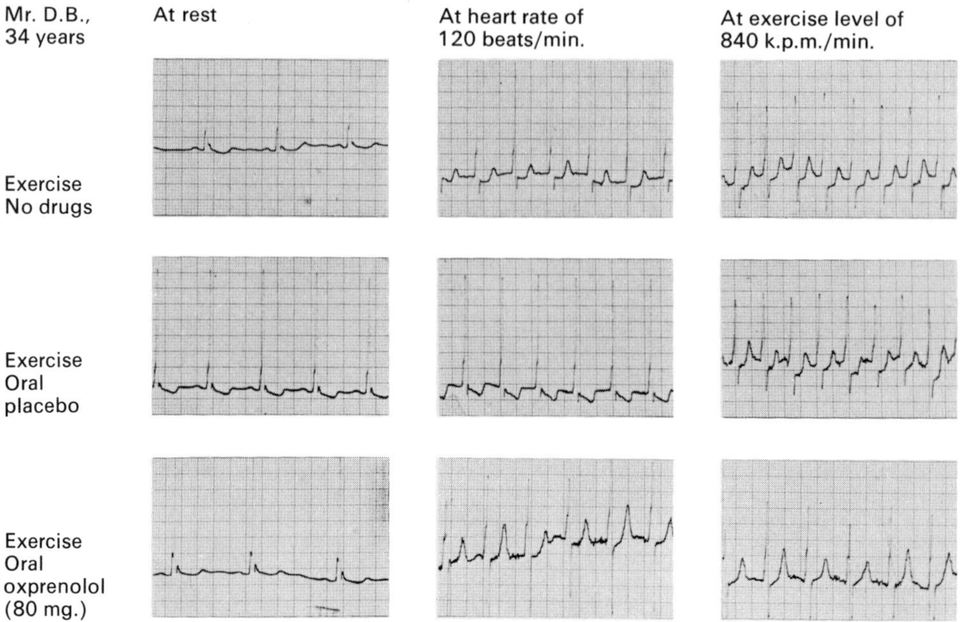

Fig. 6. S-T segment depression without angina during exercise following placebo, but not following oxprenolol, in a case of symptomless hypertrophic cardiomyopathy; k.p.m. = kilopond metres.

symptom-free, exercise after placebo produced a considerable degree of S-T segment depression. The patient had no angina or syncope at any time, and the S-T segment depression was largely abolished by oxprenolol. This effect of the drug was apparent at the same work loads and heart rates as those recorded during exercise following placebo medication.

Another conduction abnormality observed was inversion of T waves in patients exercising without taking any drugs or after placebo. In these patients, however, the T waves remained erect during the exercise test following beta-blockers. At no time did any of the patients showing conduction abnormalities experience angina during the exercise test.

Acute trial results

The mean results for the three groups are presented in Table 3. The group with mild (early) disease all exercised up to 840 k.p.m. per minute, and in the intermediate group all the patients achieved the maximum work load of 630 k.p.m. per minute (these work loads were exceeded by some patients, but, since they were not exceeded by all of them, the relevant data have not been included in this table). The patients in the severe group had a very greatly reduced exercise tolerance and could not manage multiple work loads.

The total work performed by each of the groups is presented in Table 4, from which the marked limitations of the severe group can be clearly seen.

In both the mild and the intermediate group the mean resting heart rate and blood pressure were usually the same after all three procedures, but at higher

236

Table 3. Acute study. Effect of exercise on heart rate and blood pressure following three different procedures in patients with hypertrophic cardiomyopathy of varying degrees of severity.

Work load (k.p.m./min.)		None (patient at rest)	210	420	630	840
Mild (early) group						
Heart rate (beats/min.)	No drugs	72	128	147	159	185
	Placebo	95	129	142	159	185
	Oxprenolol (80 mg.)	79	101	111	122	132
Blood pressure (mm. Hg)	No drugs	120/82	153/79	164/85	171/87	160/84
	Placebo	125/81	146/83	155/84	165/89	154/83
	Oxprenolol (80 mg.)	117/78	127/78	134/93	146/87	141/81
Intermediate group						
Heart rate (beats/min.)	No drugs	78	154	187	197	–
	Placebo	114	173	187	193	–
	Oxprenolol (80 mg.)	84	155	164	172	–
Blood pressure (mm. Hg)	No drugs	123/82	135/90	143/92	153/84	–
	Placebo	115/72	145/85	152/92	160/102	–
	Oxprenolol (80 mg.)	103/72	127/90	135/90	145/90	–
Severe (late) group						
Heart rate (beats/min.)	No drugs	84	–	–	–	–
	Placebo	78	–	–	–	–
	Oxprenolol (80 mg.)	63	–	–	–	–
Blood pressure (mm. Hg)	No drugs	105/65	–	–	–	–
	Placebo	115/65	–	–	–	–
	Oxprenolol (80 mg.)	100/65	–	–	–	–

k.p.m. = kilopond metres

work loads heart rate and blood pressure were consistently lower when the patients had received oxprenolol than when they had received placebo or no drugs at all. In the mild group the reduction in heart rate was approximately 20% at all work loads, while in the intermediate group it was 12–16%.

The diastolic blood pressure, either at rest or at any exercise level, differed very little irrespective of whether the patients had received no drugs, placebo, or oxprenolol. The systolic pressure recorded during exercise, however, was consistently lower following oxprenolol, the extent of the reduction amounting to 12–15%.

Despite these findings, the mild and intermediate group performed less exercise after having taken placebo than when they had taken no drugs at all and considerably less exercise after oxprenolol (cf. Table 4). This was particularly noticeable in the mild (early) group, who performed 12% less exercise after

Table 4. Total work performed by each of the groups following the three different procedures.

Group	Total work done (k.p.m.)		
	No drugs	Placebo	Oxprenolol
Mild (early)	5,653	5,565	4,900
Intermediate	4,077	3,535	3,290
Severe (late)	833	728	763

k.p.m. = kilopond metres

oxprenolol than after placebo. Although it is possible that the reduced amount of exercise performed after medication was related to the design of the trial rather than to the effect of the medication, this seems unlikely, because in the case of the mild group the total work performed following placebo was very similar to that performed without any drugs at all, and also because in the severe group exercise performance following oxprenolol was slightly better than that achieved after placebo. Nevertheless, the total work done in this latter group following oxprenolol was still about 10% less than the amount performed following no drugs at all.

It is noteworthy that with all three procedures the systolic blood pressure in the mild group was lower following the highest work load than following the second highest; this also applies to individual patients in the intermediate group who achieved work loads of more than 630 k.p.m./min. Of the eight patients in the mild group and of the five in the intermediate group who completed this part of the trial, three and one, respectively, stated that they found the exercise appreciably less disturbing, in terms both of dyspnoea and of awareness of the heart's action, when they had taken oxprenolol. However, in each of these cases total work performed was less following treatment with oxprenolol.

Prolonged "double-blind" oral trial

Seventeen patients started this part of the investigation, but seven (41%) had to be withdrawn because they developed serious complications when placebo was substituted for the beta-blocker which they had been taking therapeutically. Three patients (two from the intermediate and one from the severe group) experienced increased syncope, and four patients had an increase in the severity and frequency of their angina; one patient complained of frequent distressing palpitation. All these patients were successfully treated with oxprenolol. In addition, two patients complained of increased tiredness while they were taking placebo instead of the propranolol with which they had previously been treated; these symptoms disappeared when they were started on oxprenolol. Thus, in nine (53%) of the 17 patients the symptoms became worse on placebo, and in seven (41%) the deterioration was serious enough to require the urgent resumption of beta-blockade.

Table 5. Results obtained in exercise tests performed after medication with placebo and oxprenolol, each given for a period of two weeks (N = 10).

	Placebo	Oxprenolol (40 mg. t.i.d.)	Change
Total work done (k.p.m.)	3,530	3,630	+2.8%
Maximum heart rate on effort (beats/min.)	157	127	−19%
Blood pressure on effort (mm. Hg)	157/86	150/82	−4.5%

Table 6. Results obtained in exercise tests performed following medication with two different beta-blockers (N = 6).

	Propranolol	Oxprenolol	Change
Total work done (k.p.m.)	2,850	2,765	−3.5%
Maximum heart rate on effort (beats/min.)	119	117	−1.8%
Blood pressure on effort (mm. Hg)	144/78	151/83	+4.9%

In all, ten patients completed the full "double-blind" trial, including two weeks of placebo medication; two of them were from the severe group, five from the intermediate group, and three from the mild group. Of the ten patients, four preferred the placebo, five preferred oxprenolol, and one could not detect any difference between the two medications. All the patients who preferred placebo claimed to feel much better, less tired, less languid, and more mentally alert while on this medication. Two patients reported less dyspnoea compared to the oxprenolol period. The five patients who preferred oxprenolol gave as their reasons a decrease in dyspnoea (two patients), less angina (two patients), or a reduction in palpitation (two patients).

The results – in terms of total work performed, maximum heart rate during exercise, and blood pressure at the maximum work load reached – are presented in Table 5. The data on all ten patients have been combined, because the number of cases is too small to justify separate analysis according to the groups to which the patients belonged.

The heart rate was significantly lower following oxprenolol, but no appreciable difference was found between the two medications as regards their effect on blood pressure or total work performed.

Results are also available for six patients who had previously been treated with propranolol; one of them did not complete a placebo period. These results are compared in Table 6 with those obtained following treatment with oxprenolol in an equivalent dosage. No significant difference was found between the two drugs.

Symptoms

In the acute study exercise was limited by dyspnoea or by fatigue of the leg muscles in all cases (except for two patients from the severe group); this limitation was observed both before and even after beta-blockade in most of the patients who normally experienced angina. In the two patients with severe disease, in whom exercise without drugs, following placebo, and following oxprenolol was limited by angina, there was no improvement in effort tolerance. Three patients from the mild group and one from the intermediate group claimed that the exercise was easier and dyspnoea less marked after treatment with oxprenolol, but none of these four patients actually performed more work. Though not giving rise to symptoms, the ischaemia revealed by E.C.G. tracings was reduced following oxprenolol in three patients who did not experience angina.

During the prolonged "double-blind" study, exercise in the placebo period was limited by dyspnoea in seven patients; only two of them reported that this symptom was improved by the active drug, and two stated that their dyspnoea was unchanged. The remaining three thought it was worse. Overall, these findings are in keeping with previous reports that treatment with beta-blocking drugs does not afford much relief from the subjective sensation of dyspnoea. In particular, the patients who reported improvement in their dyspnoea were in the mild or intermediate groups, and patients with severe disease experienced no such benefit.

Discussion

CHERIAN et al.[1] and SLOMAN[11] reported that beta-blockade provided symptomatic relief in patients with hypertrophic cardiomyopathy, but GOODWIN[6] found that dyspnoea was usually not improved. In a double-blind trial of propranolol and practolol compared with placebo in 16 patients[8] it was observed that dyspnoea improved only in patients whose effort tolerance was markedly reduced, and the improvement occurred only in response to propranolol; propranolol was also superior (in a daily dosage of 320 mg.) to practolol (800 mg. daily) in the relief of angina and palpitation. However, two of the eight patients studied in an acebutolol trial by LEWIS et al.[10] had a markedly reduced effort tolerance and were not improved in terms of reduction of left-ventricular end-diastolic pressure or A wave after beta-blockade. The present findings bear out previous reports to the effect that the symptom of dyspnoea is poorly relieved by beta-blockade, despite the reduction in exercising heart rate and blood pressure, but do not confirm the observation of HUBNER et al.[8] that benefit is achieved in severely limited patients.

The reduction in left-ventricular outflow gradient following intravenous propranolol during cardiac catheterisation, including blockade of the isoprenaline-induced augmentation, has been demonstrated on many occasions[5, 12]. It is of interest that the reduction in left-ventricular end-diastolic pressure produced by acute beta-blockade[10, 12, 13] has been shown in the present study to persist over several years. Despite this, the patients feel better during placebo treatment and are not relieved of their dyspnoea by oxprenolol, nor is their exercise tolerance improved. The situation is further confused by the finding that syncope and angina are improved by oxprenolol, even in some patients who feel better during placebo treatment. It appears that acute haemodynamic changes are an unreliable guide to symptomatic improvement.

The response to exercise of patients with hypertrophic cardiomyopathy was found by EDWARDS et al.[3] to include undue tachycardia, and the results in the present study are to some extent in keeping with this finding. The heart rates of the intermediate group were appreciably higher at low work loads than those of patients with mild disease (cf. Table 3). The reduction in effort tolerance after beta-adrenergic blockade with oxprenolol in this study is similar to that which occurs in normal subjects treated with propranolol[4]. The exercise study reported by EDWARDS et al.[3] involved only one exercise level in some patients, and therefore no assessment of total work is available for comparison. However, the authors note that there was a significant reduction in cardiac work after intravenous, though not after oral, propranolol. This is in contrast with the present finding that exercise performance after oral oxprenolol was reduced.

Although nine of our 20 patients reported experiencing angina from time to time, this was not the limiting factor during the exercise tests in most patients. However, the marked increase in the severity and frequency of angina in four patients when placebo tablets were substituted for their propranolol provides strong support for the view that beta-blockade is effective in relieving this symptom. In all four cases the pain was reduced by oxprenolol. In both the acute oral and the prolonged oral trials, all patients except those whose condition was very severe (and who were in any case extremely limited) performed an appreciably lower total amount of exercise during the period of treatment with an active drug. Alleviation of angina after treadmill exercise was also reported by COHEN and BRAUNWALD[2] in a single-blind trial of propranolol in patients with hypertrophic cardiomyopathy. VAN DER WALL[12] likewise reported successful relief of angina with oral propranolol.

A consistent finding in the present study was the reduction in blood pressure at the maximal work load (cf. Figure 4). This can be explained as being due to the increase in gradient known to be produced by exercise in patients with cardiomyopathy or to a fall in stroke volume secondary to undue tachycardia[5, 13]. EDWARDS et al.[3] also found that the rise in systolic brachial artery pressure was less than in normal subjects, although the design of their study did not show a reduction in blood pressure at the maximal work load as compared with the value measured at the previous level of exercise.

Six of our 20 patients reported syncope, and there was a considerable increase in this symptom in four patients when placebo was substituted for the propranolol which they had previously been receiving. In all cases this symptom was relieved when oxprenolol was substituted for placebo. The presumptive association between syncopal episodes and sudden death from arrhythmia in cases of hypertrophic cardiomyopathy lends added clinical importance to these findings. Although patients may feel better on placebo, or when they are taking no drugs at all, the results obtained in this study support the view expressed by GOODWIN and OAKLEY[7] that all patients suffering from hypertrophic cardiomyopathy should be treated with beta-blockers. The only contra-indications at present appear to be the presence or provocation of heart failure and the occurrence of asystole, giving rise to syncopal attacks, as documented by JOSEPH et al.[9].

In the five patients in the intermediate group who were switched from propranolol to oxprenolol there was no significant difference in the maximum exercising heart rate, in total work performed, or in systolic or diastolic blood pressure on exercise (or, indeed, at rest). Furthermore, there was no difference in the patients' preference for either drug; on the other hand, the two beta-blockers were preferred to placebo. Only one patient had previously been treated with practolol: he experienced an increase in both angina and dyspnoea during two weeks of placebo medication, but he could not detect any difference in symptoms or exercise tolerance while taking oxprenolol. It seems clear that treatment with beta-blockers is the important step, rather than the choice of a particular drug.

Conclusions

1. This study confirms previously reported findings that the left-ventricular outflow gradient and its augmentation by intravenous isoprenaline are reduced or abolished by intravenous propranolol at rest.
2. The left-ventricular end-diastolic pressure, both "pre-A wave" and "post-A wave", can be reduced by intravenous beta-blockade, and this effect can be demonstrated over many years.
3. There was no evidence that dyspnoea was improved by beta-blockade in the group as a whole, although individual patients reported some improvement; other patients reported an increase in dyspnoea while taking oxprenolol.
4. Blood pressure at rest was unaffected by oral oxprenolol, but during exercise it was rather lower when the patients had taken oxprenolol than when they had taken placebo or no drugs at all.
5. Acute haemodynamic changes are a poor guide by which to judge symptomatic improvement produced by beta-blockers.
6. Treatment with beta-blocking drugs tends to reduce maximum effort tolerance.
7. Substitution of placebo for beta-blocking drugs resulted in severe deterioration of symptoms, particularly syncope and angina, in about half the patients.
8. It is recommended that all patients with hypertrophic cardiomyopathy who experience syncope, palpitation, or tachycardia should be treated with beta-blocking drugs unless bradycardia or asystole is implicated in their symptoms.

Acknowledgments

This investigation would not have been possible without the invaluable technical assistance of Mrs. G. Lockton. Thanks are due to Miss G. Alton of the Pharmacy, Guy's Hospital, who gave great help, and to Dr. D. Deuchar for allowing patients under his care to be included in the study. We would particularly like to thank the patients for their cooperation. Dr. D. M. Burley of ciba laboratories, Horsham, kindly provided the Trasicor and placebo tablets.

References

1 Cherian, G., Brockington, I. F., Shah, P. M., Oakley, C. M., Goodwin, J. F.: Beta-adrenergic blockade in hypertrophic obstructive cardiomyopathy. Brit. med. J. *i*, 895 (1966)
2 Cohen, L. S., Braunwald, E.: Amelioration of angina pectoris in idiopathic hypertrophic subaortic stenosis with beta-adrenergic blockade. Circulation *35*, 847 (1967)
3 Edwards, R. H. T., Kristinsson, A., Warrell, D. A., Goodwin, J. F.: Effects of propranolol on response to exercise in hypertrophic obstructive cardiomyopathy. Brit. Heart J. *32*, 219 (1970)
4 Epstein, S. E., Robinson, B. F., Kahler, R. L., Braunwald, E.: Effects of beta-adrenergic blockade on the cardiac response to maximal and submaximal exercise in man. J. clin. Invest. *44*, 1745 (1965)
5 Flamm, M. D., Harrison, D. C., Hancock, E. W.: Muscular subaortic stenosis. Prevention of outflow obstruction with propranolol. Circulation *38*, 846 (1968)
6 Goodwin, J. F.: Congestive and hypertrophic cardiomyopathies. A decade of study. Lancet *i*, 731 (1970)
7 Goodwin, J. F., Oakley, C. M.: The cardiomyopathies. Brit. Heart J. *34*, 545 (1972)
8 Hubner, P. J. B., Ziady, G. M., Lane, G. K., Hardarson, T., Scales, B., Oakley, C. M., Goodwin, J. F.: Double-blind trial of propranolol and practolol in hypertrophic cardiomyopathy. Brit. Heart J. *35*, 1116 (1973)
9 Joseph, S., Balcon, R., McDonald, L.: Syncope in hypertrophic obstructive cardiomyopathy due to asystole. Brit. Heart J. *34*, 974 (1972)
10 Lewis, B. S., Mitha, B. S., Bakst, A., Purdon, K., Gotsman, M. S.: Haemodynamic effects of beta blockade in hypertrophic cardiomyopathy using Sectral (acebutolol; M & B 17803A). Cardiovasc. Res. *8*, 249 (1974)
11 Sloman, G.: Propranolol in management of muscular subaortic stenosis. Brit. Heart J. *29*, 783 (1967)
12 Wall, E. van der: Hypertrophic obstructive cardiomyopathy. Evaluation of treatment by invasive and non-invasive methods. Thesis, Groningen 1972
13 Webb-Peploe, M.: Management of hypertrophic obstructive cardiomyopathy by beta-blockade. In Wolstenholme, G. E. W., O'Connor, M. (Editors): Hypertrophic obstructive cardiomyopathy, Ciba Found. Study Group No. 37, p. 103 (Churchill, London 1971)

Discussion

W. Schweizer: Dr. Sowton's excellent paper prompts me to raise one question right away. Should beta-blockers be given to all patients with hypertrophic cardiomyopathy, irrespective of the pressure gradient or of the presence of angina? I am very curious to hear the views of others on this point.

R. C. Tarazi: Since it has been suggested that beta-blockers are suitable for use in all patients with idiopathic subaortic stenosis, I'd like to ask Dr. Sowton about his experience in this connection. Is there any particular relationship between hypertension and the incidence of idiopathic subaortic stenosis? I ask this because both may be associated with a markedly hyperkinetic circulation and with left-ventricular hypertrophy, and because beta-blockers have been found to be of benefit in both conditions. I was also interested to hear Dr. Sowton say that propranolol led to a reduction in left-ventricular end-diastolic pressure when given intravenously. We made the same observation in a patient with hypertension who had pronounced left-ventricular hypertrophy; when he was given a vasodilator, his cardiac output increased markedly and his pulmonary artery and wedge pressure also rose. In this patient, both the systemic blood pressure and the pulmonary blood pressure fell when cardiac output had been reduced and the circulation quietened down by adding propranolol.

E. Sowton: I'm not sure, Dr. Tarazi, that we are talking about the same disease. You were referring to idiopathic subaortic stenosis, but the patients I was discussing didn't necessarily have stenosis beneath the aortic valve; they included patients with gross thickening of the left-ventricular wall, but with no outflow stenosis. Leaving this question aside, I agree that hypertension certainly can occur simultaneously in a patient with hypertrophic cardiomyopathy. Though I don't think that the two diseases are the same, it is not easy to distinguish between them in the early stages. If a patient presents with hypertension, but also has hypertrophic cardiomyopathy, the hypertrophic myopathy may easily be missed. I therefore regard it as essential to have left-ventricular angiograms and coronary arteriograms as an aid in establishing the diagnosis. Apart from that, I would agree with you that the effects of intravenous beta-blockade are very similar in the early stages.

J. Brod: With reference to the problem of diagnosis, Dr. Sowton, the symptoms as you described them could of course lead to a mistaken diagnosis of aortic stenosis. The only easily ascertainable features distinguishing the two conditions are the full pulse and the high cardiac output. Can you suggest any other clinical criteria by which to distinguish these two conditions from each other in a mild case?

E. Sowton: The main clinical guideline, of course, is the carotid pulse, which is sharp-rising and jerky in hypertrophic cardiomyopathy and, if anything, reduced and sustained in mild aortic stenosis. The way in which the diagnosis was established in these patients was by left-ventricular angiography and catheterisation.

D. W. Gau: May I also ask a very elementary question? In your paper, Dr. Sowton, you said that the S-T depression in your cases was not in fact due to coronary artery disease. How does this tie up with what Dr. Taggart told us in his paper? Though your patients did not have coronary artery disease, couldn't it have been ischaemia or something else that was causing the S-T depression?

E. Sowton: I believe that the presence of S-T depression in a patient not receiving any drugs at all indicates left-ventricular ischaemia. I am not at all convinced that, if you manage to correct the S-T depression, this means that you have in fact improved the left-ventricular ischaemia. There is also no doubt that you can have S-T depression of the ischaemic type with clear coronary arteries. All the patients in the series I reported upon had had coronary arteriography to establish that their symptoms were not due to coronary artery disease, and it's a typical finding in hypertrophic cardiomyopathy that the coronary arteries are unusually clear and large, i.e. pipe-stem arteries. Despite this, the patients suffer from angina and they show S-T depression in the E.C.G. Presumably, the reason is either small-vessel disease, obstruction at the end of the line due to myocardial hypertrophy, or possibly some metabolic cause.

J. SOMERVILLE: I am a little concerned about Dr. SOWTON's grading of this disease on the basis of angiocardiography. Perhaps I misunderstood what he said, but I don't think you can grade the disease from the appearance of the left-ventricular angiocardiogram, because you can alter its appearance by giving, for example, a little isoprenaline; you may even get an alteration if the patient is mildly excited. Another comment I'd like to make is that hypertrophic cardiomyopathy doesn't necessarily become more severe with increasing age, as Dr. SOWTON's Table 1 so beautifully indicated. I suspect that the reverse is true and that the most severe cases are to be found in the younger age groups. Finally, Dr. SOWTON, did you try to stimulate gradients in each of your cases? I ask this because I'd like to know whether the S-T changes on effort which improved in response to the beta-blocker occurred in those patients who had or did not have a gradient. Or was the improvement perhaps uniform and unrelated to gradient? Did the occurrence of angina and syncope differ, depending on whether the patients were ones who had a gradient at rest or only in response to stimulation?

E. SOWTON: We did try to stimulate gradients in all the patients. The S-T changes on exercise were not related to the presence or absence of a gradient. Nor were angina and syncope related to the presence of a gradient, whether spontaneous or induced by isoprenaline. As for your comments on the method of grading, Mrs. SOMERVILLE, the severity of the disease was graded on the basis of the patient's symptomatology and of the angiographic findings in the unstimulated state. I accept that it is not possible to grade the actual progression of a disease in this way, but I didn't claim to be able to do that. If the patients are graded incorrectly into the groups, the conclusions are not changed by shifting them into other groups. Finally, the table of mine to which you referred did not show that the disease became more severe in the same patients as they got older. It simply showed that, on the average, the patients with a lot of thick, hypertrophied muscle had higher gradients and were older. Except for one patient who has had several catheterisations, I have no data on the progression of hypertrophic cardiomyopathy. I agree that many young patients have a severe form of the disease. Perhaps they die sooner, so that the disease does not necessarily get better as they get older. This would mean that, if you take an elderly group of patients, a lot of the severe cases will already have died. I think we simply do not know how the thing works out in terms of progression of the disease.

J. SOMERVILLE: I am still not entirely satisfied, Dr. SOWTON, because I am concerned about the justification for using beta-blockers in the absence of angina and syncope and without gradients. Is it simply that the S-T changes really look better on the E.C.G. after treatment with a beta-blocker?

E. SOWTON: Where a patient develops syncope, it may, of course, be the first and only attack he ever has. In the absence of drugs, S-T segment depression on exercise must, I think, be interpreted as indicative of ischaemia. This, together with other abnormalities on exercise, is so common that you have to look for a reason not to use beta-blockers rather than for reasons to justify their use.

E. K. CHUNG: I have just a brief comment to make on your Figure 6, Dr. SOWTON. The S-T changes in the first strip were a lot different from those in the second strip. At the same heart rate of 120 beats/min., the S-T changes with no drug differ markedly from the changes following an oral placebo. How do you explain this?

E. SOWTON: I agree that the S-T segment changes in this instance were different, but it's quite a common finding that the conduction and the S-T segment changes vary from time to time in one and the same patient. They are not consistently repeatable in every exercise test.

J. SOMERVILLE: A number of these patients with hypertrophic cardiomyopathy – particularly the young ones and, of course, also some of the middle-aged ones – end up by entering the congestive phase. In view of this, there may be some patients – who don't have angina or syncope and who have no gradient of any importance on catheterisation – whom it would be wrong to treat blindly with beta-blockers just in order to improve the look of their electrocardiograms. This is the main point that I have been trying to get across to Dr. SOWTON.

E. SOWTON: I think that these patients could be treated with a beta-blocker when they are in the hypertrophic obstructive phase, but of course I wouldn't give them this treatment if and when they enter the congestive phase. I don't think anybody in this room would support the view that the treatment you start today is necessarily going to be the treatment you give five or ten years later.

J. SOMERVILLE: But what I am saying is that you may possibly hasten the onset of this congestive phase by the unnecessary administration of beta-blockers.

E. SOWTON: Yes, it is true that you might possibly do that. It is also possible that you will delay the onset of the congestive phase. Since neither of us has the requisite evidence, and since I at least have shown that the left-ventricular end-diastolic pressure drops, I prefer my interpretation to yours, Mrs. SOMERVILLE.

P. TAGGART: In the United States, a fall-off in systolic pressure is regarded as being probably the most important of all indications for stopping an exercise test. This is taken as a sign that the left ventricle is blowing up, and that the patient is heading for syncope at the best or for something a lot more dangerous at the worst. Admittedly, this applies to people suspected of suffering from coronary heart disease. I noticed, Dr. SOWTON, that you carried on with the test for quite a while after the fall in systolic pressure had started. Did those patients of yours who were subject to syncope suffer any ill-effects immediately after the test, or was it in fact quite safe to press on?

E. SOWTON: Firstly, I already knew that these patients did not have coronary artery disease, because we had done coronary arteriograms on them. Secondly, we didn't progress beyond angina in any of them. Thirdly, none of them did in fact have syncope. This problem worried me, too, at first, but I'm glad to say we had no trouble at all.

H. BRICAUD: In our research unit we have been making a comparative study of patients treated with beta-blockers and patients subjected to surgery. In this investigation we have been taking account not only of haemodynamic data, but also of data on heart work and cardiac volume. Although the study has now been in progress for 18 months, I must say that we are still extremely uncertain as to whether beta-blockers deserve preference over surgery. We have to some extent gained the impression that surgery yields better results, but we have no statistical evidence to substantiate this impression. I might add that, according to our figures, the incidence of sudden untoward episodes has not been reduced by the use of beta-blockers.

E. SOWTON: In this presentation of mine I excluded the patients who had had surgery. Among the other patients with hypertrophic cardiomyopathy that we have studied there have been quite an appreciable number who were operated upon after beta-blockade had failed to afford them relief from syncope or particularly from angina. On the whole, the results of surgery have been disappointing, except in cases where there was a very high outflow gradient. We haven't enough long-term follow-up information to say anything about the late results. One point which seems to emerge from our results is that surgery is most successful when left bundle-branch block has been produced at operation.

W. SCHWEIZER: Thank you, Dr. SOWTON. I propose that we now pass on to the next question: If you decide to employ beta-blockers in hypertrophic cardiomyopathy, which drug do you choose, and in what dosage do you give it? Are there any differences between the various beta-blockers in this indication, and is it true that a very high dosage has to be used?

E. SOWTON: The only firm data I have on this question of differences between the beta-blockers are those from a comparison between six patients who first had propranolol, were subsequently given oxprenolol (®Trasicor), and then went back to propranolol. There were no differences in the exercise tests, nor did the patients express a preference for one of the two drugs. The one patient who had been receiving practolol before being switched to oxprenolol couldn't detect any difference either. I have no information on other beta-blockers. As for the question of dosage, most of our patients have been on daily doses around the 240–320 mg. level. Both from reports in the literature and from our own experience, I have the strong impression that a remarkable degree of tolerance is liable to occur in this type of patient when higher doses are given, but we don't find these doses necessary.

W. SCHWEIZER: Now a further question: How can we or the general practitioner assess response to treatment? Must this be done solely by reference to the symptoms, or are non-invasive methods available which indicate whether or not the treatment is proving effective?

E. SOWTON: Since we don't understand the basic nature of the disease, I believe that the symptoms offer the best guide. Regarding non-invasive methods, perhaps the electrocardiogram is a help, and possibly chest X-rays too. I don't think that the apex cardiogram has proved helpful. In the hands of the group at the Hammersmith Hospital and the Harefield Hospital echocardiography and ultrasound studies certainly seem to be proving very encouraging.

R. C. TARAZI: From the literature and from what I have also heard at various meetings, I gather that marked changes in systemic blood pressure may occur as the gradient – in those patients who have a gradient – disappears and as the full force of ventricular contraction is then transmitted to the arterial side. I have not seen this in the few patients I have had. Have you, Dr. SOWTON, encountered this at all?

E. SOWTON: No, we have not met with this. We have seen tolerance develop, in which the outflow gradient disappeared at catheterisation; and we have seen the murmur vanish after intravenous propranolol. But, in spite of oral propranolol, the murmur was present again 48 hours later; it was abolished by an increase in the dose, but it returned once again despite maintenance of the bigger dose.

Some metabolic effects of beta-blockade on temperature regulation and in the presence of trauma

by M. E. CARRUTHERS*, P. TAGGART**, P. D. SALPEKAR***, and J. A. GATT***

The potency of beta-blocking compounds in decreasing the direct action of excessive sympathetic drive on the heart has tended to divert attention from their other physiological and biochemical effects. Even apart from their influence on the bronchi, few would claim that the beta-blockers are cardio-specific. Moreover, it is from consideration of the so-called "side effects" of these compounds that some of their most interesting future clinical applications may emerge. As examples, we should like to describe how beta-blockers may affect temperature regulation and the metabolic responses to trauma.

1. Temperature regulation

Introduction

In accounts of the mechanisms which maintain body temperature at a relatively constant level in widely varying hostile thermal environments[1], the emphasis is usually placed on regulation of the processes of heat loss by sweating and by circulatory changes in the skin, rather than on alterations in heat production due to increases in muscular activity and variations in metabolic rate. It therefore seemed appropriate to extend our previous studies of the biochemical effects of thermal stress[5] by examining the sequence of events occurring when healthy adults were exposed to low temperatures.

Method

Three moderately intense temperature-lowering situations were studied. Eleven healthy young males from the Hampstead Swimming Club, who enjoy swimming in their local pond all the year round, had blood samples taken before and after their pre-Christmas swim, the water temperature averaging 7°C. Readings of blood pressure were made at the same time, and the electrocardiogram recorded throughout the five-minute swim by means of a radio-electrocardiographic transmitter enclosed in a waterproof polythene urine-collection bag tucked into the bather's swimming trunks. Reception remained good, even though the transmitter was submerged for most of the time.

In a second experiment, five healthy young male subjects, wearing only shirts and shorts, were seated in the wind tunnel of the Medical Research Council

---

* Department of Chemical Pathology (Research), St. Mary's Hospital, London, England.
** Department of Cardiology, The Middlesex Hospital, London, England.
*** Multiple Injury Unit, Royal Infirmary, Preston, England.

Department of Human Physiology in Hampstead for two and a half hours. Blood samples were taken before and after this exposure to moving air at an ambient temperature of 10 °C.

In the third experiment, after having been given placebo tablets, six healthy, fasting, young male subjects had their pulse, blood pressure, and oral temperatures taken before and at the end of a 30-minute period which they spent sitting similarly lightly clad in a laboratory "cold room" at 4 °C. This experiment was repeated an hour and a half later after an oral dose of 40 mg. oxprenolol (®Trasicor).

Results

During the swimming periods, no significant changes in heart rate or E.C.G. pattern were observed, although in some subjects a brief decrease in heart rate occurred within the first few seconds of immersion. Both the systolic and diastolic pressures rose markedly in all subjects, maximum levels of up to 170/120 mm. Hg being recorded in subjects who were normotensive at rest (Table 1).

The noradrenaline level, though already relatively high in the pre-swim samples, rose well above the normal resting range. The adrenaline also in-

Table 1. Heart rate, blood pressure, plasma catecholamines, free fatty acid levels, and glucose concentrations in 11 volunteers before and after swimming for five minutes in cold water.

Variable	Time	N	Mean	S.E.	t	P
Heart rate (beats/min.)	Before	11	89.1	3.43	1.41	Not significant
	After	11	92.3	4.01		
Systolic blood pressure (mm. Hg)	Before	11	133	5.70	7.73	<0.001
	After	11	157	5.57		
Diastolic blood pressure (mm. Hg)	Before	11	84.6	3.12	7.42	<0.001
	After	11	99.1	3.98		
Total catecholamines (mcg./litre)	Before	11	0.95	0.08	7.63	<0.001
	After	11	1.79	0.16		
Noradrenaline (mcg./litre)	Before	11	0.91	0.10	3.13	<0.01
	After	11	1.43	0.08		
Adrenaline (mcg./litre)	Before	11	0.11	0.03	2.26	<0.05
	After	11	0.22	0.03		
Free fatty acids (μmol./litre)	Before	11	786	98.0	2.55	<0.05
	After	11	599	74.1		
Glucose (mg./100 ml.)	Before	11	83.2	3.52	1.01	Not significant
	After	11	80.8	4.40		

Table 2. Plasma catecholamines, free fatty acid levels, and glucose concentrations in five volunteers before and after two and a half hours spent in a wind tunnel.

Variable	Time	N	Mean	S.E.	t	P
Total catecholamines	Before	5	0.65	0.04		
(mcg./litre)	After	5	1.07	0.08	3.71	<0.025
Noradrenaline	Before	5	0.60	0.04		
(mcg./litre)	After	5	0.93	0.04	4.76	<0.01
Adrenaline	Before	5	0.06	0.01		
(mcg./litre)	After	5	0.14	0.05	1.58	Not significant
Free fatty acids	Before	5	796	46.3		
(μmol./litre)	After	5	1,362	180.7	2.89	<0.05
Glucose	Before	5	61.6	3.31		
(mg./100 ml.)	After	5	62.0	2.53	0.34	Not significant

Table 3. Heart rate, blood pressure, plasma catecholamines, free fatty acids, and glucose concentrations in six placebo-treated volunteers before and after 30 minutes spent in a "cold room".

Variable	Time	N	Mean	S.E.	t	P
Heart rate	Before	6	73.7	2.88		
(beats/min.)	After	6	60.5	1.76	9.56	<0.001
Systolic blood pressure	Before	6	117	6.0		
(mm. Hg)	After	6	134	13.9	2.82	<0.05
Diastolic blood pressure	Before	6	77.5	7.58		
(mm. Hg)	After	6	96.7	12.91	3.14	<0.05
Total catecholamines	Before	6	0.84	0.08		
(mcg./litre)	After	6	1.32	0.17	6.19	<0.005
Noradrenaline	Before	6	0.78	0.10		
(mcg./litre)	After	6	1.26	0.14	6.73	<0.005
Adrenaline	Before	6	0.06	0.04		
(mcg./litre)	After	6	0.06	0.05	0.13	Not significant
Free fatty acids	Before	6	422	71.9		
(μmol./litre)	After	6	545	98.7	2.49	<0.05
Glucose	Before	6	80.7	4.51		
(mg./100 ml.)	After	6	84.7	6.97	0.61	Not significant
Body temperature	Before	6	36.6	0.09		
(°C.)	After	6	36.0	0.14	9.46	<0.001

Table 4. Heart rate, blood pressure, plasma catecholamines, free fatty acid levels, and glucose concentrations in the same six volunteers as in Table 3 when the "cold room" experiment was repeated with oxprenolol instead of placebo.

Variable	Time	N	Mean	S.E.	t	P
Heart rate (beats/min.)	Before After	6 6	62.7 50.3	3.20 2.34	7.62	<0.001
Systolic blood pressure (mm. Hg)	Before After	6 6	111 112	7.53 5.16	0.18	Not significant
Diastolic blood pressure (mm. Hg)	Before After	6 6	69 81	7.03 8.01	2.76	<0.05
Total catecholamines (mcg./litre)	Before After	6 6	0.76 1.93	0.04 0.64	4.45	<0.01
Noradrenaline (mcg./litre)	Before After	6 6	0.71 1.76	0.10 0.55	4.69	<0.01
Adrenaline (mcg./litre)	Before After	6 6	0.08 0.18	0.08 0.16	1.33	Not significant
Free fatty acids (μmol./litre)	Before After	6 6	398 343	41.7 34.0	2.54	<0.05
Glucose (mg./100 ml.)	Before After	6 6	82.8 77.3	2.83 7.65	0.77	Not significant
Body temperature (°C.)	Before After	6 6	36.4 34.8	0.13 0.24	13.9	<0.001

creased but was still within the normal range. The level of the free fatty acids fell, whereas cholesterol increased. No changes in the triglyceride or glucose concentrations occurred.

The subjects in the wind tunnel showed increases in plasma noradrenaline and free fatty acids, but no other significant biochemical changes (Table 2).

In the "cold room" experiment, the volunteers' temperatures decreased more when they had been given oxprenolol, despite a greater increase in noradrenaline. During the placebo period, free fatty acid levels rose, as in the "wind tunnel" experiment, and fell slightly from a low level during the oxprenolol period. The increase in systolic blood pressure in particular was greater in the placebo period (Tables 3 and 4).

Discussion

The deliberately limited amount of muscular activity occurring while the swimmers were in the water appeared insufficient to produce the required increase in heat production. It is therefore suggested that the rapid mobilisation of free fatty acids by noradrenaline and their metabolism at various sites, including especially the liver and muscles, were responsible for supplying a major proportion of the abruptly raised thermal demands. Such a switch from carbohydrate to fat utilisation can greatly increase the body's metabolic rate, since fat has more than twice the caloric value of carbohydrates.

That the catecholamine response to cold consisted mainly of an increase in noradrenaline secretion was shown by the rise in both systolic and diastolic blood pressure, which was unaccompanied by any appreciable acceleration of the heart rate. This would seem to indicate that open-air swimming in countries with a climate like that of England might well be inadvisable for hypertensive subjects.

The tests in the wind tunnel and those with unblocked subjects in the "cold room" revealed the pattern of longer-term adaptation to more moderate cooling. The noradrenaline levels were again raised, but only to about 75% of those recorded in the swimmers, and in both these more prolonged tests the increase in free fatty acids suggests that mobilisation had exceeded utilisation. The constancy of the blood sugar concentrations in all three groups is a further indication of the relatively minor part played by alterations in carbohydrate metabolism in the regulation of body temperature in cold surroundings.

From the results of these experiments, and from similar biochemical changes reported in other studies on people swimming in cold water[3], it would appear that the thermogenic responses mediated by noradrenaline – which result in the release of free fatty acids from adipose tissue and in their increased utilisation – may persist into adult life.

The "cold room" experiment showed that, as in emotional stress[4], oxprenolol abolishes the rise in free fatty acids and also suppresses the "cold pressor" response.

Further investigations of the relationships between catecholamine secretion rates and body temperature are in progress, and the influence of beta-blocking compounds on diurnal rhythms is being studied. It is suggested that the "cold room" experiment described is a safe, convenient, and reproducible means of testing the efficacy of beta-blockers in high-noradrenaline states and may also provide an explanation for the reduced tolerance of cold conditions experienced by some subjects.

2. Trauma

Introduction

HALMÁGYI et al.[2], while investigating hypovolaemic shock in pigs, found that beta-blockade improved survival in some groups, apparently by making the animals poikilothermic and by preventing the hyperpyrexia which killed a proportion of the unblocked animals even after reinfusion of lost blood. This interesting observation was highlighted by the findings of a group led by Mr. P. D. SALPEKAR working at the Multiple Injury Unit of the Royal Infirmary in Preston. During the period 1964–1973 metabolic responses were studied in 50 severely injured patients consisting mainly of road accident cases, including particularly those requiring external fixation of the chest wall and tracheostomy. Clinically, these patients appeared to reach a "crisis point" some 24–48 hours after the accident. At that stage the blood pressure, which had already been restored by transfusions and various anti-shock measures, began to oscillate around the new level. Various ventricular arrhythmias also developed, and these cardiac irregularities were evidently one of the most serious of the multiple organ failures.

Biochemically, there was evidence of gross sympathetic overactivity. This first presented as glycosuria, which occurred in the majority of these seriously injured cases. From being normal on admission, the blood glucose frequently rose on the second and third days to levels between 150 and 250 mg./100 ml. At the same time, the free fatty acid levels also showed increases. As part of the "sick-cell syndrome", plasma sodium fell and potassium rose, whereas the reverse changes occurred in the red-cell electrolytes. Periods of hypoxic hypoxia, tissue hypoxia, respiratory acidosis, and then metabolic alkalosis further complicated the picture. On the fourth or fifth day, this combination of events frequently led to death from ventricular arrhythmias, or from one of the other multiple organ failures.

As evidence mounted that these deaths were associated with high levels of sympathetic activity, even when the patients were heavily sedated, it was decided two years ago by the Preston team to try out the beta-blocker practolol in cases of this kind. The compound was administered intramuscularly or intravenously, starting from the time when the blood pressure had been restored, i.e. usually 24–48 hours after the accident. The dosage was adjusted on the basis of clinical observation of the heart rate and the number of ectopic beats, together with the blood pressure readings. It was soon found that the death rate was more than halved from 13% to 6% (i.e. from 20 out of 150 cases not treated with the beta-blocker to three out of 53 in treated cases), mainly because of a reduction in the number of arrhythmias. Also, the metabolic evidence of sympathetic overactivity, such as elevated blood glucose and free fatty acid levels, was much less pronounced in patients treated with the beta-blocker. Though the evidence indicative of excessive and probably harmful sympathetic drive in these accident cases was strong, it was felt that the picture needed to be completed by performing plasma catecholamine estimations (at St. Mary's Hospital), as well as by comparing the action of practolol with that of another beta-blocking compound, oxprenolol.

Method
This experiment, which is still in progress, was designed to study plasma catecholamine and other metabolic indices in 12 patients admitted to the Multiple Injury Unit at Preston Royal Infirmary. Practolol and oxprenolol are being used in alternate cases.

Results
Of the three cases so far tested, each has shown a peak of plasma catecholamines – largely noradrenaline – at the previously observed crisis point occurring

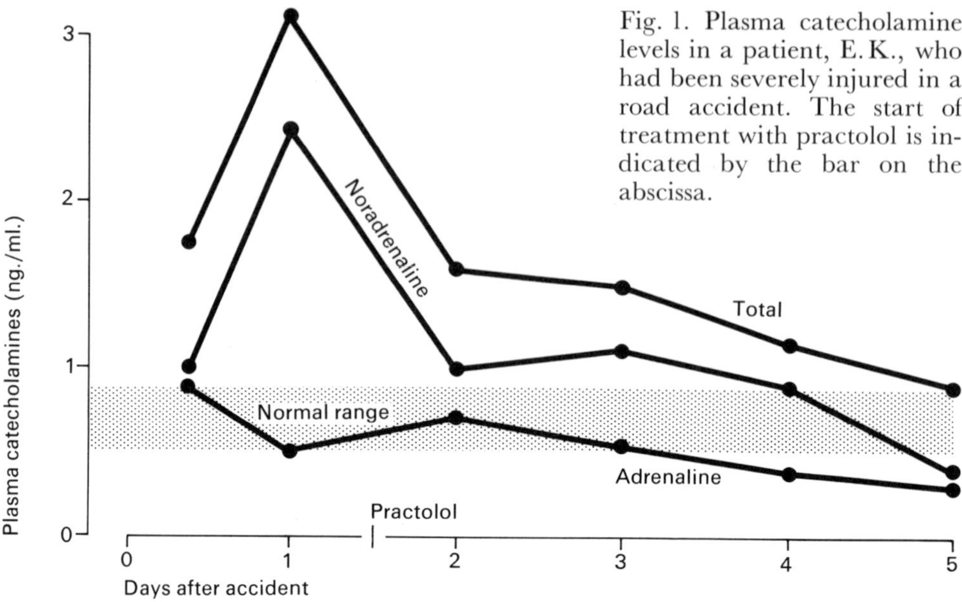

Fig. 1. Plasma catecholamine levels in a patient, E. K., who had been severely injured in a road accident. The start of treatment with practolol is indicated by the bar on the abscissa.

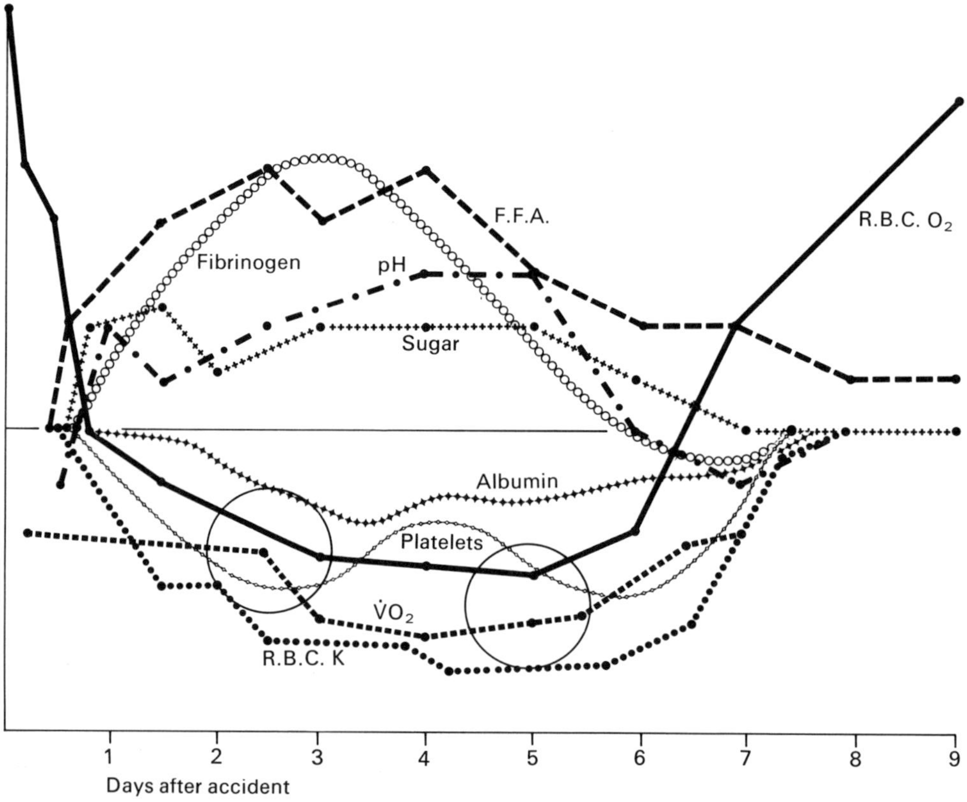

Fig. 2. Metabolic changes following severe injury in 50 cases not treated with beta-blockers. The first circle indicates a period of hypoxic hypoxia and the second a period of tissue hypoxia.

254

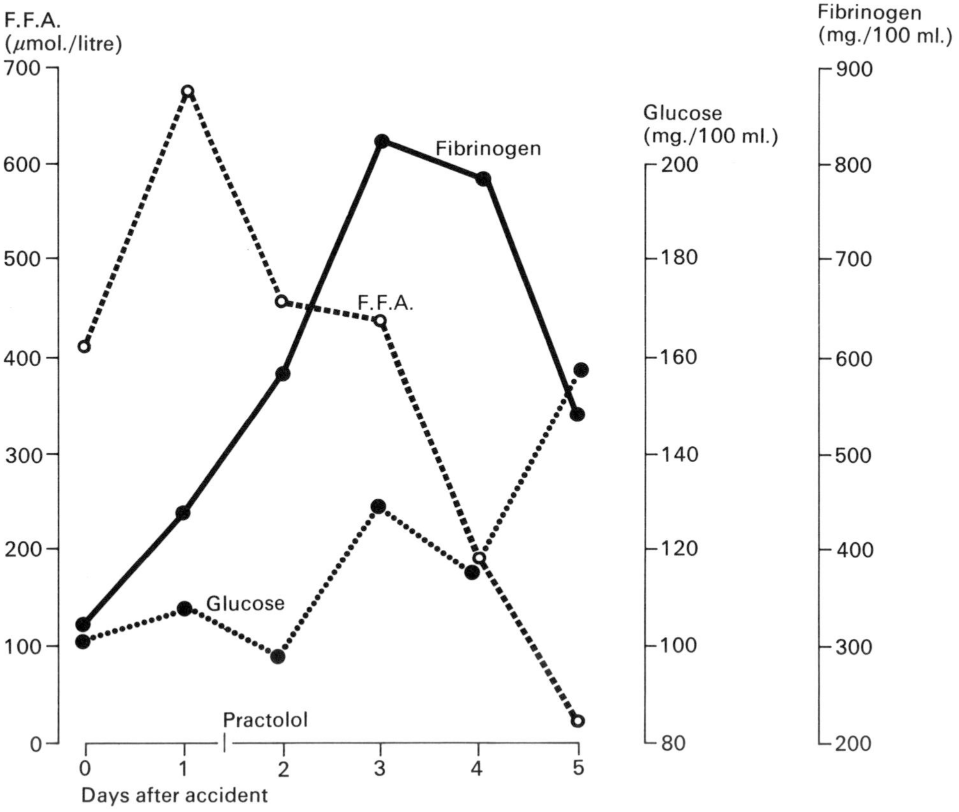

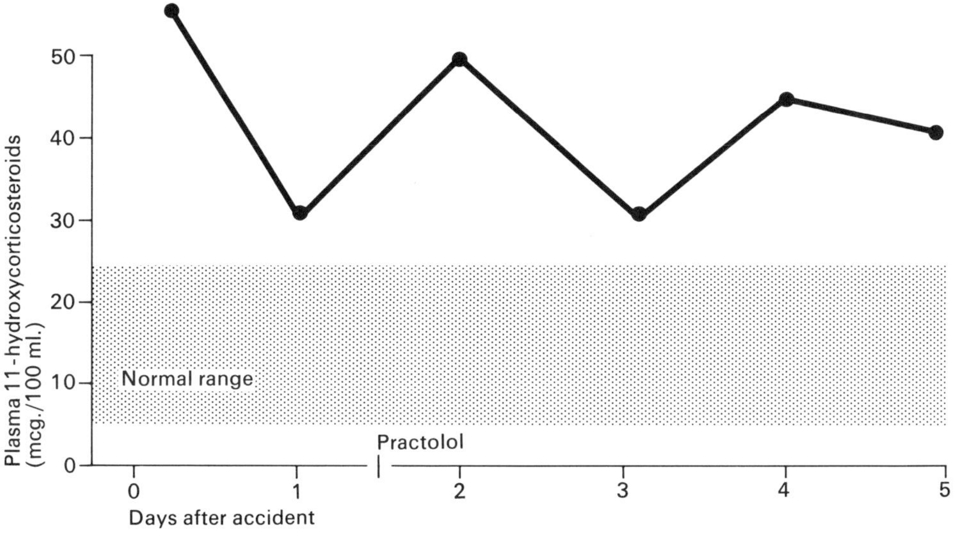

Fig. 3. Changes in free fatty acids, glucose, and fibrinogen in Patient E. K. following treatment with practolol.

Fig. 4. Plasma cortisol levels in Patient E. K.

between the first and second day (Figure 1). The levels were similar to those seen in a series of post-thoracic surgery cases investigated at the Middlesex Hospital, a report on which is to be published shortly. Many of the complex interrelated metabolic changes recorded in unblocked subjects (Figure 2), including in particular the elevations in free fatty acids (Figure 3), were abolished in both the present and previous cases treated with practolol, as well as in the single case treated with oxprenolol. The increases in plasma glucose were only partially suppressed, perhaps owing to the raised cortisol levels (Figure 4), and the elevations in the fibrinogen level appeared unaltered (Figure 3). Reductions in body temperature were also noted when beta-blockade was started.

Discussion

The clinical benefits derived from introducing treatment with beta-blockers at the crisis point in these severely injured patients were sufficiently great to render an unblocked control group ethically impossible at the present time in the Preston Unit. Although the series has yet to be completed, previous observations in cases treated with practolol, and the first results obtained in the new group, make it appear highly probable that beta-blocking compounds are extremely effective in suppressing the potentially lethal effects of sympathetic overactivity.

These drugs were originally used in largely empirical fashion in order to suppress cardiac arrhythmias and angina, but were then found to have a far wider and more useful spectrum of clinical and metabolic effects. New fields of application for such compounds now appear to be opening up, so that the surgeon may perhaps shortly be using them to produce hypothermia, for instance, during operations on the brain and heart, as well as for suppression of the metabolic responses to trauma. Beta-blockade is no longer the secret weapon of the physician!

References

1 CLOTHIER, J.G.: Work in hostile climates. Ann. occup. Hyg. *14*, 197 (1971)
2 HALMÁGYI, D.F.J., KENNEDY, M., VARGA, D.: Combined adrenergic receptor blockade and circulating catecholamines in haemorrhagic shock. Europ. surg. Res. *3*, 378 (1971)
3 STONE, E.A.: Swim-stress-induced inactivity: relation to body temperature and brain norepinephrine, and effects of d-amphetamine. Psychosom. Med. *32*, 51 (1970)
4 TAGGART, P., CARRUTHERS, M.: Suppression by oxprenolol of adrenergic responses to stress. Lancet *ii*, 256 (1972)
5 TAGGART, P., PARKINSON, P., CARRUTHERS, M.: Cardiac responses to thermal, physical, and emotional stress. Brit. med. J. *iii*, 71 (1972)

Discussion

S. H. Taylor: Could I ask Dr. Carruthers how much practolol and oxprenolol (®Trasicor) was given and by what route?

M. E. Carruthers: Initially, practolol – the drug used over the past three years – was given on clinical demand and by intravenous injection. The swings in the various metabolic parameters, however, suggest that this may not have been the best method of administration and that it was somewhat too erratic. We are looking now at the possibility of trying to ensure a more even form of beta-blockade which might prove more beneficial; this could perhaps be done by introducing the beta-blocking compound into the intravenous fluids which these subjects are on, or possibly by resorting to intramuscular injections.

G. Muiesan: What importance, Dr. Carruthers, do you ascribe to the evaluation of adrenosympathetic activity by estimating the concentrations of catecholamines in the plasma or in the urine? We have been estimating the plasma catecholamines – i. e. noradrenaline and adrenaline separately – in ischaemic patients before and after treatment with oxprenolol and have obtained results which are at least in partial agreement with those that you have presented. In an ischaemic patient who has not received oxprenolol the plasma noradrenaline and adrenaline levels rise with the onset of anginal pain. We found that, when we pre-treated our patients with oxprenolol, their effort tolerance was increased, and the intensity and duration of the angina were reduced, but the plasma adrenaline, and especially the plasma noradrenaline, levels became very much higher than before. It was our experience, however, that, if you induce angina in an ischaemic patient by means of atrial pacing, the noradrenaline and adrenaline concentrations are not increased, despite the fact that the patients feel anginal pain.

J. Hodler: Why, Dr. Carruthers, was practolol chosen for use in the experiments you have described? If I had had to make such a choice, I should definitely not have taken practolol – for one thing, because it doesn't inhibit lipolysis to anything like the same extent as certain other beta-blockers.

M. E. Carruthers: I entirely agree. I appeared on the scene only six months ago, and, when I found this gold-mine of clinical information and metabolic results, I suggested that a comparison should be made with different beta-blockers, and particularly with oxprenolol. As was mentioned in our paper, we are in fact now engaged on such a comparative study.

J. Brod: It was suggested by Nickerson* some 20 years ago that irreversible shock after severe trauma is due to exaggerated vasoconstriction in various organs, associated with a breakdown in their metabolism. Nickerson proposed that these patients should be treated with alpha-blockers, which he claimed were capable of improving the prognosis. You, Dr. Carruthers, accept the idea of Halmágyi, i. e. that the irreversibility of shock is due to adrenaline poisoning, and you therefore advocate that these metabolic consequences of trauma be combated with beta-blockers. Does this mean that people who are in grave shock should be given both alpha-blockers and beta-blockers? Or would this be dangerous?

M. E. Carruthers: During the initial stage of shock, when the blood pressure is excessively low, beta-blockers are obviously not indicated. The time to use beta-blockers seems to be when the blood pressure has been restored, and when a clinical crisis picture supervenes with a normal or even raised blood pressure and with ventricular arrhythmias. I don't think that adrenaline poisoning is the sole cause of this condition. I

---

* Nickerson, M.: Factors of vasoconstriction and vasodilatation in shock. J. Mich. med. Soc. *54*, 45 (1955)

believe that a combination of effects – possibly acting indirectly – is involved; it's not just adrenaline toxicity, but perhaps more particularly noradrenaline toxicity, with excessive levels of free fatty acids and a very wide range of metabolic changes.

A. Zanchetti: You know, Dr. Carruthers, that a number of patients receiving beta-blockers complain of cold hands and feet. Do you think this is due to interference with the thermoregulatory mechanisms you studied, or is it perhaps due to a relative preponderance of alpha-vasoconstrictor tone?

M. E. Carruthers: The work we have done seems to suggest that the normal mechanism responsible for increasing heat production when people are exposed to cold conditions is put out of action by beta-blockers. But this mechanism – or its absence – only appears to become important when the patients are exposed to fairly severe temperature-lowering conditions. In the patients who were being treated for severe trauma, and who were in fact lying fairly exposed in the intensive care unit, we also found that, particularly during the winter months, there tended to be some lowering of body temperature when beta-blockade was started.

W. Schweizer: Looking at it from the ethical viewpoint, Dr. Carruthers, would you now suggest that a heavily injured man should receive beta-blockade?

M. E. Carruthers: Yes. Assuming that administration of the beta-blocker were correctly timed, I would. I would like to be given a beta-blocker myself if I were in that position.

S. H. Taylor: But surely a word of caution should be voiced here. The very survival of a patient who is severely injured is probably critically dependent upon his having full cardiac function, especially if he is hypovolaemic and in shock. I would debate whether under these conditions beta-receptor blockade of the heart is justifiable or even ethical. That is why I initially asked Dr. Carruthers what dose of beta-blocker he is using. I think this is terribly important. If he is merely trying to inhibit the effect of adrenaline on the heart, then this could be achieved with very small doses of beta-blockers. I feel that this must be made quite clear. Can Dr. Carruthers please tell us exactly how much oxprenolol was given, how frequently, and by what route? I'd also like the same information on any other beta-receptor antagonist that was being used.

M. E. Carruthers: These compounds are not being given during the hypovolaemic stage, but only after the blood pressure and circulating volume have been restored. Up to now, the clinical use of these compounds has been based on a close, hour-by-hour assessment of the patient's overall condition. Besides electrocardiographic monitoring, the team even have emergency facilities for determining free fatty acids, and they shortly hope to add emergency catecholamine estimations to their biochemical back-up in this intensive care situation. I myself am not directly concerned with the clinical care of the patients; but the people dealing with this side are experts in this field, and they are so impressed with the results obtained with beta-blockers that they regard it as not only ethical to use them but as unethical not to – given, of course, what they consider to be the correct clinical indications. I can only defer to their clinical judgment on the matter. As regards the question of dosage, Dr. Taylor, the dose was 1–2 mg. practolol, given intravenously as and when the clinical condition appeared to warrant it. Oxprenolol was given in a higher dosage, i.e. in an equivalent dose worked out with help from ciba laboratories in Horsham. Incidentally, may I emphasise once again that we are not simply blocking the effects of adrenaline on the heart but are also trying to block some of the other metabolic changes which may be potentially unfavourable.

K. D. Bock: Could you explain, Dr. Carruthers, precisely what kind of complication or disease you believe that you are able to prevent in these cases, if it is not shock?

M. E. CARRUTHERS: From my clinical colleagues I gather that they are trying to prevent the syndrome of what they call "multiple organ failure", in which the cardiac manifestations are the most important but in which renal manifestations also play a part. This is the ultimate aim of using beta-blockers in these cases and their use appears in fact to be justified by the reduction in the mortality rate that was achieved in quite a large series. The mortality dropped from 20 out of 150 before the use of beta-blockers to three out of 53, even after only intermittent dosage with practolol had been introduced.

Metabolic effects of beta-blockers

by J. HODLER*

A very thorough review of the influence exerted by the sympathico-adrenal system on metabolic processes has been given by HIMMS-HAGEN [24] in particular. This influence manifests itself in an extremely wide variety of ways. Active substances belonging to the sympathico-adrenal system affect carbohydrate, fat, and – to a lesser extent – protein metabolism as well. In addition, these substances produce effects on a number of other metabolic variables, ranging from feelings of hunger and satiety to gastric secretion and electrolyte metabolism; and, albeit indirectly, they also influence such physiological mechanisms as blood coagulation. Particularly important in this connection is the role played by catecholamines in controlling the secretion or the activity of various hormones, e. g. insulin, growth hormone, glucagon, and calcitonin.

It is by no means easy to provide an exact description of the metabolic effects of the catecholamines. Questions of dosage, variations in the specificity of differing organs and animal species, and the fact that inhibitors employed in analysing the metabolic processes under study may also exert intrinsic effects of their own have resulted in the publication of many contradictory findings in the literature. In response to an alteration in sympathetic tone or in catecholamine concentration, the metabolic processes often undergo changes in some organs that are diametrically opposed to the changes occurring in other organs; consequently, studies on the organism as a whole yield net results which cannot be accurately interpreted until the changes taking place in the various individual organs have been analysed *in vitro* or *in vivo*. The fact that both alpha and beta-adrenergic receptors participate in the effects which the sympathico-adrenal system exerts on metabolism also frequently adds to the difficulties of interpretation.

In view of the major influence which the sympathico-adrenal system has on metabolic processes throughout the body, a clinician would have been justified – at the time when substances capable of blocking the sympathetic beta-receptors were first introduced – in fearing that these drugs might provoke a variety of possibly undesirable effects on metabolism. Fortunately, such fears as may have been entertained on this score have since proved unfounded. So far as is known at present, severe metabolic disturbances seldom occur in response to beta-blockade. Although, in a recently published study on 368 hospitalised patients treated with propranolol, it is stated that signs of disturbed cardiac function developed in as many as 25 cases [18], the incidence of serious side effects demonstrably attributable to the metabolic repercussions of beta-

    * Medizinische Poliklinik, University of Berne, Berne, Switzerland.

blockade has, according to figures contained in the literature, been extremely low; this, for example, applies both to hypoglycaemia[1, 15, 23, 24, 28, 56, 68], which – as has been well known for several years – may occur as a side effect not only in diabetics receiving treatment with insulin but also occasionally in other patients, as well as to hyperglycaemic, non-ketotic coma in untreated diabetics[45-47], a side effect which is even more uncommon and has, in fact, only been encountered in a few isolated cases.

To illustrate the changes occurring in carbohydrate metabolism following beta-blockade, I should like to refer here to a simple investigation we carried out some years ago on ten young subjects at the time when oxprenolol (®Trasicor) was undergoing human-pharmacology studies (Figure 1). This investigation of ours took the form of a double-blind trial in which glucose tolerance tests (100 g. glucose by mouth) and tolbutamide tests (1 g. tolbutamide intra-

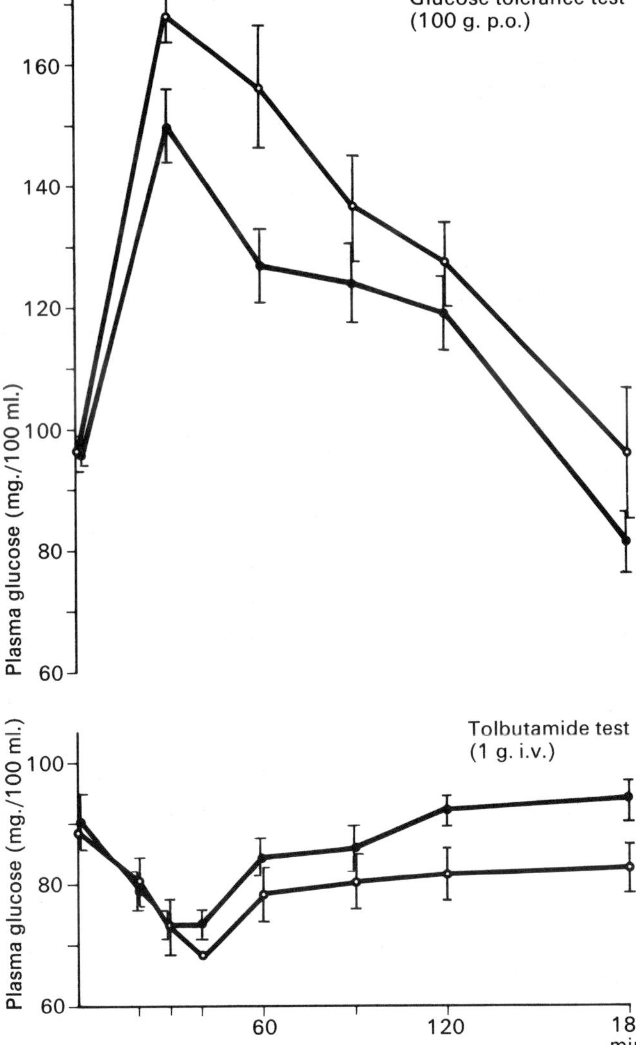

Fig. 1. Effect of oxprenolol (o——o) as compared with placebo (●——●) on the glucose tolerance and tolbutamide tests in ten healthy students. Ordinate: mean values (±S.E.) for the plasma glucose concentrations (mg./100 ml.); abscissa: time in minutes. Beta-blockade produced an upward shift in the glucose tolerance curve and also prolonged the duration of hypoglycaemia in the tolbutamide test. For details of the dosage and comments on the results obtained, see text.

venously) were performed with and without pre-treatment with oxprenolol (40 mg. three times by mouth + 5 mg. intravenously). A marked deterioration in glucose tolerance was observed following pre-treatment with oxprenolol, i.e. the blood glucose concentration rose to hyperglycaemic levels. In the tolbutamide tests, on the other hand, although the primary decrease in the blood sugar levels was the same with and without prior beta-blockade, the recovery phase was markedly prolonged under the influence of oxprenolol.

Tests of this kind have meanwhile been repeated by numerous other authors, who at the same time have also determined the effect of beta-blockade on the insulin levels and various other parameters. In these studies, the glucose concentrations have been found to behave differently, in both the glucose tolerance and tolbutamide tests, depending on the type of procedure adopted. This variability is attributable to the interplay of at least four major factors: in the first place, beta-blockers inhibit one component involved in the secretion of insulin from the pancreas in response to the stimulus provided by glucose[12, 33, 50, 55]; in addition, however, the activity of circulating insulin becomes enhanced owing not only to direct and indirect inhibition of peripheral lipolysis[24, 44] and hepatic phosphorylase[59], but probably also to the facilitation of glucose uptake in the periphery[3, 58]. The situation is further complicated by the fact that in various respects the beta-blockers themselves display certain differences between one another[3, 8]. Whether the beta-blockers also inhibit insulin secretion induced by tolbutamide is a question on which opinions still differ. According to WIDSTRÖM and CERASI[65], no such inhibition occurs in man – at least not in response to propranolol. On the other hand, beta-blockade does have the effect of prolonging and potentiating the activity of circulating insulin in the tolbutamide test as well[60]. These biochemical findings offer an explanation for the hypoglycaemia that occasionally occurs as a complication where beta-blockers are employed in patients under treatment with insulin. In rare instances, hypoglycaemia may also result from potentiation of endogenous insulin activity. Furthermore, in exceptional cases, the combination of blocked lipolysis together with very pronounced inhibition of endogenous insulin secretion may give rise in an untreated diabetic to hyperosmolar, nonketotic coma[45-47]. Attempts to stabilise carbohydrate metabolism in diabetics by treating them with a beta-blocker are not generally viewed with much optimism[52]. Of theoretical interest in this connection, however, is the finding that in certain diabetic patients the pancreas, though not responding properly to glucose, does react in the normal way to isoprenaline[53, 54]. Factors contributing to the changes in carbohydrate metabolism following beta-blockade may also include promotion of the secretion of growth hormone[20, 25, 27, 63] and inhibition of glucagon secretion[21, 31].

Whether, apart from the above-mentioned complications occurring in diabetics, beta-blockade may also have other repercussions on carbohydrate metabolism that are of any real clinical significance, cannot yet be stated with certainty. It is conceivable that, for example, when given in high doses over a prolonged period of time, beta-blockers might in particular affect muscle and energy metabolism. The results of experiments with beta-blockers in animals suggest that they may reduce the working capacity of the muscles. As for the well-known cardiac effects of beta-blockade, no reliable estimate can be made as to the extent to which these are also ascribable to changes in carbohydrate metabolism.

In contrast to the influence of beta-blockers on carbohydrate metabolism, there are a number of other metabolic effects of beta-blockade which, at least when these drugs are employed on a short-term basis, give rise to no clinical signs and symptoms of any significance. If plans to administer long-term treatment with beta-blockers for the prevention of myocardial infarction were to be put into practice, however, these other metabolic effects might well also come to assume considerable importance; fortunately, several of them open up promising therapeutic prospects. One example in point is the effect exerted by beta-blockers on fat metabolism. These drugs inhibit catecholamine-induced lipolysis in fatty tissue; in other words, they prevent the catecholamines from producing a rise in free fatty acid levels in the blood [24, 25, 27, 37, 39, 44, 63]. Less clear-cut, on the other hand, is the nature of their effect on the blood concentrations of triglycerides, phospholipids, and cholesterol, which show a delayed increase in response to the infusion of adrenaline, this delay being probably attributable to the time taken for resynthesis of the fatty acids to occur in the liver [24]. Attempts to reduce the blood lipid levels in patients with hyperlipaemia by treating them with beta-blockers have yielded inconsistent results [34, 35]. Nor has beta-blockade been found to elicit any marked decrease in alimentary, i.e. post-prandial, lipaemia [4]; and in patients with pre-existing hyperlipaemia of Type IV, alimentary lipaemia is even reported to have been increased by treatment with a beta-blocker [5]. But contrasting impressively with these findings are the results of studies showing that beta-blockade suppresses the rise in free fatty acid levels occurring in response to everyday situations of stress [11]. The possibility thus remains that, by diminishing fluctuations in the free fatty acid levels, prolonged treatment with beta-blockers might serve to control lipid metabolism and minimise its atherogenic potential. To investigate this possibility in man would require long-term studies, and no such studies have yet been carried out. Meanwhile, animal experiments have revealed that atheromatosis of the aorta provoked by feeding a high-cholesterol diet to rabbits can be reduced by administering propranolol [64], and that propranolol also affords protection against isoprenaline-induced myocardial necrosis in rats.

Among other metabolic effects of beta-adrenergic blockade which might likewise help to minimise vascular damage in hypertensives are those which beta-blockers exert on platelet aggregation, fibrinolysis, and red-cell oxygen transport. It has been known for quite some time that adrenaline increases platelet adhesiveness [24, 57, 69] – an effect which propranolol is capable of inhibiting. This inhibitory action has been ascribed partly to a decrease in the concentration of free fatty acids, which promote platelet aggregation, and also partly to a direct effect which propranolol itself has on platelet adhesiveness and platelet mobility [39] – although it should, incidentally, be noted that alpha-blockers are alleged to be, if anything, even more effective against catecholamine-induced platelet aggregation than beta-blockers [57]. In patients with angina pectoris or other coronary artery diseases, the enhanced susceptibility to aggregation which their blood platelets display in response to adenosine diphosphate can be appreciably reduced with propranolol[16, 19, 49]. Beta-blockade also prevents the increase in certain coagulation factors which adrenaline induces[26]. In addition, fibrinolysis is reported to be activated by oxprenolol in particular, as well as by the two optical isomers of N-isopropyl-p-nitrophenylethanol-amine, but not by other beta-blockers[48]. Finally, another side effect of pro-pranolol deserving mention in this connection is the rightward shift it produces in the haemoglobin oxygen-dissociation curve, a shift which promotes the release of oxygen to the tissues[43]. This effect is believed to be due to the fact that propranolol liberates a fraction of the 2,3-diphosphoglycerate which is bound to the red-cell membrane, thereby raising the intracellular concentration of unbound 2,3-diphosphoglycerate[42]. Both these findings, however, are dis-puted[7]. In experiments which we ourselves have performed, we have not yet succeeded in demonstrating a difference in the 2,3-diphosphoglycerate content of whole and membrane-free haemolysates. In a double-blind study carried out in 12 healthy volunteers, who were given propranolol (80 mg. three times by mouth), CGP 2175 (100 mg. three times by mouth), and oxprenolol (80 mg. three times by mouth), we therefore confined ourselves to an investigation of the cell-free supernatant. Here, after periods of one, two, and 24 hours, we were unable to discern any differences in the 2,3-diphosphoglycerate concen-trations as compared with those in placebo experiments; in the case of ox-prenolol, we did admittedly observe some increase in the concentration, but this was not statistically significant.

To conclude this brief review, I should now like to refer to one further aspect of beta-blockade which, like the other effects just mentioned, might also prove of importance in cases where beta-blockers are employed on a long-term basis for the treatment of hypertension or in an attempt to prevent myocardial infarction. The fact that renal secretion of renin can be inhibited by beta-blockers – and, though probably to a lesser degree, also by alpha-blockers – has recently been attracting widespread attention[14, 21, 29, 35–37, 39, 61, 65, 67]. According to BÜHLER et al.[10], this inhibitory effect makes a major contribution to the antihypertensive activity of the beta-blockers. BRUNNER et al.[9] likewise empha-sise the significant bearing which elevated renin concentrations have on the prognosis in cases of hypertension. In their studies, BRUNNER et al. found evidence that beta-blockade also has an antihypopotassaemic effect.

After it had been discovered in experiments on dogs that hypopotassaemia occurring in response to diuretics plus a low-potassium diet could be markedly attenuated by beta-blockade (BRUNNER, H., unpublished results), we decided to investigate this phenomenon ourselves in metabolic studies on six healthy students. In these studies, which were carried out on a double-blind basis, the students were hospitalised in a metabolic unit and given oxprenolol and a placebo for one week each. Throughout both weeks they received an identical diet containing an average of 41 mEq. potassium per day. After two control days, during which treatment either with the beta-blocker (80 mg. three times daily by mouth) or with the placebo had already been commenced, we ad-

ministered intravenous injections of 40 mg. furosemide daily (at 8 a.m.) from
the third day onwards. The subjects' urine was collected in two portions, i.e.
one during the diuretic phase (8 a.m.–11 a.m.) and the other during the antidiuretic phase (11 a.m.–8 a.m.), and the electrolyte content of the two portions determined separately. To calculate the overall potassium balance we
likewise measured the excretion of potassium in the stools. In addition, we also
recorded total body potassium at the beginning and end of both weeks.
The main findings obtained are outlined in Figures 2–6. Except on the day of
the first furosemide injection, beta-blockade exerted very little influence on

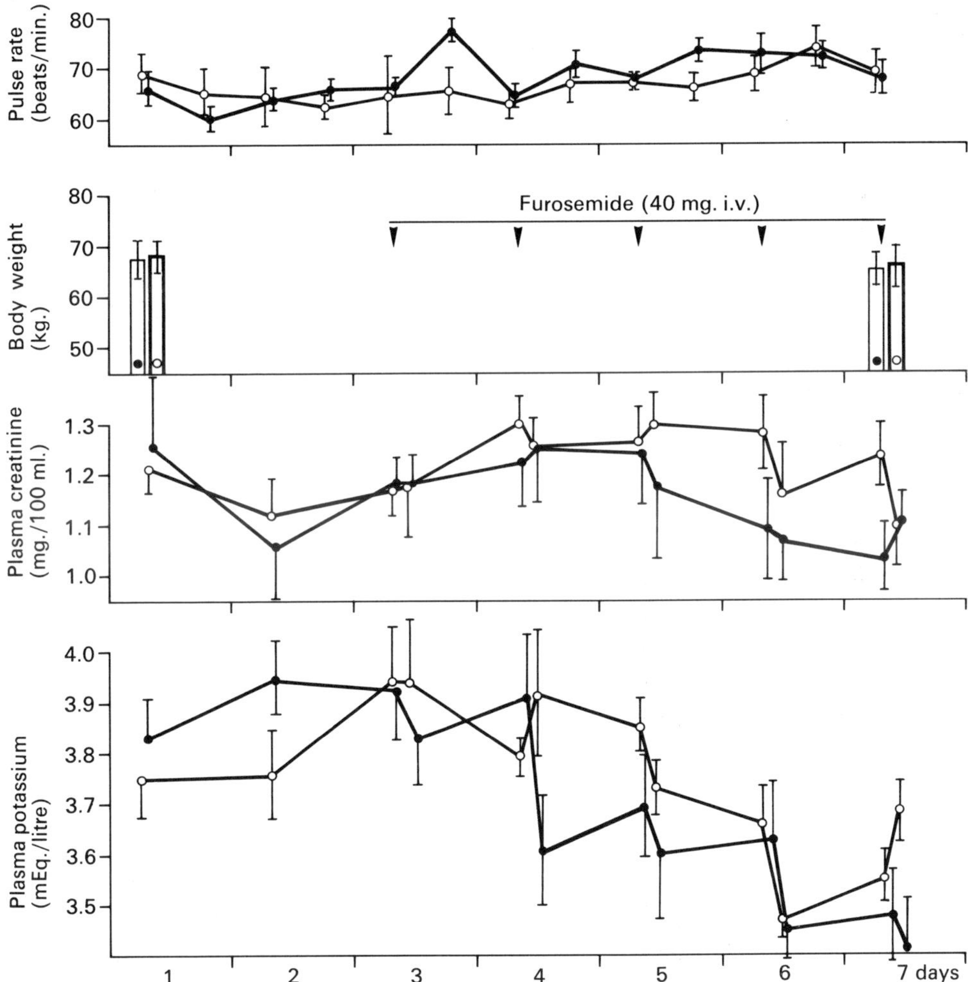

Fig. 2. Pulse rate, body weight, plasma creatinine, and plasma potassium ($\bar{x} \pm$ S.E.)
in six healthy students during a double-blind study lasting two weeks in all, the purpose
of which was to compare the influence of oxprenolol (○———○) in a dosage of 80 mg.
three times daily by mouth with that of a placebo (●———●) on the effects produced,
after two control days, by repeated injections of furosemide (40 mg. i.v. daily). For
further details, see text. Beta-blockade attenuated the hypopotassaemia which these
subjects developed in response to furosemide plus a low-potassium diet (P < 0.01).

Table 1. Potassium balance and total body potassium ($\bar{x} \pm$ S.E.) in six healthy students during a double-blind study lasting two weeks, in which oxprenolol was administered for one week and placebo for one week. For further details, see text.

Dietary intake of potassium per week	Oxprenolol + diuretic 289 mEq.	Placebo + diuretic 289 mEq.
Potassium excretion		
– Urine	363.15 ± 15.6 mEq.	343.31 ± 13.7 mEq.
– Faeces	55.3 ± 6.79 mEq.	50.8 ± 6.05 mEq.
Balance	−129.3 ± 25 mEq.	−105.1 ± 19.4 mEq.
Total body potassium		
– 2nd day	66.4 ± 2.3 mEq./kg.	67.2 ± 3.9 mEq./kg.
– 7th day	64.01 ± 2.8 mEq./kg.	65.7 ± 1.8 mEq./kg.

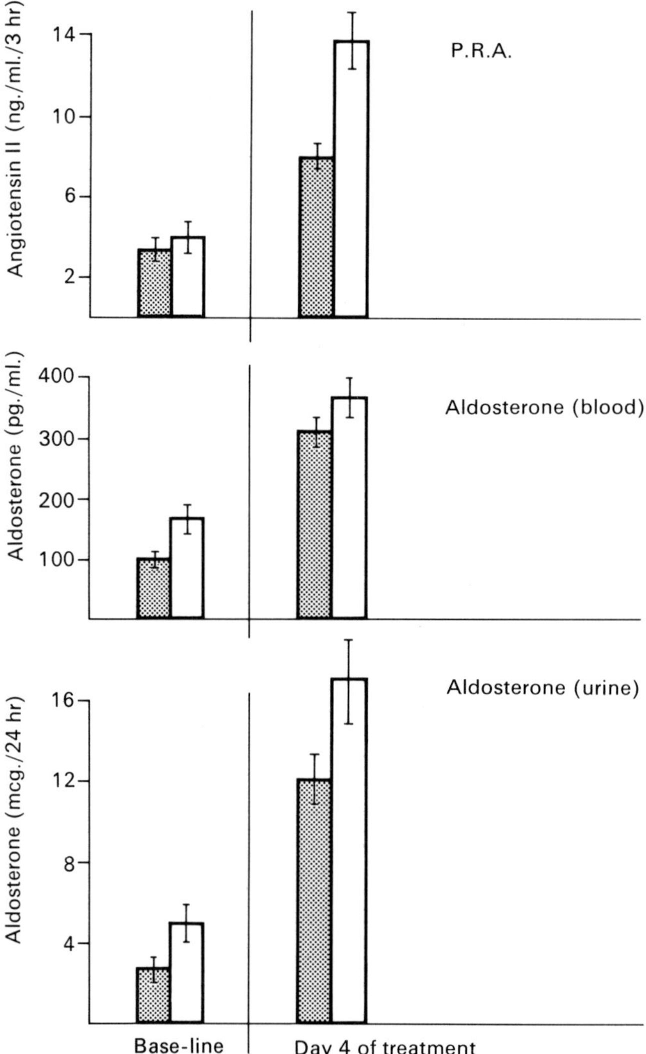

Fig. 3. Influence of oxprenolol on changes in peripheral plasma renin activity (P.R.A.) occurring in response to furosemide injections, as well as on blood and urinary aldosterone levels ($\bar{x} \pm$ S.E.), in six healthy students. Investigations performed under the same conditions as in Figure 2. The increase in P.R.A. induced by the diuretic medication was significantly inhibited by pre-treatment with oxprenolol ($P < 0.05$); but the increase in the blood and urinary aldosterone levels was not significantly diminished by beta-blockade. The author is indebted to the Renin Laboratory of the Medical Polyclinic (Drs VETTER and ARMBRUSTER), Cantonal Hospital, Zurich, for performing the determinations of the P.R.A. and of the blood and urine aldosterone levels.

the pulse rate in these hospitalised subjects. Neither their body weight nor their plasma creatinine levels showed any significant change in response to the oxprenolol (Figure 2). During the placebo week, as well as during the week in which oxprenolol was given, the diuretic medication together with the relatively low potassium diet led to a decrease in the plasma potassium concentrations. Under treatment with the beta-blocker, however, this decrease was significantly less pronounced (P < 0.01), despite the fact that measurement of the potassium balance and of total body potassium revealed no significant difference in the potassium loss occurring in the treated as compared with the untreated state (Table 1). The less pronounced decrease in plasma potassium levels during oxprenolol medication must therefore have been attributable to exchange processes taking place within the body. Beta-blockade significantly inhibited the increase in plasma renin activity provoked by diuretic therapy (P < 0.05) and also reduced, but not to a statistically significant extent, the increase in aldosterone (Figure 3). Urinary excretion of the individual electro-

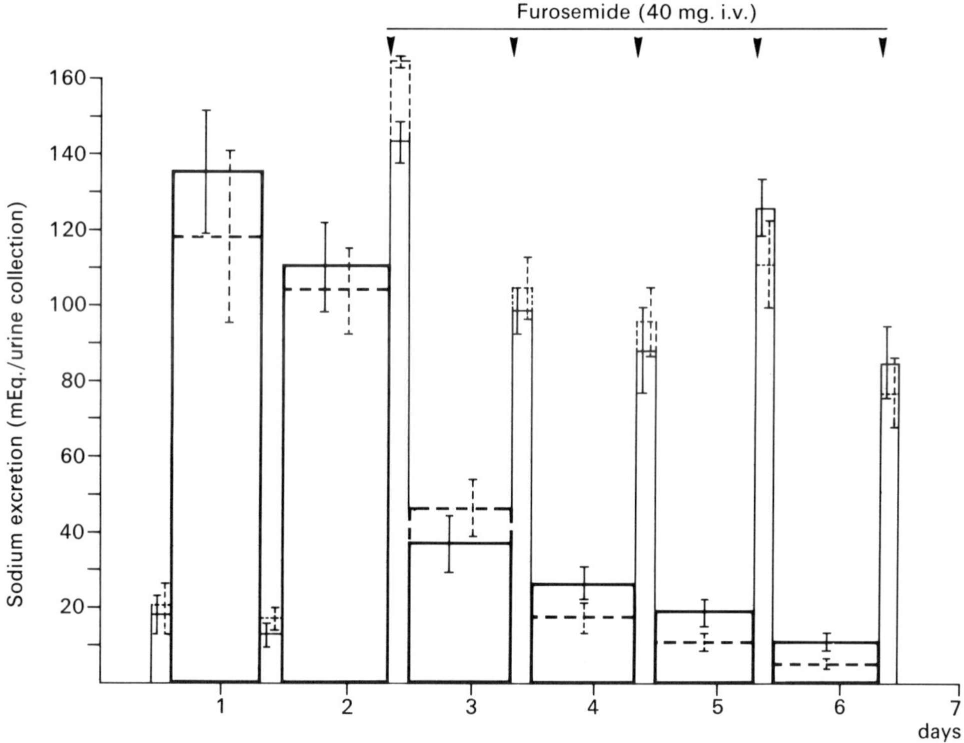

Fig. 4. Influence of oxprenolol (——) as compared with placebo (– – –) on sodium excretion following injections of furosemide. Investigations performed under the same conditions as in Figure 2. The sodium excretion values ($\bar{x} \pm$ S.E.) were obtained by separately measuring the sodium content of two urine collections made between 8 a.m. and 11 a.m. (narrow columns) and between 11 a.m. and 8 a.m. (wide columns). Beta-blockade initially produced a non-significant inhibition of sodium excretion during the diuretic phase following the furosemide injection (8 a.m.–11 a.m.), whereas towards the end of the week it led to significantly higher sodium excretion during the anti-diuretic phase (P < 0.01).

lytes varied during the two weeks of the study. In the case of sodium and chloride, the diuretic medication induced clear-cut periods of diuresis and antidiuresis from the third day onwards, the losses of sodium chloride occurring during the diuretic phase being followed by a compensatory decrease in sodium and chloride excretion during the antidiuretic phase (Figures 4 and 5). By contrast, the urinary excretion of potassium (Figure 6) in the antidiuretic phase – i. e. after the diuretic effect of the furosemide had worn off – remained at roughly the same level as in response to placebo during the same period of time (11 a.m.–8 a.m.), although the rise in potassium excretion during the diuretic phase (8 a.m.–11 a.m.) was still of high statistical significance. Whereas the beta-blocker exerted no significant influence on sodium-chloride excretion during the first few days on which the diuretic was administered, it did very significantly diminish the kidneys' capacity to excrete urine of low sodium-chloride content during the antidiuretic phase (P < 0.01). Owing possibly to the presence of more elevated blood potassium concentrations, potassium

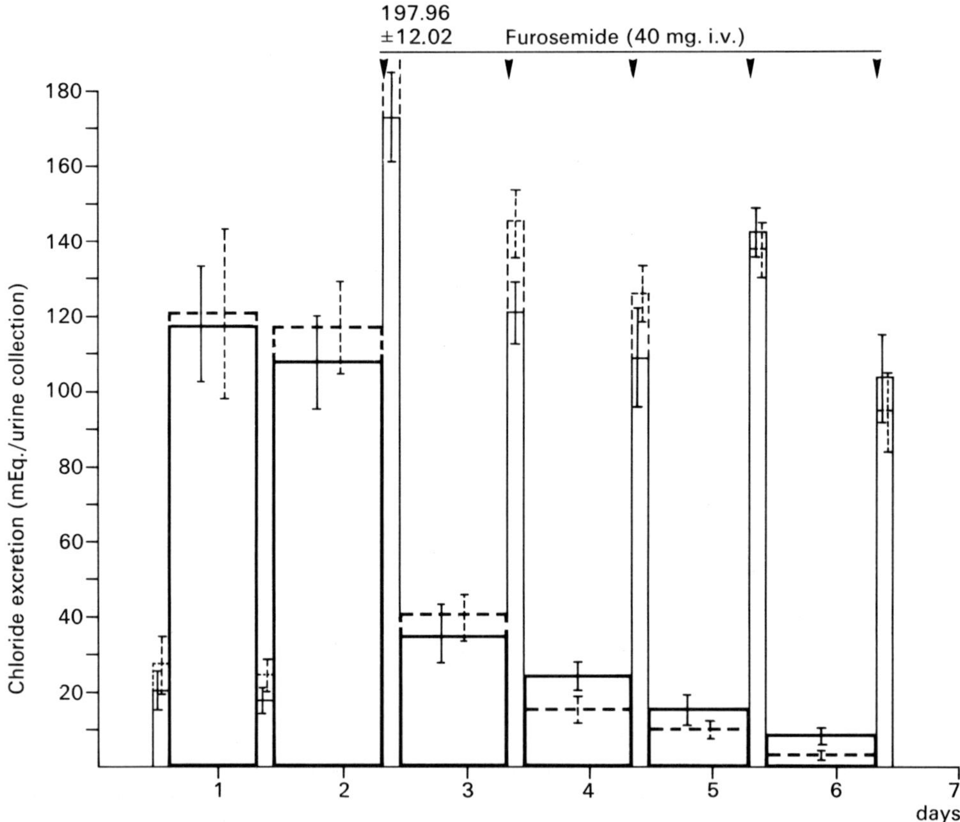

Fig. 5. Influence of oxprenolol (——) as compared with placebo (– – –) on chloride excretion following furosemide. Investigations performed under the same conditions as in Figure 2. As in the case of sodium, beta-blockade also initially inhibited the excretion of chloride following the furosemide injection, whereas at the end of the week it markedly increased chloride excretion during the antidiuretic phase.

268

excretion towards the end of the week finally reached higher levels under treatment with oxprenolol than when the placebo was given.

From the results of these investigations of ours it cannot be stated with certainty whether the differences in the excretion of electrolytes during the phases of furosemide-induced diuresis and during the antidiuretic phases must be ascribed directly to the beta-blocker or indirectly to its influence on the renin-angiotensin-aldosterone system. Both possibilities deserve consideration. Although the results of needle biopsy in dogs indicated that neither beta nor alpha-adrenergic stimulation has any significant effect on proximal sodium reabsorption[6], in man pre-treatment with propranolol does delay the excretion of hyperosmotic saline infusions[29], whereas in rats propranolol steps up the urinary excretion of sodium and chloride[32]. The fact that beta-blockade tends to raise blood potassium levels in man – a fact which, so far as I am aware,

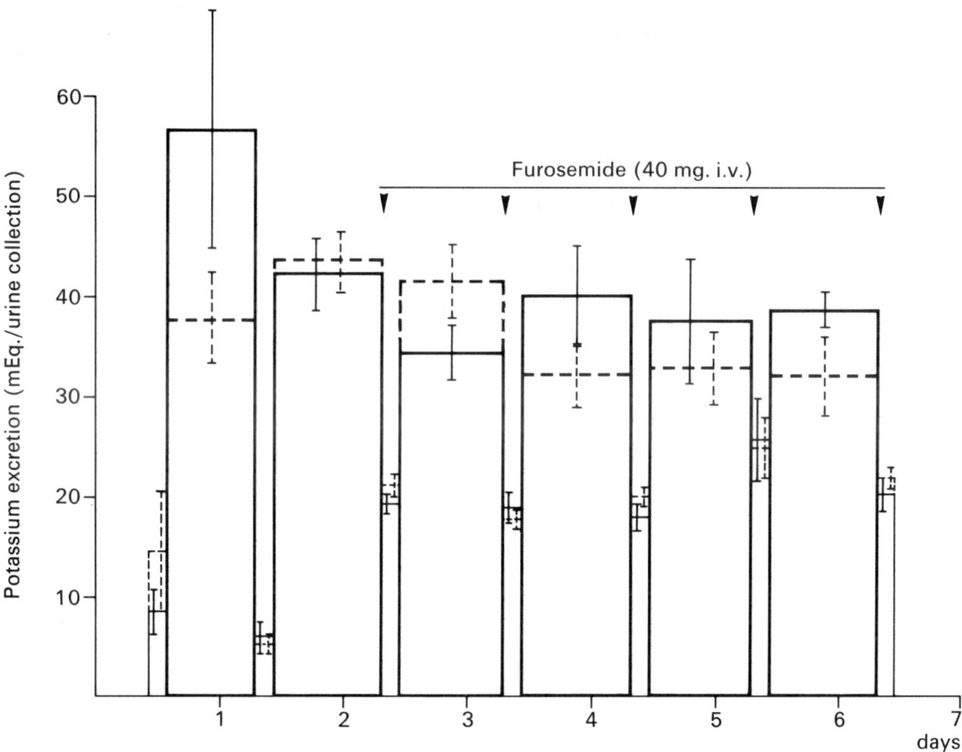

Fig. 6. Influence of oxprenolol (——) as compared with placebo (–––) on potassium excretion following furosemide. Investigations performed under the same conditions as in Figure 2. From the figures for the two portions of urine collected between 8 a.m. and 11 a.m. (narrow columns) and between 11 a.m. and 8 a.m. (wide columns) it can be seen that, during the diuretic phase, furosemide caused a significant increase in potassium excretion which was unaffected by beta-blockade. In contrast to sodium and chloride excretion, potassium excretion in the antidiuretic phase, too, remained at roughly the same level as that recorded during the same period (11 a.m.–8 a.m.) following placebo; at the end of the week it was actually somewhat higher than in response to placebo, a possible reason for this being the presence of more elevated blood potassium concentrations. For further details, see text.

our own studies on the potassium balance have been the first to demonstrate –
is of interest in view of various earlier observations on changes in the potassium balance occurring in response to adrenaline infusions. It has long been
known[17, 37] that catecholamines give rise in dogs not only to hyperglycaemia
but also to a biphasic fluctuation in the blood potassium concentration: the
initial rise in blood potassium is attributed to a release of potassium from the
liver, and the hypopotassaemia – which occurs afterwards and can be inhibited
by beta-blockade – to an uptake of potassium into the skeletal muscles[17]. Of
more recent date is the discovery that propranolol strongly reduces the potassium content of the erythrocytes, a finding which can be partly related to the
drug's effect on the oxygen-binding capacity of haemoglobin[2]. Changes in the
potassium balance have also been reported in the myocardium under the influence of propranolol[13, 61].

All the metabolic side effects of beta-blockade to which I have referred here
are admittedly relatively minor ones when considered individually. But, in view
of the necessity for undertaking long-term trials on large numbers of hypertensive patients in order to study the possible value of beta-blockers in the
prevention of myocardial infarction, even such minor effects of these drugs
might come to assume some importance in the future. It is therefore all the more
fortunate that, so far as is known at present, the beta-blockers seldom provoke
more serious, potentially harmful, metabolic side effects and that many of their
comparatively insignificant side effects might, on the contrary, actually prove
therapeutically beneficial.

References

1 ABRAMSON, E.A., ARKY, R.A., WOEBER, K.A.: Effects of propranolol on the
  hormonal and metabolic responses to insulin-induced hypoglycaemia. Lancet *ii*,
  1386 (1966)
2 AGOSTINI, A., BERFASCONI, C., GERLI, G.C.: Oxygen affinity and electrolyte distribution of human blood: changes induced by propranolol. Science *182*, 300 (1973)
3 ASMAL, A.C., COLEMAN, A.J., LEARY, W.P.: The effects of adrenergic β-blockade
  with oxprenolol on peripheral metabolism. In: Proc. VIIth Ann. Congr. S.Afr.
  Pharmacol. Soc., Johannesburg 1973
4 BARBORIAK, J.J., FRIEDBERG, H.D.: Propranolol and alimentary lipemia in man.
  J. Lab. clin. Med. *78*, 986 (1971); abstract of paper
5 BARBORIAK, J.J., FRIEDBERG, H.D.: Propranolol and hypertriglyceridemia. Atherosclerosis *17*, 31 (1973)
6 BLENDIS, L.M., AULD, R.B., ALEXANDER, E.A., LEVINSKY, N.G.: Effect of renal
  beta- and alpha-adrenergic stimulation on proximal sodium reabsorption in dogs.
  Clin. Sci. *43*, 569 (1972)
7 BRAIN, M.C., CARD, R.T., KANE, J., LYONNAIS, J., DOLLERY, C.T.: Acute effects
  of varying doses of propranolol upon oxygen haemoglobin affinity in man. Brit.
  J. clin. Pharmacol. *1*, 67 (1974)
8 BRUNNER, H., HEDWALL, P.R., MAIER, R., MEIER, M.: Pharmacological aspects
  of oxprenolol. Postgrad. med.J. *46*, Suppl. (Nov.): 5 (1970)
9 BRUNNER, H.R., LARAGH, J.H., BAER, L., NEWTON, M.A., GOODWIN, F.T., KRAKOFF, L.R., BARD, R.H., BÜHLER, F.R.: Essential hypertension: renin and aldosterone, heart attack and stroke. New Engl.J. Med. *286*, 441 (1972)

10 BÜHLER, F. R., LARAGH, J. H., BAER, L., VAUGHAN, E. D., Jr., BRUNNER, H. R.: Propranolol inhibition of renin secretion. A specific approach to diagnosis and treatment of renin-dependent hypertensive diseases. New Engl. J. Med. *287*, 1209 (1972)

11 CARRUTHERS, M., TAGGART, P.: Vagotonicity of violence: biochemical and cardiac responses to violent films and television programmes. Brit. med. J. *iii*, 384 (1973)

12 CERASI, E., LUFT, E., EFENDIĆ, S.: Effect of adrenergic blocking agents on insulin response to glucose infusion in man. Acta endocr. (Kbh.) *69*, 335 (1972)

13 CHOI, S. J., ROBERTS, J., KELLIHER, G. J.: Modification of the action of propranolol and quinidine on ^{22}Na- and ^{42}K-exchange by reserpine. Europ. J. Pharmacol. *20*, 22 (1972)

14 CHOKSHI, D. S., YEH, B. K., SAMET, P.: Effects of dopamine and isoproterenol on renin secretion. Circulation *44*, Suppl. II: 151 (1971); abstract of paper

15 FELLER, J. M.: Danger of hypoglycaemia with use of propranolol. Med. J. Aust. *60/ii*, 92 (1973); corresp.

16 FRISHMAN, W., WEKSLER, B., CHRISTODOULOU, J., SMITHEN, C., KILLIP, T.: Platelet aggregation in angina pectoris. Response to oral propranolol. Circulation *48*, Suppl. IV: 57 (1973); abstract of paper

17 GRASSI, A. O., LEW, W. F. DE, CINGOLANI, H. E., BLESA, E. S.: Adrenergic beta blockade and changes in plasma potassium following epinephrine administration. Europ. J. Pharmacol. *15*, 209 (1971)

18 GREENBLATT, D. J., KOCH-WESER, J.: Adverse reactions to propranolol in hospitalized medical patients: a report from the Boston Collaborative Drug Surveillance Program. Amer. Heart J. *86*, 478 (1973)

19 HAFT, J. I., FANI, K., ALCORTA, C., TOOR, M.: Effect of propranolol on stress induced intravascular platelet aggregation in the heart. Circulation *48*, Suppl. IV: 57 (1973); abstract of paper

20 HANSEN, A. P.: The effect of adrenergic receptor blockade on the exercise-induced serum growth hormone rise in normals and juvenile diabetics. J. clin. Endocr. *33*, 807 (1971)

21 HARVEY, W., FALOONA, G., UNGER, R.: Effect of adrenergic blockade on exercise-induced hyperglucagonemia. J. clin. Invest. *52*, 38a (1973); abstract of paper

22 HAUGER-KLEVENE, J. H.: Actinomycin D effect on the adrenergic receptor mediation of renin release. Clin. Res. *19*, 687 (1971); abstract of paper

23 HESSE, B., PEDERSEN, J. T.: Hypoglycaemia after propranolol in children. Acta med. scand. *193*, 551 (1973)

24 HIMMS-HAGEN, J.: Sympathetic regulation of metabolism. Pharmacol. Rev. *19*, 367 (1967)

25 IMURA, H., KATO, Y., IKEDA, M., MORIMOTO, M., YAWATA, M.: Effect of adrenergic-blocking or -stimulating agents on plasma growth hormone, immunoreactive insulin, and blood free fatty acid levels in man. J. clin. Invest. *50*, 1069 (1971)

26 INGRAM, G. I. C., JONES, R. V.: The rise in clotting factor VIII induced in man by adrenaline: effect of *a*- and *β*-blockers. J. Physiol. (Lond.) *187*, 447 (1966)

27 JENKINS, D. J. A.: Propranolol and hypoglycaemia. Lancet *i*, 164 (1967)

28 KOTLER, M. N., BERMAN, L., RUBENSTEIN, A. H.: Hypoglycaemia precipitated by propranolol. Lancet *ii*, 1389 (1966)

29 KRAUSS, H., SCHALEKAMP, M. A. D., KOLSTERS, G., ZAAL, G. A., BIRKENHÄGER, W. H.: Effects of chronic beta-adrenergic blockade on systemic and renal haemodynamic responses to hyperosmotic saline in hypertensive patients. Clin. Sci. *43*, 385 (1972)

30 KRUMLOVSKY, E., GRECO, F. DEL: Effects of adrenergic stimulators on renin levels in hypertension. Clin. Res. *19*, 351 (1971); abstract of paper

31 LAWRENCE, A. M., HAGEN, T. C., KIRSTEINS, L.: Propranolol hypoglycemia and impaired glucagon secretion. Clin. Res. *20*, 770 (1972); abstract of paper

32 LEES, P.: The influence of *β*-adrenoceptive receptor blocking agents on urinary function in the rat. Brit. J. Pharmacol. *34*, 429 (1968)

33 LERNER, R. L., PORTE, D., Jr.: Epinephrine: selective inhibition of the acute insulin response to glucose. J. clin. Invest. *50*, 2453 (1971)

34 LIPSON, M. J., NAIMI, S., PROGER, S.: Effect of a combination of propranolol and nicotinic acid on the inhibition of norepinephrine-induced elevations of serum triglyceride, serum glycerol, and plasma free fatty acids in the dog. Metabolism *20*, 580 (1971)

35 LLOYD-MOSTYN, R. H., KRIKLER, D. M.: Raised serum-lipids unaffected by beta-blockade. Lancet *ii*, 1040 (1970)

36 LOEFFLER, J. R., STOCKIGT, J. R., GANONG, W. F.: Effect of alpha- and beta-adrenergic blocking agents on the increase in renin secretion produced by stimulation of the renal nerves. Neuroendocrinology *10*, 129 (1972)

37 MAYER, S., MORAN, N. C., FAIN, J.: The effect of adrenergic blocking agents on some metabolic actions of catecholamines. J. Pharmacol. exp. Ther. *134*, 18 (1961)

38 McALLISTER, R. G., Jr., MICHELAKIS, A. M.: Effect of chronic adrenergic receptor blockade on renin. Clin. Res. *19*, 353 (1971); abstract of paper

39 MEESTER, W. D., SEKHAR, N. C.: The effect of epinephrine-induced elevation of plasma free fatty acids on platelet aggregation in man. Blood *34*, 537 (1969); abstract of paper

40 MEURER, K.-A.: Die Bedeutung des sympathico-adrenalen Systems für die Renin-freisetzung. Klin. Wschr. *49*, 1001 (1971)

41 MEYER, D. K., PESKAR, B., TAUCHMANN, U., HERTTING, G.: Potentiation and abolition of the increase in plasma renin activity seen after hypotensive drugs in rats. Europ. J. Pharmacol. *16*, 278 (1971)

42 OSKI, F. A., MILLER, L. D., DELIVORIA-PAPADOPOULOS, M., MANCHESTER, J. H., SHELBURNE, J. C.: Oxygen affinity in red cells: changes induced in vivo by pro-pranolol. Science *175*, 1372 (1972)

43 PENDLETON, R. G., NEWMAN, D. J., SHERMAN, S. S., BRANN, E. G., MAYA, W. E.: Effect of propranolol upon the hemoglobin-oxygen dissociation curve. J. Pharmacol. exp. Ther. *180*, 647 (1972)

44 PINTER, E. J., PATTEE, C. J.: Effect of β-adrenergic blockade on resting and stimulated fat mobilization. J. clin. Endocr. *27*, 1441 (1967)

45 PODOLSKY, S.: A possible role of propranolol in recurrent nonketotic hyperosmolar diabetic coma. Diabetes *19*, Suppl. 1: 398 (1970)

46 PODOLSKY, S.: Recurrent hyperosmolar nonketotic diabetic coma (HNC) caused by propranolol therapy. Clin. Res. *19*, 354 (1971); abstract of paper

47 PODOLSKY, S., PATTAVINA, C. G.: Hyperosmolar nonketotic diabetic coma: a complication of propranolol therapy. Metabolism *22*, 685 (1973)

48 PONARI, O., CIVARDI, E., POTÌ, R.: Action of some β-blockers on plasma fibrinolysis in vitro and in vivo in man. Arzneimittel-Forsch. (Drug Res.) *22*, 629 (1972)

49 POPŁAWSKY, A., SKORULSKA, M., NIEWIAROWSKI, S.: Increased platelet adhesiveness in hypertensive cardiovascular disease. J. Atheroscler. Res. *8*, 721 (1968)

50 PORTE, D., Jr.: Sympathetic regulation of insulin secretion. Its relation to diabetes mellitus. Arch. intern. Med. *123*, 252 (1969)

51 PURI, P. S., MAMMEN, E. F.: The effect of β-adrenergic blockade (propranolol) on the coagulability of the blood across the human heart. Med. Pharmacol. exp. (Basle) *17*, 239 (1967)

52 REVENO, W. S., ROSENBAUM, H.: Propranolol and hypoglycaemia. Lancet *i*, 920 (1968)

53 ROBERTSON, R. P., PORTE, D., Jr.: Beta adrenergic mediated insulin responses in normal and diabetic men. Clin. Res. *20*, 193 (1972); abstract of paper

54 ROBERTSON, R. P., PORTE, D., Jr.: The glucose receptor: a defective mechanism in diabetes mellitus distinct from the β-adrenergic receptor. Clin. Res. *20*, 555 (1972); abstract of paper

55 ROBERTSON, R. P., PORTE, D., Jr.: Adrenergic modulation of basal insulin secretion in man. Diabetes *22*, 1 (1973)

56 Ross, G.: Propranolol in myocardial infarction. Lancet *i*, 396 (1967)

57 Schnetzer, G.W.: Platelets and thrombogenesis – current concepts. Amer. Heart J. *83*, 552 (1972)

58 Scholhamer, C., Felig, P., Hendler, R.G.: Propranolol hypoglycemia: increased peripheral effectiveness of insulin. Clin. Res. *19*, 736 (1971); abstract of paper

59 Sherline, P., Lynch, A., Glinsman, W.H.: Cyclic AMP and adrenergic receptor control of rat liver glucogen metabolism. Endocrinology *91*, 680 (1972)

60 Specchia, G., Fratino, P., Gamba, G.: Influenze del propranololo sull'azione della tolbutamide. Boll. Soc. ital. Biol. sper. *46*, 586 (1970)

61 Tarr, M., Luckstead, E.F., Jurewicz, P.A., Haas, H.G.: Effect of propranolol on the fast inward sodium current in frog atrial muscle. J. Pharmacol. exp. Ther. *184*, 599 (1973)

62 Tobert, J.A., Slater, J.D.H., Fogelman, F., Lightman, S.L., Kurtz, A.B., Payne, N.N.: The effect in man of (+)-propranolol and racemic propranolol on renin secretion stimulated by orthostatic stress. Clin. Sci. *44*, 291 (1973)

63 Werrbach, J.H., Gale, C.C., Goodner, C.J., Conway, M.J.: Effects of autonomic blocking agents on growth hormone, insulin, free fatty acids and glucose in baboons. Clin. Res. *17*, 111 (1969); abstract of paper

64 Whittington-Coleman, P.J., Carrier, O., Jr., Douglas, B.H.: The effects of propranolol on cholesterol-induced atheromatous lesions. Atherosclerosis *18*, 337 (1973)

65 Widström, A., Cerasi, E.: On the action of tolbutamide in normal man. I. Role of adrenergic mechanisms in tolbutamide-induced insulin release during normoglycaemia and induced hypoglycaemia. Acta endocr. (Kbh.) *72*, 506 (1973)

66 Winer, N.: Effects of adrenergic antagonists on renin secretion induced by adrenalectomy, sodium depletion, haemorrhage and renal artery constriction. Clin. Res. *19*, 690 (1971); abstract of paper

67 Winer, N., Chokshi, D.S., Walkenhorst, W.G.: Effects of cyclic AMP, sympathomimetic amines, and adrenergic receptor antagonists on renin secretion. Circulat. Res. *29*, 239 (1971)

68 Wray, R., Sutcliffe, S.B.J.: Propranolol-induced hypoglycaemia and myocardial infarction. Brit. med. J. *ii*, 592 (1972); corresp.

69 Yamazaki, H., Sano, T., Odakura, T., Takeuchi, K., Matsumura, T., Hosaki, S., Shimamoto, T.: Appearance of thrombogenic tendency induced by adrenaline and its prevention by β-adrenergic blocking agent, nialamide and pyridinolcarbamate. Thrombos. Diathes. haemorrh. (Stuttg.) *26*, 251 (1971)

Discussion

J. Brod: First of all, a word about the decrease in sodium reabsorption during beta-blocker medication. What your test subjects had, Dr. Hodler, was really secondary hyperaldosteronism induced by furosemide. It is easy to imagine that blocking the release of renin will result in a reduction in aldosterone secretion, and this would explain your observations.

I also have a question concerning the protective action of beta-blockers against the effects of stress, a subject that has already been discussed during this symposium. Selye has been claiming for many years that the effects of stress on the myocardium and blood vessels can be reduced or even prevented altogether by the administration of potassium. Do you consider that the changes in potassium you mentioned could perhaps be the factor mediating this effect of the beta-blockers?

J. Hodler: The inhibition of this secondary hyperaldosteronism, Dr. Brod, lies of course at the root of the whole problem. It was with this consideration in mind that we designed our experimental procedure. The inhibition of secondary hyperaldosteronism may possibly also account for the increased loss of sodium during the antidiuretic phase. This hypothesis, however, doesn't provide a satisfactory explanation for the behaviour of potassium, because no decrease in urinary potassium loss could be demonstrated, with the result that the cause of the apparent antihypopotassaemic effect of beta-blockade remains obscure.

As for the second question you raised, I feel that the "antihypopotassaemic" effect of beta-blockade is still too unclear for us to be able to postulate that it might be responsible for protecting the myocardium from stress. Other effects of beta-blockade (for example, on energy consumption) are more likely to be involved in this protective action.

H. Brunner: With regard to sugar metabolism you said, Dr. Hodler, that glucose uptake in the periphery is facilitated by beta-blockade. Is this a secondary effect, due perhaps to the reduction in the concentration of free fatty acids, or is it a direct effect?

J. Hodler: Both possibilities have been discussed. I know of only two studies dealing in detail with this question*. In both, the conclusion was reached that the increase in glucose utilisation in the periphery is probably brought about less by a change in fatty acid metabolism than by a direct effect on the membrane and on glucose uptake by the peripheral cells.

F. Gross: Dr. Hodler, you have demonstrated, on the one hand, that oxprenolol (®Trasicor) increases the diuretic-induced excretion of potassium and, on the other, that it attenuates the hypopotassaemia provoked by the diuretic plus a low-potassium diet. Consequently, there must probably be an internal shift in potassium.

J. Hodler: Yes, that's right. There is actually a higher loss of potassium following beta-blockade, but the difference is not significant.

F. Gross: But how are we to interpret this shift? You have tried to account for it by ascribing it to an effect on the kidney and to a change in aldosterone secretion, but there must doubtless be still another mechanism. We don't know how catecholamine excretion changes, or whether the catecholamines may also possibly play a role. And there may perhaps be additional factors that influence the potassium balance.

---

* Asmal, A.C., Coleman, A.J., Leary, W.P.: The effects of adrenergic β-blockade with oxprenolol on peripheral metabolism. In: Proc. VIIth Ann. Congr. S. Afr. Pharmacol. Soc., Johannesburg 1973

Scholhamer, C., Felig, P., Hendler, R.G.: Propranolol hypoglycemia: increased peripheral effectiveness of insulin. Clin. Res. *19*, 736 (1971); abstract of paper

J. Hodler: Our experiments have yielded two findings: firstly, beta-blockade leads to an inhibition of aldosterone and prevents the organism from retaining a maximum amount of sodium and chloride. This, however, sheds no light on the behaviour of potassium – except, possibly, indirectly. On the other hand, we know that the displacement of potassium from the intracellular to the extracellular space, such as I have described, does occur following adrenaline injections. It is conceivable that the secondary re-uptake of potassium is modified by beta-blockade. This is why I was so interested to know whether anybody could say how much potassium can be released by the erythrocytes.

H. Brunner: The results of tests we have conducted on dogs might help to clarify this question. The animals were treated with furosemide and oxprenolol daily for four days (corresponding controls being run with furosemide alone), and the blood potassium levels were determined. In the animals receiving the beta-blocker the blood potassium levels were invariably higher than in the controls *before* administration of the diuretic; during the diuretic phase they then decreased to the values recorded in the control animals, and during the recovery phase until the next dose of diuretic on the following day they rose again. Hence, the effect of beta-blockade is most marked at the end of the relatively long phase during which the electrolyte balance is recovering, i. e. at the end of the phase in which a reduction in aldosterone would be most likely to occur.

J. Hodler: But we found a negative potassium balance.

F. R. Bühler: Doesn't it worry you, Dr. Hodler, that your test subjects were losing increased amounts of potassium even though the potassium-excreting factor, aldosterone, was relatively reduced?

J. Hodler: The test subjects had deliberately been placed on a low-potassium diet, and they were receiving a diuretic. Hence the negative potassium balance. This was precisely the situation we wanted to produce and investigate. The slight increase in renal potassium excretion during beta-blockade (cf. Table 1 in my paper) may possibly have been due to an increased blood potassium level.

The metabolism of beta-blockers in relation to their pharmacokinetic and pharmacodynamic behaviour

by W. Riess, S. Brechbühler, L. Brunner, P. R. Imhof, and D. B. Jack*

The various beta-blockers known to date not only feature structural elements that are common to them all but, when broken down by processes of biotransformation, they also undergo reactions which are typical of their structure. A study of the known representatives from this category of active substances, however, reveals that the extent of the metabolic changes to which they are subject, and the degree to which certain metabolites predominate in the mixture of compounds finally eliminated from the body, differ very considerably from one preparation to another. Although the existing documentation on the metabolism of some of these preparations is still very incomplete and heterogeneous, it is nevertheless already possible at this stage to make an attempt at presenting a systematic outline of the metabolism and kinetics of the betablockers.

If we start with the metabolites of propranolol and oxprenolol (®Trasicor), which in our opinion are at present the two beta-blockers that have been most thoroughly investigated, we find certain basic metabolic reactions of the same type in the case of both preparations (Figure 1):

1. Glucuronidation of the active substance, which otherwise remains unchanged.
2. N-dealkylation.

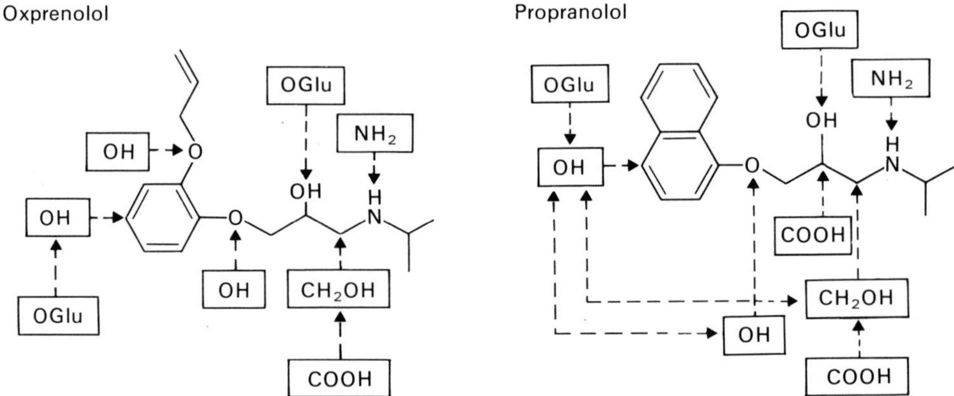

Fig. 1. Biotransformation of oxprenolol[5, 15, 16, 23] and propranolol[2, 3, 22, 26, 30, 31]. The boxes and arrows indicate groups introduced by metabolism.

* Research Department, Pharmaceuticals Division, CIBA-GEIGY LIMITED, Basle, Switzerland.

3. Oxidative deamination of the carbon atom in the *a*-position to the nitrogen atom, resulting initially in formation of the corresponding aldehyde, which is partially reduced to the carbinol and partially undergoes further oxidation to the carboxylic acid.
4. Oxidation of the carbinol, present in the original structure, to carboxylic acid.
5. Splitting of the ether bond to yield the corresponding phenol.
6. Aromatic hydroxylation, as well as combination of aromatic hydroxylation with oxidation of the carbon atom in the *a*-position to the nitrogen atom and with ether-splitting.
7. Conjugation of the inserted phenolic group with polar ligands such as glucuronic acid.

Propranolol and oxprenolol, however, also exhibit certain fundamental differences: in man, oxprenolol is eliminated via the kidneys largely in glucuronide but otherwise unchanged form, whereas propranolol is excreted in the form of metabolites consisting predominantly of hydroxycarboxylic acid resulting from oxidative deamination.

When considering other representatives from this class of substances, one is obliged in most instances to confine oneself to a comparison of such data as can be found on their total renal excretion and on the portion of the total that is accounted for by unchanged substance (Table 1).

But even these simple data reveal tentative evidence of an interesting systematic pattern. Assessed in terms of their total renal excretion, the substances listed in Table 1 show hardly any differences. Propranolol, bunolol, alprenolol, and oxprenolol nevertheless do differ from pindolol, sotalol, and practolol: whereas sotalol and practolol are excreted in the urine in unchanged form to the extent of approximately 90% of the dose administered, and pindolol to the extent of 40%, propranolol, bunolol, alprenolol, and oxprenolol are almost entirely metabolised, one peculiar feature of oxprenolol being the fact that, as already mentioned, it is largely excreted in a form which differs from the original structure only insofar as it is a glucuronide. From such data as are available on the half-life of the plasma concentrations in man, practolol appears to have the highest value, amounting to approximately ten hours. The extent to which these preparations become broken down into metabolites is thus evidently not dependent *per se* on the length of time during which the original substances remain in the body.

One property of these substances frequently taken into consideration when comparing various pharmacological and chemical aspects is their liposolubility, the relative degree of which is determined by reference to their respective partition coefficients. There does in fact also seem to be some connection between the relative liposolubility of beta-blockers and the extent to which they are metabolised in the body. According to the partition coefficients as listed in Table 2, propranolol, bunolol, and alprenolol – which are broken down to a large degree by oxidation – are readily soluble in fats; oxprenolol, on the other hand, with its distinctive metabolic behaviour, is moderately

Table 1. Data currently available on the renal excretion, plasma half-life, and liposolubility of seven different beta-blockers.

		Propranolol	Bunolol	Alprenolol	Oxprenolol	Pindolol	Sotalol	Practolol
I. Totally eliminated in urine (metabolised and unchanged)	Human	>90% after 48 hr[22]		>90% after 48 hr. Enzyme saturation[13, 14]	95% after 48 hr[23]	~95%[20]		
	Animal		Up to 75% after 72 hr (dog)[18]					
II. Unchanged drug in urine (% of dose)	Human	5%[22]		≤ 2%[13, 14]	<5%[23, 30]	40%[8]		85% after 24 hr[4]
	Animal		1 % (dog)[18]				90% after 72 hr (dog)[19]	>85% (rat[4] and dog)[24]
III. Half-life	Human	2.3 hr i.v. 3.2 hr p.o.[6, 26]		1.7–2.2 hr[1]	1.3–1.5 hr*	3.1 hr i.v. 3.6 hr p.o.[8, 11]		10 ± 2 hr[4]
	Animal							6–8 hr (rat and dog)[24]
IV. Liposolubility		High[9, 10, 29]	High[29]	High[9, 13]	Medium[9]	Low[9, 10]	Very low[9, 19, 29]	Very low[9, 10]

* Data from the present paper

278

Table 2. Partition coefficients of beta-blockers in n-octanol-phosphate buffer (0.16 M) at pH 7 and pH 8 and in chloroform-water.

Drug	Partition coefficients in n-octanol-phosphate buffer[9]		Partition coefficients in CHCl$_3$-water[29]
	pH 7	pH 8	
Propranolol	5.39	52.5	15.67
Bunolol	–	–	10.90
Alprenolol	3.27	27.8	–
Oxprenolol	0.43	4.11	8.09
Pindolol	0.12	1.55	–
Sotalol	0.011	0.031	0.06
Practolol	0.009	0.053	–

liposoluble, whereas pindolol is sparingly and sotalol and practolol only very sparingly soluble in fats. It would appear that the degree to which beta-blockers are metabolised decreases as their hydrophilic character increases, but that their half-life values do not conform to this same pattern. There would certainly be justification for supposing that compounds showing greater hydrophilism than practolol would also be correspondingly more rapidly eliminated. Whether such structural homologues would still possess beta-blocking properties, however, remains a moot point.

When factors of this kind are considered in relation to the clinical use of beta-blockers, it might at first sight seem quite irrelevant whether the preparation administered is inactivated by biotransformation, i.e. metabolised, or whether it is directly eliminated in unchanged form. This would be true, were it not for the fact, firstly, that biotransformation is known to be capable of giving rise to biologically active metabolites and, secondly, that enzymatic degradation processes can reach a saturation point due to the limited supply of enzymes upon which they depend; owing to this latter limiting factor, the bio-availability of the active substance may be dependent on the size of the dose administered and on the route of administration.

In the case of propranolol, for example, it has been reported[3] that, following an oral dose of the drug, pharmacological activity equivalent to twice that produced by an intravenous dose can be recorded over a brief period, despite the fact that *the concentration of unchanged substance in the serum is in both instances the same.* The explanation given for this phenomenon is the initial formation of 4-hydroxypropranolol during the first passage of the oral dose through the liver.

Where lipophilic beta-blockers are employed by the oral route, particular clinical importance attaches to the first passage of the dose through the liver, regardless of whether or not active metabolites are produced in the process. It is in the question of the bio-availability of orally administered doses that this clinical significance lies. The term "bio-availability" may be defined as the transport rate and the percentage of the dose that can be detected in the general circulation, i.e. in the arterial and venous blood which conveys the

279

active substance to the tissues (excluding the hepatoportal blood during the phase of absorption).

The bio-availability of an oral dose may be far less than that of an intravenous dose *of the same size;* the difference in the case of one and the same substance may depend very much on the size of the dose, and the extent of this dose-dependence may in turn differ drastically from one substance to another.

As can be convincingly demonstrated by theoretical deduction, it is possible to obtain a relative indication of that portion of the dose which is measurable in the general circulation by reference to the area beneath the curve plotted for the time course of the concentration of active substance in the blood or plasma. Figure 2 illustrates how the concentration integral for propranolol in the plasma reportedly depends on the size of orally and intravenously administered doses[27]. Following intravenous administration, the specific increase in the area beneath the curve – i.e. the increase in the number of area units per dose unit – remains constant within the range measured:

$$\left[\frac{\Delta \text{AREA}}{\Delta \text{DOSE}}\right]_{\substack{\text{Prop.}\\ \text{i.v.}}} = 17 \, \frac{(\text{ng.} \cdot \text{hr/ml.})}{(\text{mg.})}$$

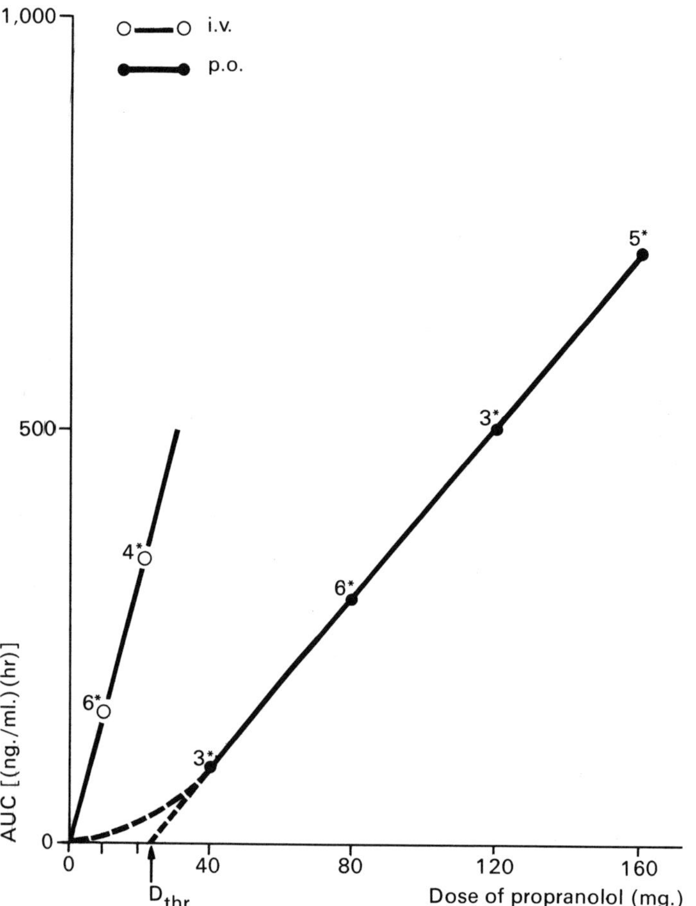

Fig. 2. Area under the plasma concentration curve in man in relation to the dose of propranolol administered either intravenously or orally (as reported by SHAND and RANGNO[27]). The figures followed by an asterisk indicate the number of subjects involved; $D_{\text{thr.}}$ = threshold dose.

When oral doses are administered in progressively larger amounts, the specific increase occurring in the area under the curve is found to be strongly dose-dependent up to a level of 40 mg., but becomes constant at doses of between 40 and 160 mg.:

$$\boxed{\frac{\varDelta \text{AREA}}{\varDelta \text{DOSE}}} \quad \frac{\text{Prop.}}{> 40 \text{ p.o.}} = 5\,\frac{(\text{ng.} \cdot \text{hr/ml.})}{(\text{mg.})}$$

In other words, for each unit by which the oral dose is raised above the 40 mg. level, the area beneath the curve increases by an amount equivalent to only about one-third of the increase recorded when the intravenous dose is raised by one unit.

This difference in the specific increase which the area beneath the curve undergoes following an oral as compared with an intravenous dose is a phenomenon that GIBALDI et al.[7] have interpreted as the so-called "first-pass effect". As revealed by the explanation outlined in Figure 3, this first-pass effect is of no significance under circumstances where the elimination or inactivation constant is very much smaller than the rate constant of transition from the hepato-portal system to the central compartment.

The model interpretation outlined here, however, only holds good in cases where the increase in the area beneath the curve per oral dose unit remains constant.

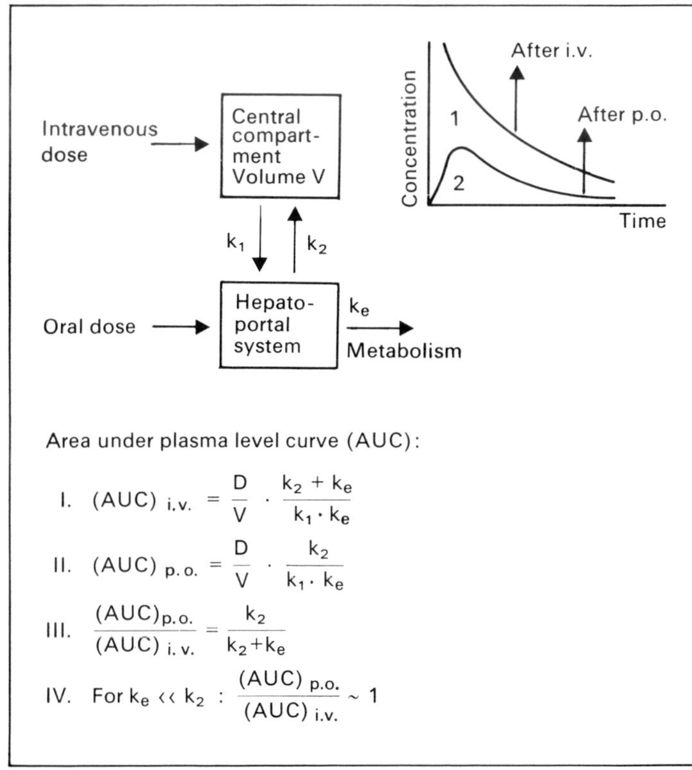

$$\text{I.} \quad (\text{AUC})_{\text{i.v.}} = \frac{D}{V} \cdot \frac{k_2 + k_e}{k_1 \cdot k_e}$$

$$\text{II.} \quad (\text{AUC})_{\text{p.o.}} = \frac{D}{V} \cdot \frac{k_2}{k_1 \cdot k_e}$$

$$\text{III.} \quad \frac{(\text{AUC})_{\text{p.o.}}}{(\text{AUC})_{\text{i.v.}}} = \frac{k_2}{k_2 + k_e}$$

$$\text{IV.} \quad \text{For } k_e \ll k_2 : \quad \frac{(\text{AUC})_{\text{p.o.}}}{(\text{AUC})_{\text{i.v.}}} \sim 1$$

Fig. 3. Theoretical explanation for the so-called "first-pass effect", as a result of which the bio-availability of an oral dose is less than that of an equally large intravenous dose. (Adapted from: GIBALDI et al.[7]).

To account for the dose-dependence observed at dosage levels of below 40 mg. propranolol, SHAND and RANGNO[27] have introduced the concept of a "threshold dose" ($D_{thr.}$) and have modified Equation III in Figure 3 as follows:

$$\frac{(AUC)_{p.o.}}{(AUC)_{i.v.}} = \frac{D - D_{thr.}}{D} \cdot \frac{k_2}{k_2 + k_e}$$

This modified equation, however, is merely equivalent to a linear extrapolation of the area-dose dependence to the abscissa in Figure 2 and, as such, is meaningless for $D < D_{thr.}$. It is, in fact, inadequate in the case of substances which also exhibit a dose-dependent first-pass effect at higher dosage levels. According to ÅBLAD et al.[1], one beta-blocker whose specific increase in the area beneath the curve displays a particularly marked dose-dependence, even at therapeutic dose levels, is alprenolol (Figure 4). The degree of ascent of the tangents to the dose-dependent plot for the area under the curves indicates the specific increase in this area at each dosage level. If one were to plot a steady curve through the experimental points, the maximum degree of ascent would be attained at a dose of 200 mg., this maximum only reaching a value of 5 area units/mg.:

$$\left[\frac{\Delta AREA}{\Delta DOSE}\right] Alpr.\ 200\ p.o. = 5\ \frac{(ng. \cdot hr/ml.)}{(mg.)}$$

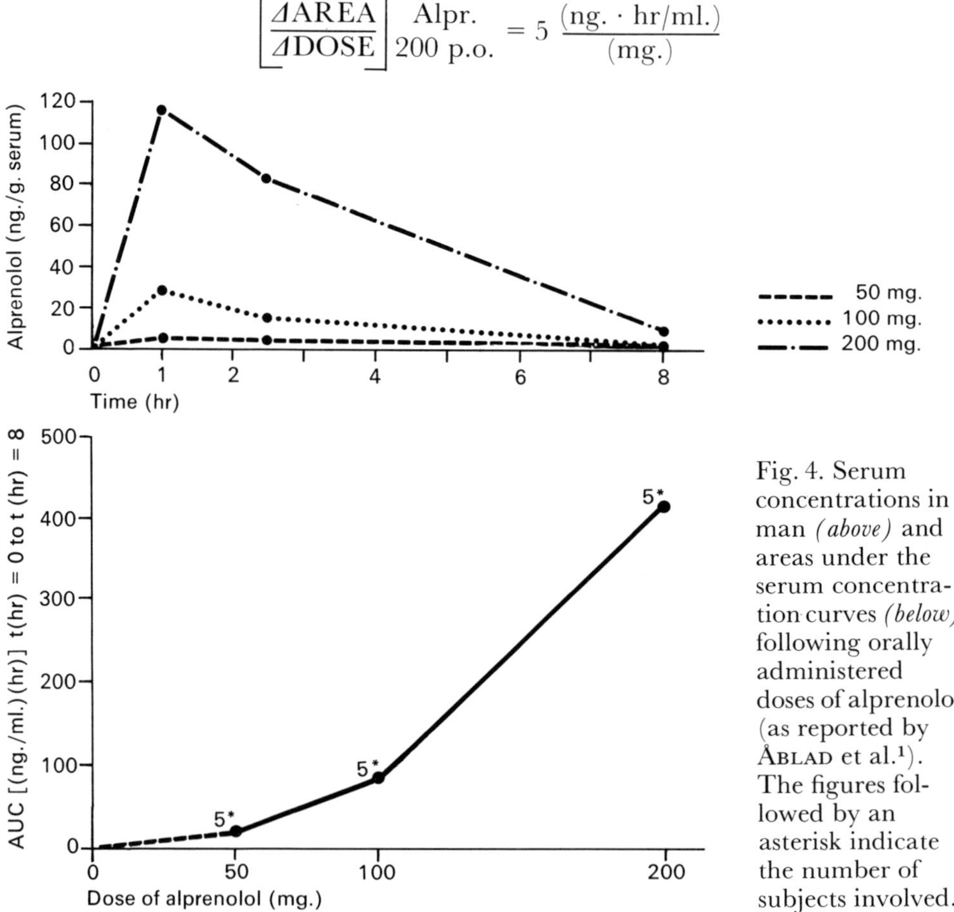

Fig. 4. Serum concentrations in man *(above)* and areas under the serum concentration curves *(below)* following orally administered doses of alprenolol (as reported by ÅBLAD et al.[1]). The figures followed by an asterisk indicate the number of subjects involved.

Table 3. Plasma concentrations (ng./ml.) of oxprenolol and areas under the plasma concentration curves (AUC) (ng./ml. · hr) after single oral doses of 40 mg., 80 mg., and 160 mg. administered to each of the same seven subjects.

Dose (mg.)	Subject	Time (hr)								(AUC) $\frac{t\,(hr)=8}{t\,(hr)=0}$
		0	0.5	1	2	2.5	4	6	8	
40	B	<10	188	169	125	100	50	18	<10	548
	C	<10	238	200	113	62	20	<10	<10	450
	D	<10	375	181	125	94	44	15	<10	645
	E	<10	269	156	100	106	50	<10	<10	539
	F	<10	94	150	125	100	53	34	<10	535
	G	<10	268	245	94	75	20	<10	<10	476
	H	<10	356	212	144	84	50	20	<10	680
	Mean	<10	255	188	118	89	41	12	<10	533
80	B	<10	376	362	200	175	75	10	10	945
	C	<10	625	375	200	138	69	18	<10	1,061
	D	<10	860	400	150	150	63	20	<10	1,180
	E	<10	425	575	175	162	20	<10	<10	928
	F	<10	400	330	144	131	92	52	22	977
	G	<10	663	262	100	88	25	<10	<10	780
	H	<10	600	380	250	188	75	20	<10	1,214
	Mean	<10	564	383	174	147	60	17	<10	1,012
160	B	<10	514	525	350	200	137	56	37	1,659
	C	<10	876	1,010	600	420	200	60	10	2,543
	D	<10	1,090	660	388	326	140	43	10	2,069
	E	<10	338	500	615	350	150	94	49	1,758
	F	<10	625	338	200	187	150	69	20	1,353
	G	<10	825	569	340	270	162	47	20	1,873
	H	<10	625	750	400	288	123	50	20	1,914
	Mean	<10	699	622	413	292	152	60	24	1,881

If we ignore the difference between the molecular weights of propranolol and alprenolol, we find that it is not until given in a dose of 200 mg. that alprenolol attains the specific area value which in the case of propranolol is already reached at a dose of 40 mg.

In view of the fact that adequate pharmacokinetic data are available on ox-prenolol, this preparation is suitable for purposes of comparison. Listed in Table 3 are the individual and mean plasma concentrations of oxprenolol, as well as the areas under the plasma concentration curves, following oral doses of 40 mg., 80 mg., and 160 mg. administered to the same seven subjects. The doses were given randomly with an interval of one week between them, and the plasma concentrations were determined by gas-liquid chromatog-raphy[12].

Shown in Figure 5 are the resultant mean plasma concentration curves for oxprenolol and – derived therefrom – the dose-dependence of the curve inte-gral. In contrast to propranolol, and particularly to alprenolol, the mean area-dose dependence in these seven subjects is linear, the line passing through

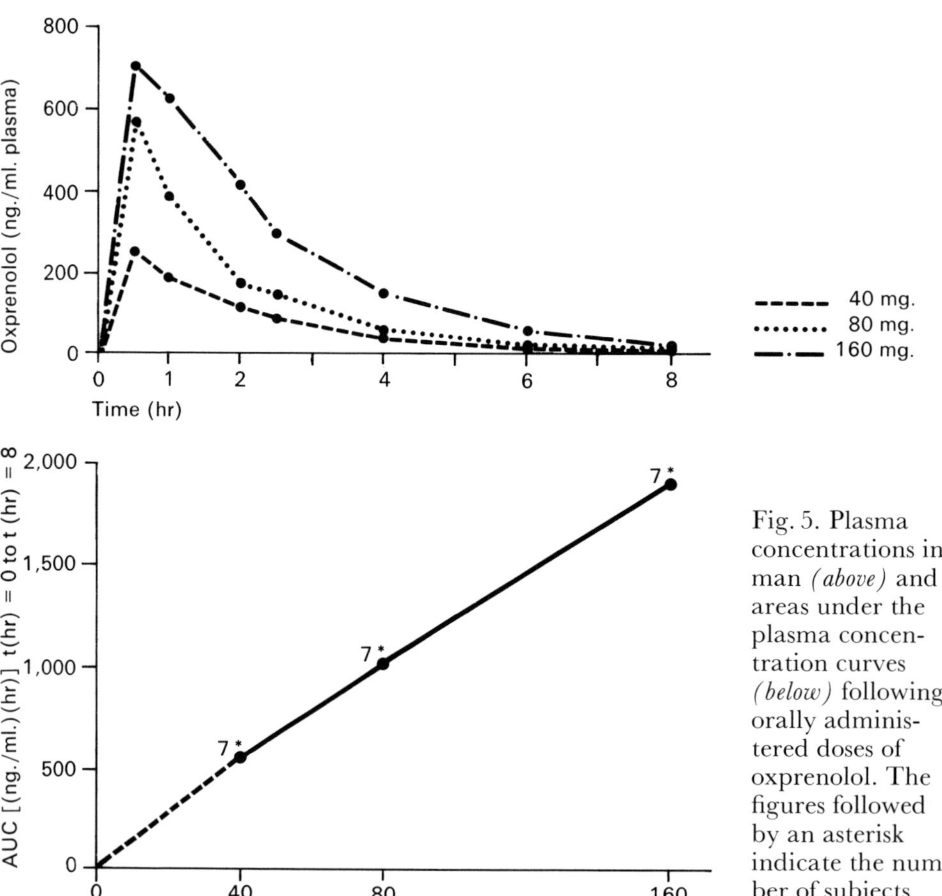

Fig. 5. Plasma concentrations in man *(above)* and areas under the plasma concentration curves *(below)* following orally adminis-tered doses of oxprenolol. The figures followed by an asterisk indicate the number of subjects involved.

the zero point when extrapolated backwards. The specific area under the curve,
i. e. the increase in the area per dose unit,

$$\left[\frac{\varDelta \text{AREA}}{\varDelta \text{DOSE}}\right]\frac{\text{Oxpr.}}{\text{p.o.}} \sim 12 \; \frac{(\text{ng.} \cdot \text{hr/ml.})}{(\text{mg.})}$$

is constant over the entire dosage range and exceeds, by a factor of roughly
two, the specific increase for propranolol at oral dosage levels of 40–160 mg.
and the maximal specific increase observed with alprenolol at an oral dosage
level of 200 mg. Oxprenolol is thus completely devoid of any dose-dependent
first-pass effect. To what extent the dose-independent specific area under the
curve depends on the distribution volume, or on a dose-independent first-pass
effect as postulated by GIBALDI et al.[7], can only be determined by comparisons
with the specific area under the curve after intravenous administration.
In the experiment in which the dose-dependence of the plasma concentrations was measured (Figure 6), we also recorded the following circulatory
variables: heart rate (Figure 7) and systolic blood pressure (Figure 8) in the
supine and upright positions as well as during exercise on the ergometer; in
addition, we measured the systolic time interval (WEISSLER et al.[32]), including
in particular the pre-ejection period (Figure 9), both while the subjects were
lying and after they had been passively tilted upwards.

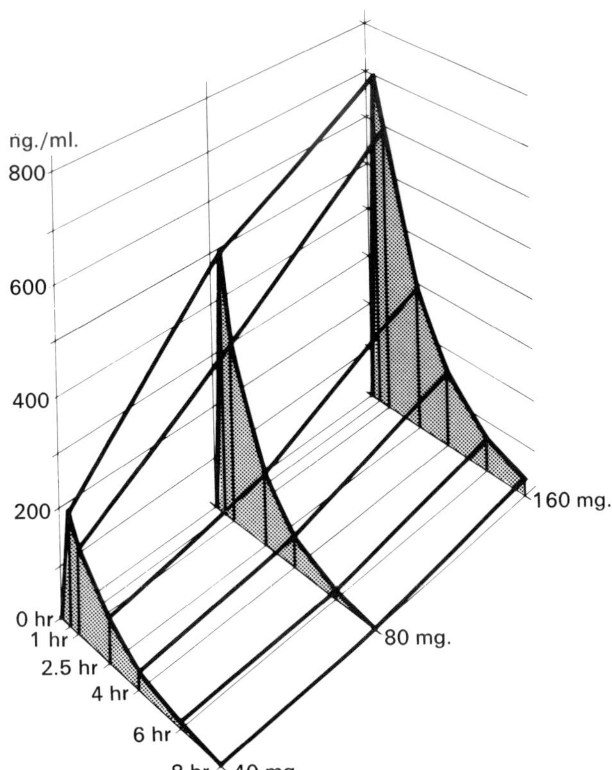

Fig. 6. Concentra-
tions of oxprenolol
measured in the
plasma of the same
seven subjects follow-
ing oral administra-
tion of 40 mg.,
80 mg., and 160 mg.

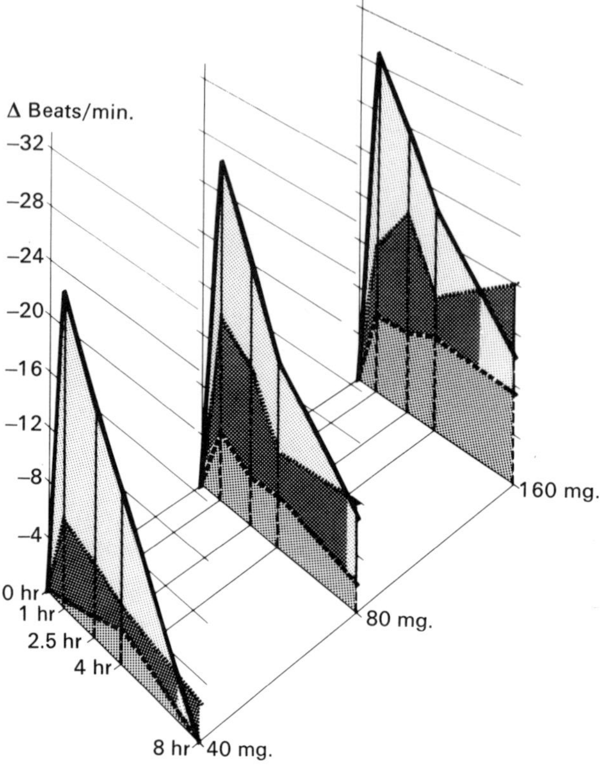

Fig. 7. Decrease in heart rate recorded in the same seven subjects following oral administration of 40 mg., 80 mg., and 160 mg. oxprenolol.

– – – – – measured recumbent
⋯⋯⋯⋯ measured standing
——— measured during the fourth and fifth minute of exercise (ergometer, 120 watts)

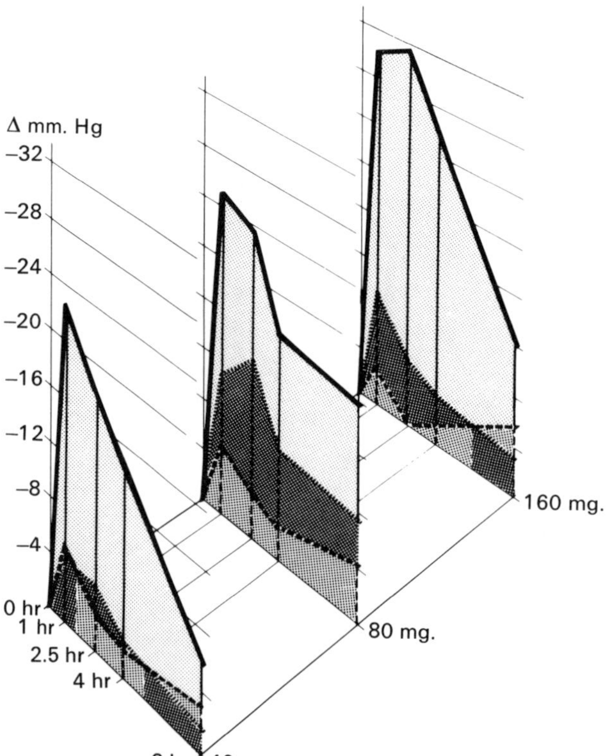

Fig. 8. Decrease in systolic blood pressure in the same seven subjects following oral administration of 40 mg., 80 mg., and 160 mg. oxprenolol.

– – – – – measured recumbent
⋯⋯⋯⋯ measured standing
——— measured during the fourth and fifth minute of exercise (ergometer, 120 watts)

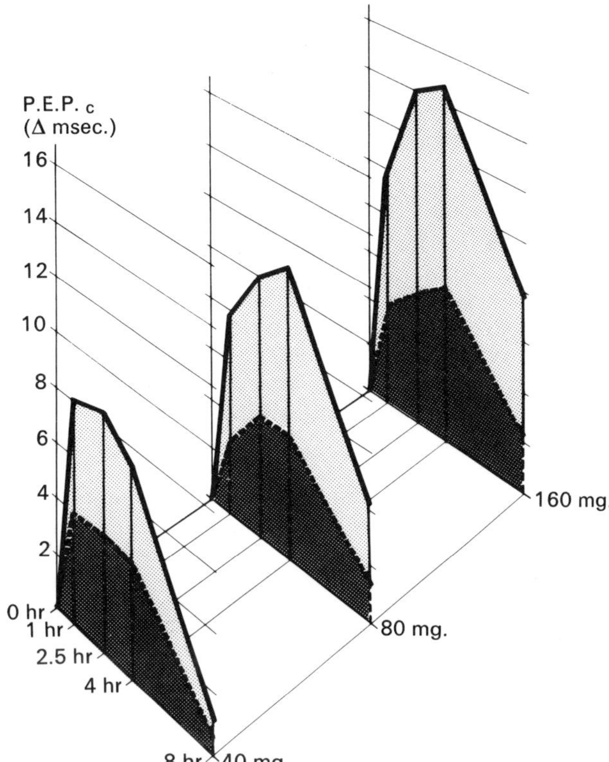

Fig. 9. Prolongation of frequency-corrected pre-ejection period (P.E.P.$_c$) in the same seven subjects following oral administration of 40 mg., 80 mg., and 160 mg. oxprenolol.

- - - - - - measured recumbent
———— measured during the third minute of passive upward tilting

A comparison between the plasma concentrations measured (cf. Figure 6) and the values recorded for these various circulatory parameters (cf. Figures 7, 8, and 9) reveals that the peak concentrations already appear in the flat upper portion of the concentration-response curve, which, as was to be expected, is of sigmoid configuration, whereas the concentrations and responses recorded after eight hours are to be sought in the steeply ascending portion.

Further human-pharmacology studies of this same type will have to be undertaken with the most important representatives of the beta-blocking group of substances in order to ensure that more systematic comparisons can be made between the concentration-response curves and the pharmacokinetic and physical parameters than are at present possible with such fragmentary documentation as is now available on the various preparations in this category.

References

1 ÅBLAD, B., ERVIK, M., HALLGREN, J., JOHNSSON, G., SÖLVELL, L.: Pharmacological effects and serum levels of orally administered alprenolol in man. Europ. J. clin. Pharmacol. *5*, 44 (1972)

2 BOND, P.A.: Metabolism of propranolol ("Inderal"), a potent, specific β-adrenergic receptor blocking agent. Nature (Lond.) *213*, 721 (1967)

3 DOLLERY, C.T., DAVIES, D.S., CONOLLY, M.E.: Differences in the metabolism of drugs depending upon their route of administration. Arch. int. Pharmacodyn. *192*, Suppl.: 214 (1971)

4 FITZGERALD, J.D., SCALES, B.: Effect of a new adrenergic beta-blocking agent (ICI 50,172) on heart rate in relation to its blood levels. Int. J. clin. Pharmacol. Ther. Toxicol. *1*, 467 (1968)

5 GARTEIZ, D.A.: Metabolism of a beta blocking drug, oxprenolol. J. Pharmacol. exp. Ther. *179*, 354 (1971)

6 GEORGE, C. F., FENYVESI, T., CONOLLY, M.E., DOLLERY, C.T.: Pharmacokinetics of dextro-, laevo- and racemic propranolol in man. Europ. J. clin. Pharmacol. *4*, 74 (1972)

7 GIBALDI, M., BOYES, R. N., FELDMAN, S.: Influence of first-pass effect on availability of drugs on oral administration. J. pharm. Sci. *60*, 1338 (1971)

8 GUGLER, R., HEROLD, W., DENGLER, H.J.: Pharmacokinetics of pindolol in man. Europ. J. clin. Pharmacol. *7*, 17 (1974)

9 HELLENBRECHT, D., LEMMER, B., WIETHOLD, G., GROBECKER, H.: Measurement of hydrophobicity, surface activity, local anaesthesia and myocardial conduction velocity as quantitative parameters of the non-specific membrane affinity of nine β-adrenergic blocking agents. Naunyn-Schmiedeberg's Arch. Pharmacol. *277*, 211 (1973)

10 HICKS, C. D.: The buccal absorption of some β-adrenoceptor blocking drugs. Brit. J. Pharmacol. *47*, 680P (1973); abstract of paper

11 HICKS, D. C., ARBAB, A.G., TURNER, P., HILLS, M.: A comparison of intravenous pindolol and propranolol in normal man. J. clin. Pharmacol. *12*, 212 (1972)

12 JACK, D.B., RIESS, W.: Determination of low levels of oxprenolol in blood or plasma by gas-liquid chromatography. J. Chromatogr. *88*, 173 (1974)

13 JOHANSSON, R., REGÅRDH, C.G., SJÖGREN, J.: Absorption of alprenolol in man from tablets with different rates of release. Acta pharm. suecica *8*, 59 (1971)

14 JOHNSSON, G., REGÅRDH, C.G., SÖLVELL, L.: Lack of biological interaction of alprenolol and salicylate in man. Europ. J. clin. Pharmacol. *6*, 9 (1973)

15 LEESON, G.A., GARTEIZ, D.A., KNAPP, W.C., WRIGHT, G.J.: N-methylation, a newly identified pathway in the dog for the metabolism of oxprenolol, a β-receptor blocking agent. Drug Metab. Disposition *1*, 565 (1973)

16 LEESON, G.A., GARTEIZ, D.A., KNAPP, W.C., WRIGHT, G.J.: Oxprenolol quantitation and metabolic identification. Fed. Proc. *32*, 747, Abstr. 3002 (1973)

17 LEINWEBER, F.-J., GREENOUGH, R. C., SCHWENDER, C.F., HAYNES, L.J., DI CARLO, F.J.: Bunolol metabolism by dogs: isolation and identification of two acidic metabolites. J. pharm. Sci. *60*, 1516 (1971)

18 LEINWEBER, F.-J., HAYNES, L.J., CREW, M.C., DI CARLO, F.J.: Absorption, tissue distribution and excretion of bunolol-^{14}C by dogs. J. pharm. Sci. *60*, 1512 (1971)

19 LISH, P.M., SHELANSKI, M.V., LABUDDE, J.A., WILLIAMS, W.R.: Inhibition of cardiac chronotropic action of isoproterenol by sotalol (MJ 1999) in rat, dog and man. Curr. ther. Res. *9*, 311 (1967)

20 OHNHAUS, E.E., NÜESCH, E., MEIER, J., KALBERER, F.: Pharmacokinetics of unlabelled and ^{14}C-labelled pindolol in uraemia. Europ. J. clin. Pharmacol. *7*, 25 (1974)

21 PACHA, W.L.: A method for the fluorimetric determination of 4-(2-hydroxy-3-isopropylaminopropoxy)-indole (LB 46), a β-blocking agent, in plasma and urine. Experientia (Basle) *25*, 802 (1969)

22 PATERSON, J.W., CONOLLY, M.E., DOLLERY, C.T., HAYES, A., COOPER, R.G.: The pharmacodynamics and metabolism of propranolol in man. Pharmacol. clin. *2*, 127 (1970)

23 RIESS, W., HÜRZELER, H., RASCHDORF, F.: The metabolites of oxprenolol (Trasicor) in man. Xenobiotica *4*, 365 (1974)

24 SCALES, B., COSGROVE, M.B.: The metabolism and distribution of the selective adrenergic beta blocking agent, practolol. J. Pharmacol. exp. Ther. *175*, 338 (1970)

25 SHAND, D.G., EVANS, G.H., NIES, A.S.: The almost complete hepatic extraction of propranolol during intravenous administration in the dog. Life Sci. *10/I*, 1417 (1971)

26 SHAND, D.G., NUCKOLLS, E.M., OATES, J.A.: Plasma propranolol levels in adults. With observations in four children. Clin. Pharmacol. Ther. *11*, 112 (1970)

27 SHAND, D.G., RANGNO, R.E.: The disposition of propranolol. I. Elimination during oral absorption in man. Pharmacology (Basle) *7*, 159 (1972)

28 SVEDMYR, N., JAKOBSSON, B., MALMBERG, R.: Effects of sotalol and propranolol administered orally in man. Europ. J. Pharmacol. *8*, 79 (1969)

29 TRUELOVE, J.F., VAN PETTEN, G.R., WILLES, R.F.: Action of several β-adrenoceptor blocking drugs in the pregnant sheep and foetus. Brit. J. Pharmacol. *47*, 161 (1973)

30 WALLE, T., GAFFNEY, T.E.: Metabolism of propranolol in man: mass spectrometric identification of four new metabolites. Pharmacologist *13*, 221, Abstr. 165 (1971)

31 WALLE, T., GAFFNEY, T.E.: Propranolol metabolism in man and dog: mass spectrometric identification of six new metabolites. J. Pharmacol. exp. Ther. *182*, 83 (1972)

32 WEISSLER, A.M., HARRIS, W.S., SCHOENFELD, C.D.: Systolic time intervals in heart failure in man. Circulation *37*, 149 (1968)

Discussion

A. Zanchetti: In collaboration with Chidsey and his group, we in Milan recently did some pharmacokinetic work with propranolol in man*. We observed that the half-life of propranolol differs quite considerably, depending not so much on the route of administration but rather on whether the administration is acute or chronic. The difference as between intravenous and acute oral administration is thus very slight: you get the same half-life when you give the drug intravenously or orally in a single dose. But, if you resort to chronic oral administration, you get a much longer half-life. The acute half-life is roughly two hours, whereas the chronic half-life is about four hours.

With regard to the relationship between the drug's pharmacokinetic behaviour and its biological effects, this is a question on which studies recently undertaken by Morselli in rats shed some interesting light. These studies are still in progress and have not yet been published. When Morselli compared the half-life in the plasma with the half-life in various organs, he found that the half-life in the plasma, which was about 45 minutes, was very similar to that in the brain; but in the atria and in the kidney it was much longer, i.e. about 120 minutes. This may perhaps be an important finding, because the atria and kidney are probably two of the target organs involved in the mechanism by which beta-blockers reduce the heart rate and the production of renin.

W. Riess: What methodology did Morselli use?

A. Zanchetti: It was a highly selective gas-chromatographic method which makes it possible to measure very small amounts of the order of about 1 ng./ml.

W. Riess: That's fine. I wondered for a moment whether it was perhaps the old fluorimetric method.

A. Zanchetti: No. It was for this reason that we tackled the work together, i.e. because Chidsey and his group had developed this very sensitive method**.

W. Riess: Our colleagues at the Hammersmith Hospital have criticised some of the old propranolol data on the grounds that these data need revising, because they were obtained using the fluorimetric method, which is not as specific as one would wish for the purposes of cross-reference.

A. Zanchetti: Yes, that's certainly very true.

W. Riess: The second part of your comment, Dr. Zanchetti, prompts me to add a word or two about our findings with oxprenolol (®Trasicor). We have collected a fairly large amount of analytical data using different methodologies. Some of these data were obtained at a very early stage in the development of the drug, when we were using radio-isotopically labelled material; here, we were able to make very sensitive measurements down to below the 1 ng. level. One thing we observed in the low-concentration range was deviation from the ideal one-compartment kinetic behaviour. And this may possibly explain your finding of a discrepancy between the so-called half-lives recorded in the plasma and the tissues. But again I would like to stress that this is not something specific to beta-blockers.

Where the methods of measurement used are sensitive enough, such a discrepancy is a very common observation with many drugs. The ideal pharmacokinetic behaviour

---

* Morselli, P. L., Morganti, A., Bianchetti, G., Di Salle, E., Leonetti, G., Zanchetti, A., Chidsey, C.A.: Blood levels and pharmacokinetic studies of propranolol during chronic treatment in hypertensive patients. Europ. J. clin. Invest. (printing)
** Di Salle, E., Baker, K. M., Bareggi, F. R., Watkins, W. D., Chidsey, C.A., Frigerio, A., Morselli, P. L.: A sensitive gas chromatographic method for the determination of propranolol in human plasma. J. Chromatogr. 84, 347 (1973)

in terms of single-compartment open-model kinetics – is as a rule seen only in the upper portion of the concentration course, i. e. down to, say, 10 ng.

A. Zanchetti: I should perhaps emphasise again that this discrepancy, as you have termed it, is not between plasma and organs, but only between plasma and *some* organs, notably the atria and kidney; the half-life in the brain, for example, is the same as in the plasma.

W. Riess: I should expect that those organs which have an abundant blood supply and are very well perfused would also equilibrate very rapidly with the plasma; in these organs I should therefore think you would tend to find the same kinetics as recorded in plasma.

A. Zanchetti: Are you suggesting that the kidney doesn't have a very abundant blood supply?

W. Riess: No, of course not. But I think we have to regard the kidney as a metabolically active organ itself.

F. Gross: Dr. Riess, do you happen to have a curve – similar to the ones you showed us in your paper – indicating the area under the plasma concentration for practolol? Secondly, may I ask the clinicians how they account for the fact that practolol has about five times the half-life of propranolol and of other beta-adrenergic blockers? Isn't there a certain danger that, at least at the beginning of treatment, you might have cumulative effects with practolol – unless, of course, you give it in more widely spaced doses than propranolol? Someone said today, however, that practolol is given three times daily, similarly to propranolol.

W. Riess: I don't have the blood level curves for practolol with me, but they are to be found in a paper by Fitzgerald and Scales*.
Though the question you ask, Dr. Gross, about the possibility of cumulative effects with practolol was addressed to the clinicians, perhaps I may be permitted to add a word on that, too. Accumulation of a drug in the organism is often thought of as something that continues throughout treatment with repeated doses. But, if in a single-dose experiment it has been demonstrated that a compound does in fact leave the body again, then such a thing as infinite accumulation during repeated administration does not occur. There's always a state reached where the input per dose interval is equal to the amount eliminated in the same time span.

F. Gross: Yes, I agree. You arrive at a steady state for the plasma concentration, and there are various examples illustrating that this steady state may be attained either more rapidly or more slowly. But isn't it possible that the plasma concentrations reached with practolol may be relatively high as compared with those which might already be sufficient to produce a similar effect?

W. Riess: This question could only be answered if we had the same kind of data for practolol as we have for the other beta-blockers we have discussed.

J. Ménard: In connection with the studies reported on by Dr. Riess, I should like to mention some of the findings we have obtained in tests on the bio-availability of oxprenolol. In these tests, 80 mg. oxprenolol was given orally in the form of a single tablet which also contained 25 mg. dihydralazine. The purpose of these tests was to see whether oxprenolol would suppress two unwanted effects of dihydralazine – namely, tachycardia and headache. To do so, it would of course have to reach the beta-adrenergic receptors in sufficiently high concentrations before the dihydralazine had taken effect. We administered, then, oxprenolol (80 mg.) and dihydralazine (25 mg.)

---

* Fitzgerald, J. D., Scales, B.: Effect of a new adrenergic beta-blocking agent (ICI 50,172) on heart rate in relation to its blood levels. Int. J. clin. Pharmacol. Ther. Toxicol. *1*, 467 (1968)

at 8 a.m., i. e. before breakfast, to 29 patients suffering from arterial hypertension who had no past history of angina pectoris or myocardial infarction and who had been in hospital for 5–10 days. Their arterial blood pressure was determined in the lying position, as well as after they had been standing for one minute and after exercise (10 knee-bends), and their heart rates were recorded at the same time. In all the tests, the blood pressure showed a statistically significant reduction after one hour; the reduction reached its maximum after two and three hours, and was no longer statistically significant after four hours. This time course tallies well with the known effects of dihydralazine on the blood pressure. Heart rate decreased to a statistically significant extent after only one hour, the decrease disappearing after five hours. The plasma levels of oxprenolol reached their maximum after some 60–90 minutes. No unwanted cardiac effects were noted in these 29 patients, but headache was reported in ten cases. In four of these cases it was not treated, but in the remaining six, in which it persisted until the end of the test, the patients were given either acetylsalicylic acid (three cases) or clonidine (three cases). Whereas the acetylsalicylic acid had no effect, clonidine (75 mcg.) eliminated the headache within 30 minutes.

J. SCHWARTZ: It strikes me that there is some confusion concerning this question of bio-availability. The aim of bio-availability studies is to provide clinicians and pharmacologists alike with information from which they can deduce whether or not a given drug reaches a certain number of sites of action. Biological half-life, reflecting as it does the plasma kinetics of a compound, doesn't furnish an exact picture of what is actually happening to the drug in the organism. In fact, it is the pharmacological data you have given us, Dr. RIESS, that are relevant, inasmuch as they may enable us to anticipate the biochemical data which you are not yet in a position to supply.

W. RIESS: I don't claim that the data we now have cover all the points which have to be settled in order to ensure that the best, safest, and most reliable use can be made of the beta-blockers or that the most judicious selection can be made from those now at our disposal. I'd like to stress once again that we are only at the beginning of the task of sorting out our data, and we still have a long way to go before we can determine exactly what this dose-dependence of the area under the curve actually means in terms of advantages or disadvantages. It wasn't my intention to present here some sort of key that would enable you to give one compound an extra plus mark and another an extra minus mark. We are firmly convinced that, if you concentrate on only one parameter of pharmacokinetic behaviour, you will certainly not be doing justice to the extreme complexity of the system that needs to be investigated. It may well be that the strong dose-dependence of the area under the curve reflects a physico-chemical parameter which is also essential for the pharmacological effect *per se*. But we do not know whether this is indeed so. If our paper did create this impression, then I certainly want to correct it. It was not our intention to draw up, on the basis of the very limited data contained in this paper, an order of merit for the various compounds or to imply that oxprenolol is absolutely the best one currently available. True, it has no dose-dependent first-pass effect, but this paper of ours was intended only as an attempt to convey some tentative idea of the implications that can be drawn from the data now available. It was meant to encourage further investigations and, above all, to emphasise how essential it is that the types of experimental set-up used in the various countries should be much more closely coordinated.

J. SCHWARTZ: Thank you for this explanation, but I think that the opposite approach – i. e. to start with the pharmacological half-life and then, using this information as a basis, to try to elucidate the biochemical phenomena underlying the half-life – might perhaps be more fruitful.

U. C. DUBACH: Dr. RIESS, which preparation would you consider preferable – on the basis of its excretion pattern and half-life – for the long-term treatment of hypertension?

W. Riess: I don't believe that the half-life has a decisive role to play in the choice of the most suitable preparation for the treatment of hypertension, because in this indication the dosage schedule can presumably always be adjusted accordingly. I think, though, that once sufficient metabolic data on the various preparations are available, preference should be given to the one with the simplest metabolism. Oxprenolol, the bulk of which is merely glucuronised, would fulfil this prerequisite. But no final assessment of the various preparations can be made until we have more metabolic data and wider experience at our disposal.

U. C. Dubach: Speaking as a clinician, I'd like to express the wish that the pharmaceutical industry should, if possible, supply us with a preparation that doesn't have to be administered too often and is also safe. In the case where a patient has to take tablets for a period of years, or even for the rest of his life, once a day is better than twice a day and twice a day is better than three times a day.

W. Riess: In this discussion of the pharmacokinetic data on beta-blockers no mention has so far been made of tolerability. Oxprenolol displays a very wide therapeutic margin. This means that recourse can safely be had to high single doses of the drug, because such doses prolong the duration of action without leading to a corresponding potentiation of the drug's cardiac effects. Thus, safety margin and tolerability are criteria to which due consideration should certainly be given when selecting the most suitable preparation.

P. R. Imhof: One thing our investigations have very clearly shown is that, although the plasma concentration of oxprenolol has already fallen to unmeasurable or almost unmeasurable levels eight hours after the administration of a dose of 40, 80, or 160 mg., the signs of beta-blockade are still very much apparent at this point in time. In addition, it has been found that the higher the dose administered, the longer the duration of effect. It seems that the substance remains considerably longer at the receptor sites than would be supposed from its plasma half-life. Incidentally, the time quoted for the duration of action of beta-blockers varies depending on the parameter measured. Isoprenaline antagonism persists longest, followed by inhibition of emotion-induced tachycardia and tachycardia of effort. Depending on the parameter measured, the duration of action of a single oral dose of 80 mg. oxprenolol ranges from 12 to 16 hours, with the result that a twice-daily dosage schedule can quite safely be recommended.

G. Hitzenberger: I should like to revert to what Dr. Zanchetti had to say about the prolongation of the half-life of beta-blockers following repeated administration. Do you know, Dr. Riess, whether any other studies have been done in man and, if so, whether a prolongation has in fact been observed? I ask this because, if such a prolongation does occur, one would have to assume the existence of a kind of "deep compartment" to explain it.

W. Riess: I can only refer to data we have obtained with oxprenolol in test subjects given single daily doses for 14 days. Over the entire period of observation the blood concentration curves were virtually superposable on one another. We didn't see any significant increase or decrease in the apparent half-life.

P. Cottier: What is the excretion pattern of the various beta-blockers in patients with renal failure? It would be interesting to know this so that the dosage could, where necessary, be adjusted in such cases.

W. Riess: From Table 1 in my paper, in which the liposolubility of various beta-blockers is indicated in descending order, it can be seen that virtually all beta-blockers are eliminated via the kidneys. In the case of those substances which are hardly metabolised at all, a reduction in the dosage would certainly be indicated in the presence of renal failure; the single doses and the interval between them would have, in fact, to be readjusted in the light of the clinical response. In the case of the other substances,

293

whose duration of action is chiefly governed by their rate of metabolism – i.e. those substances which are inactivated by metabolism – a dose reduction does not seem to be necessary, but the metabolites certainly do then accrue to some extent.

P.R. Imhof: Under beta-blockade glomerular filtration and renal blood flow may be slightly reduced – i.e. by something like 10–15%. A report was recently published* on three cases of moderately severe uraemia which became worse in response to beta-blockade, with the result that in two of them dialysis had to be carried out earlier than had been planned. In the third patient, renal function returned to its pre-treatment status following withdrawal of the beta-blocker and no further measures were required. This report, however, has not gone unanswered, inasmuch as other authors** have emphasised precisely the fact that in patients suffering from disorders of renal function beta-blockers are associated with a low degree of risk. Our own view is that patients with impaired renal function may, in principle, be given beta-blockers, but that caution is indicated in the presence of severe uraemia.

A. Zanchetti: In our experience, the half-life of propranolol is not dramatically increased in patients with severely impaired renal function. It might be more markedly prolonged in patients undergoing haemodialysis, but so far we have only very preliminary data on this point.

P. Cottier: While we are on this subject, I should like to point out that clinical experience has shown beta-blockers to be effective in hyperkinetic syndrome associated with chronic uraemia. We have no misgivings about using these drugs, even without digitalisation, in patients undergoing dialysis. With regard to the treatment of hyperkinetic syndrome in patients with chronic uraemia who are not being dialysed, the question as to the renal excretory mechanisms of the various beta-blockers assumes of course added significance. At all events, chronic uraemia associated with hyperkinetic syndrome – with or without an increase in blood pressure – seems to me to be an important indication for this type of drug.

W. Riess: May I close this discussion by expressing the wish that, to facilitate comparisons, an attempt be made to standardise, as far as possible, all experimental procedures employed in the evaluation of beta-blockers. In addition, during cardiological assessment of the properties of beta-blockers a great deal more attention should be paid to the quantitative assay of drug concentrations in biological media.

---

* Warren, D.J., Swainson, C.P., Wright, N.: Deterioration in renal function after beta-blockade in patients with chronic renal failure and hypertension. Brit.med.J. *ii*, 193 (1974)
** Thompson, F.D., Joekes, A.M.: Beta-blockade in the presence of renal disease and hypertension. Brit.med.J. *ii*, 555 (1974); corresp.

General discussion on guidelines for the use of beta-blockers in general practice

B. GARNIER: Mr. Chairman, Ladies, and Gentlemen, the effectiveness of beta-blockers in the treatment of arrhythmias, angina pectoris, and hypertension has been amply demonstrated in the course of this symposium. Clinical experience conclusively indicates that these drugs are well tolerated, that they seldom provoke serious side effects, and that they can confidently be expected to occupy in future an important position in the management of cardiovascular diseases.

It now remains for us to draw from our discussions useful information and conclusions which will help general practitioners to employ beta-blockers in their proper indications and in such a way as to ensure that they yield optimum results. As a general practitioner myself, I have been entrusted with the task of pointing out to you the questions, doubts, and misgivings which my fellow G.P.s might have about the routine use of beta-blockers and of inviting you to provide simple guidelines, and possibly treatment schedules, for use in everyday medical practice. The fears and misgivings to which a G.P. falls prey when confronted with a new method of treatment are to be accounted for, to some extent, by the fact that in the last analysis he and only he is responsible for the well-being of the patients under his care. It is true that he receives a wealth of advice on how to employ a new drug, but, should he make a mistake, he will then find himself alone with his patient and his conscience.

To aid me in my task, two specialists in hypertension are sitting here beside me – Dr. GROSS on my left and Dr. MÉNARD on my right. They will help me to discuss with you the various questions with which the general practitioner may be faced.

1. The use of beta-blockers in the treatment of cardiac arrhythmias

B. GARNIER: Let us begin by examining the use of beta-blockers in the management of cardiac arrhythmias. Dr. CHUNG has covered this subject very fully in his paper. The first conclusion to be drawn from his remarks is that the intravenous administration of beta-blockers should be reserved for hospital use only. I think that everybody is in agreement on this point. The second conclusion emerging from Dr. CHUNG's paper is that beta-blockers should only be used once the nature of the arrhythmia has been precisely diagnosed. What implications does this have for the general practitioner?

E. K. CHUNG: When emphasising the need for precise diagnosis, I was not referring only to the use of beta-blockers. If we make a wrong diagnosis, or if we overlook possible aetiological factors, we cannot achieve good therapeutic results. The advice I always give to students, interns, and specialists is to record an E.C.G. in cases of arrhythmia before giving any drug. The only exception is an extreme emergency, where the patient appears to be dying as a result of arrhythmia. In such a situation, you can of course try, as a last resort, any particular agent you like, or any other emergency therapeutic measure, such as D.C. shock or cardiac massage.

W. SCHWEIZER: While I think that Dr. CHUNG is right, in principle, to insist on having electrocardiographic evidence of the dysrhythmia before treating it, I would say that, in cases where the patient is a young man in whom repeated attacks of paroxysmal tachycardia are the only abnormality and where you can depend on the story given by the patient, you can use beta-blockers prophylactically without first recording an E.C.G.

W. SOMERVILLE: We must consider the two most common situations. The first is that of the individual who is subject to repetitive attacks of rapid palpitation, going back for years, and who asks for treatment to stop the attacks. Under these circumstances I would have thought it pedantic to insist on an electrocardiogram before starting anti-arrhythmic treatment. The second is that of the individual who for the first time becomes aware of disturbing palpitation. If this palpitation is causing some haemodynamic upset, and particularly if it is associated with myocardial ischaemic pain, then I would think it unwise to prescribe treatment without knowing exactly what you are dealing with.

B. Garnier: May I ask Dr. Rivier what he thinks about the use of beta-blockers in young patients suffering from functional tachycardia?

J.-L. Rivier: One type of case which general practitioners are frequently called upon to deal with is that of the patient, usually a young person, who presents with troublesome sinus tachycardia of obscure origin, in which psychic or environmental factors may often play a role. In such cases, prolonged treatment with beta-blockers frequently proves most beneficial and entails no risks.

E. K. Chung: It was to this kind of case that I was also referring when I mentioned in my paper that propranolol is extremely useful as a means of slowing persistent and marked sinus tachycardia in young individuals suffering from anxiety neuroses or from hyperkinetic heart syndrome. When the practitioner is confronted with a patient of this type, he should try to determine the possible cause. The patient may be consuming too much coffee, tea, or coca-cola, etc., in which case he should simply be instructed to avoid these beverages. In the presence of a high-output state, the possibility of hyperthyroidism should be investigated. When no precise cause can be found, most of these patients will derive benefit from treatment with a beta-blocker. In the rare instances where the tachycardia persists or where frequent bouts continue to occur, the patient's E.C.G. should be re-examined, since the condition may possibly not be one of sinus tachycardia at all.

P. Kielholz: May I endorse what Dr. Rivier and Dr. Chung have just said. There is a very common tendency in cases of functional cardiac disturbances to prescribe minor tranquillisers. Since these drugs often lead to habituation after only a few weeks, they have to be given in ever-increasing doses. When they are finally withdrawn, the patient's anxiety and functional symptoms become even more pronounced. Hence, we should certainly urge that minor tranquillisers should not be employed in conditions of this type.

B. Garnier: You would therefore prefer beta-blockers to conventional minor tranquillisers. Aren't you afraid that beta-blockers, too, might in time give rise to some degree of habituation?

P. Kielholz: According to the information at our disposal today, there is no risk of habituation. Moreover, these cases of sinus tachycardia already show a good response to very low doses of beta-blockers.

2. The use of beta-blockers in the treatment of angina pectoris

2.1. General aspects

J. Ménard: Whatever the type of disorder for which beta-blockers are used, the first and foremost requirement is that the general practitioner should be clearly aware of their contra-indications. Absolute contra-indications are: a past history of asthmatic attacks, the presence of an obstructive pulmonary disease or of an A-V conduction disorder, and prolongation of the P-Q interval. Relative, or temporary, contra-indications include cardiac insufficiency, as revealed by questioning the patient, by the presence of exertional dyspnoea, and by clinical examination. In these cases the heart failure must be treated first of all. It is also important to know in which indications preference should be given to beta-blockers; among such indications are combinations of two disorders both of which are amenable to beta-blockade, as in the case of patients presenting with essential hypertension and angina pectoris or with angina pectoris and a cardiac arrhythmia likely to respond to treatment with beta-blockers.

B. GARNIER: I was very interested in the classification of angina that Dr. SOMERVILLE gave us during the general discussion after Dr. RIVIER's paper. This classification was as follows: Grade 1 – angina in response to unusual effort; Grade 2 – angina in response to usual effort; Grade 3 – angina in response to less than usual effort; and Grade 4 – angina at rest. Would Dr. SOMERVILLE perhaps tell us once again in which grade he uses beta-blockers, for how long he uses them, and at what stage he considers changing the treatment if it is not proving as efficacious as it should?

W. SOMERVILLE: The first thing to decide, when faced with a case of angina pectoris, is whether the patient can afford to wait for treatment or whether he is an emergency case. Where the latter can be excluded, the physician then has to find out the provocative causes of the angina and deal with these first. If, for instance, the patient is living an irregular life and smoking 40 cigarettes a day, it would be incorrect to begin treatment with a drug, even a beta-blocker, before bringing about certain changes in his life style. A programme of intensive treatment would take account of the following factors: the patient's occupation; his domestic situation; his body weight; his consumption of cigarettes and alcohol; the proper use of glyceryl trinitrate; and the use of beta-blockers. For a person in Grade 3 I would prescribe this programme for up to one week. If it were not successful in that period of time, then I would consider the next step, i.e. a coronary arteriogram. In Grade 2 I would allow the treatment to continue for a matter of weeks and in Grade 1 for 1–3 months. I always leave open the possibility that any person in any of these grades might at any time require coronary arteriography.

D.W. GAU: In connection with the contra-indications to the use of beta-blockers, I would like to mention two other relative contra-indications – or, at least, situations in which care must be taken. Firstly, in the discussion following the paper presented by Dr. RIESS, Dr. IMHOF has already referred to patients who are in incipient renal failure, and I think that doctors should be warned to keep an eye on the blood urea before they start treatment. Secondly, in cases where an insulin-dependent diabetic is in the habit of relying on incipient hypoglycaemia to tell him when he should take sugar, it should be borne in mind that this hypoglycaemia is sometimes inhibited by beta-blockers; I have heard of cases in which the patient went straight into coma because he had not been warned about this risk.

As regards the classification of angina, most of the cases in my practice belong to Dr. SOMERVILLE's Grade 2. The patients in Grade 1 do not seem to require treatment, because they do not take much exercise. Nine of our 14 angina patients are now on beta-blockers, and they are so delighted with these drugs that they refuse to stop taking them.

2.2. Angina pectoris associated with disturbances of conduction and rhythm

J.G. LEWIS: Could I ask whether bundle-branch block is a contra-indication to beta-blockade, because I have been using beta-blockers in this situation without any apparent trouble.

E. SOWTON: I do not regard bundle-branch block as a contra-indication at all, provided that only one of the three fascicles is blocked. In my view, it is perfectly all right to use beta-blockers, for instance, in the presence of right bundle-branch block with a normal axis. I would be worried, though, if there was evidence of right bundle-branch block and left anterior hemi-block, because in such a case the depression of conduction caused by beta-blockers might provoke sudden complete heart block.

E.K. CHUNG: May I add a few comments on the use of beta-blockers in the elderly? Many people aged over 60 years, and particularly those aged over 70 years, display some degree of sinus node disease. This may manifest itself, for example, as atrial fibrillation with a relatively slow ventricular rate. We should be extremely careful in this type of patient, because beta-blockers may slow down the ventricular rate still further and may even produce ventricular standstill.

2.3. Surgical operations and anaesthesia in patients undergoing treatment with beta-blockers

P. TAGGART: Could I just ask Dr. IMHOF one question? Supposing you find yourself committed to using beta-blockade in a given case, and that then you decide that surgery is fairly urgently necessary, for just how long do you have to withdraw the beta-blocker in order to be able to assure the surgeon that there is no beta-blocking activity left?

B. GARNIER: I think I can possibly answer that question myself. It seems that most surgeons, or anaesthetists, advise doctors to stop beta-blockers 48 hours before operation and that this is long enough for normal myocardial function to become restored. Beta-blockade is then resumed 24 or 48 hours after the operation.

R. C. TARAZI: The time at which beta-blockers should be withdrawn prior to surgery probably depends to some extent on the type of operation contemplated. The anaesthetists in our clinic, guided perhaps more by clinical observations than by the results of strictly controlled studies, require that, in the case of coronary artery bypass operations, beta-blockers be stopped two or maybe three weeks beforehand.

J.-L. RIVIER: To date, more than 200 patients have undergone coronary bypass operations in our department, and in some of them the operation had to be performed at relatively short notice. Several of these patients were receiving beta-blockers, and these drugs were withdrawn at the latest two days prior to surgery. We never had the impression that cardiac function was disturbed after the operation in these patients. As a rule, however, we ask patients under treatment with beta-blockers to stop taking the drugs four or five days before surgery in cases where the date of the operation is known in advance.

J. BROD: I can't see why the administration of beta-blockers has to be stopped before carrying out an ordinary operation. I am not thinking now of heroic operations such as coronary bypass or cardiac surgery but of the thousands of patients requiring appendicectomy or herniotomy – i.e. operations which certainly do not impose any major strain on the heart. I wonder whether it is really necessary to postpone such operations for 48 hours just because the patient has for years been receiving treatment with a beta-blocker.

E. SOWTON: We have operated on about 25 patients with acute coronary insufficiency – i.e. with the "crescendo" type of angina. They all received beta-blockers right up until the moment of surgery, and we had no trouble whatever restarting the heart after the bypass had been completed. I think the difference is that the left-ventricular muscle is completely normal in these patients. Their situation is quite different from that of patients who have had a previous infarction, who display some muscle damage and fibrosis, and who may have been receiving beta-blocking treatment for chronic angina for a period of many months.

J. G. LEWIS: In Britain, anaesthetists have an equal say in the question of drug treatment prior to surgery – in fact, they are sometimes the final judges. In the case of operations other than heart surgery, it is our practice not to withdraw beta-blockers until the day of the operation, on which no oral drugs at all are taken.

2.4. Combination of beta-blockers with other drugs

E. SOWTON: In a patient who is not receiving any other drugs the use of sublingual nitroglycerin produces one major effect which is not beneficial – i.e. reflex tachycardia.

This is sympathetically mediated and, as FITZGERALD* has shown, usually involves an increase of about 30–35 beats a minute in the heart rate. As a result, the oxygen demand of the heart will be stepped up considerably. This tachycardia can be almost entirely prevented by beta-blockade, and it is of course for this reason that FITZGERALD advocated the use of nitroglycerin as a test for establishing adequate beta-blockade. Clinically, it means that a beta-blocker and nitroglycerin represents an extremely satisfactory regimen.

As regards the question of the so-called long-acting nitrates and nitrites, I think it is now generally agreed that they are totally ineffective. Although these drugs act the first or the second time they are given, they prove ineffective after only two or three days.

P. COTTIER: I would like to ask Dr. SOWTON if there is any risk in combining nitroglycerin with a beta-blocker in patients with arteriosclerotic hypertension. I am thinking of the possibility of syncope in orthostasis.

E. SOWTON: Although it does happen that patients react to nitroglycerin with a considerable fall in blood pressure, this nevertheless seems to be a rare occurrence. I think this problem can be dealt with by carefully trying out the combined treatment under controlled conditions.

F. GROSS: Would someone now like to comment on the combined administration of beta-blockers together with digoxin or other digitalis preparations and/or diuretics in cases of angina?

E. K. CHUNG: In cases where the patient displays significant symptoms of heart failure, I would not hesitate to use digitalis in combination with a beta-blocker. But in the presence of only mild, virtually insignificant symptoms due to heart failure I try not to use digitalis as a prophylactic measure; in such instances, the combination of a diuretic with a beta-blocker could be tried first of all, and, if the symptoms of heart failure subsequently become more severe, digitalis could be added.

W. SOMERVILLE: I would point out that the term "heart failure" is a loose one. If heart failure is defined as a condition marked by an elevated venous pressure and an enlarged heart, with or without radiological equivalents of heart failure, then one must be extremely careful about advising practitioners to use beta-blockers.

2.5. Long-term effects of beta-blockers

B. GARNIER: One last point on the subject of angina. If I remember rightly, in the general discussion on the use of beta-blockers in hypertension, Dr. WERKÖ said that, on the basis of his data, treatment with beta-blockers may reduce the incidence of sudden deaths but has no effect on the incidence of myocardial infarction. Is this correct, Dr. WERKÖ?

L. WERKÖ: Yes, but I ought to say that confirmation is needed. More patients will have to be studied under controlled conditions before treatment with beta-blockers can be recommended as a means of preventing sudden death.

3. The use of beta-blockers in the treatment of hypertension

3.1. Criteria for the selection of patients and choice of treatment

J. MÉNARD: A certain number of criteria have been proposed for the selection of patients for treatment with beta-blockers. These criteria should be discussed from two

---

* FITZGERALD, J. D.: A new test on the degree of adrenergic beta receptor blockade. Int. J. clin. Pharmacol. Ther. Toxicol. *4*, 125 (1970)

angles: firstly, have they been proved to be valid? Secondly, can they be applied on a wide scale in general practice today? The following four different types of criteria have been suggested: haemodynamic criteria, criteria based on the classification of essential hypertension according to the various W.H.O. stages, criteria based on examination of the renin-angiotensin system, and, possibly, criteria based on examinations of sympathetic tone.

R. C. TARAZI: I shall confine my remarks first of all to the use of beta-blocking agents as monotherapy. We started from the premise that hypertensive patients displaying signs of a hyperkinetic circulation – i. e. either a high cardiac output or a high heart rate – might be more responsive to beta-blockers than hypertensive patients with a normokinetic or hypokinetic circulation. While this premise subsequently proved unfounded, it is nevertheless true that beta-blockers will control a hyperkinetic circulation. Whether or not they will also control the associated hypertension, cannot be predicted with certainty. Their effectiveness in this indication can therefore only be established by empirical means. On the other hand, I do not know of any haemodynamic criterion – apart from heart failure, of course – that would preclude the use of beta-blockers in hypertension.
As regards combined therapy for hypertension, beta-blockers have a valuable part to play in preventing the subjective manifestations of tachycardia or hyperkinetic circulation resulting from the administration of vasodilators. We have also found the combination of beta-blockers with diuretics to be very effective.

B. GARNIER: Dr. BÜHLER, you have spoken about the renin-sodium index, and you also said that more reliable renin kits may be just around the corner. Do you think that in the near future this renin-sodium index could be used as a criterion for selecting patients for treatment with beta-blockers in general practice?

F. R. BÜHLER: Let me approach this question from a slightly different angle. I think that data are now available to show that the majority of all patients with essential hypertension could be satisfactorily treated with one single drug. Both beta-blockers and long-acting diuretics have a key role to play as monotherapy in such cases. Not only are the beta-blockers effective and well tolerated in high-renin hypertensives and in about 50–60% of normal-renin patients, but also they suppress renin and angiotensin in such cases. What is more, they may inhibit the vasculotoxic effect of these two substances, they diminish platelet adhesiveness and possibly also viscosity, and, as Dr. WINTERHALTER has shown, they may reduce the haematocrit. All these factors will lead to an improvement in tissue perfusion and perhaps also to a diminution in coronary complications. Even if an infarct does occur, we know now that its size may be reduced as a result of either prior or immediate treatment with beta-blockers.
As for the long-acting diuretics, it may be true that diuretics are not the best therapy for hypertension and, as reported in some studies, may possibly be partly responsible for the fact that the incidence of myocardial infarction has not declined as a result of antihypertensive therapy. If this view is borne out by further investigations, there will be good grounds for combining diuretics with renin-suppressive beta-blockers or even replacing them with vasodilators. For the time being, however, the first priority is to lower blood pressure at the cost of as few side effects as possible.
Although long-acting diuretics, such as chlortalidone (®Hygroton), have been shown to be most effective in low-renin patients, I don't think that we necessarily have to subclassify our patients in terms of their renin-sodium index. On the other hand, we can use this index in order to define different types of hypertension from the pharmacological angle and thus to ensure that antihypertensive agents are used in a more systematic manner.

A. ZANCHETTI: We have had considerable success so far in treating essential hypertension by means of a completely empirical approach, i. e. simply by endeavouring to reduce the elevated blood pressure. But, in doing so, we have not really learnt much about the causes of this very complex disease. Being a pathophysiologist myself, I would be the last to discourage people from trying to acquire a better understanding of the physiopathology of hypertension. At the same time, I would urge that we should not attempt to translate physiopathological hypotheses into recommendations for the G.P. Hence, any suggestion that diuretics can already at this stage be replaced by beta-blockers – let alone by vasodilators which, of course, exert an extremely marked sodium-retaining effect – leaves me very doubtful.

K. D. Bock: I think that, when considering the use of beta-blockers by general practitioners, we have to start from three premises: 1. The G.P. must make sure that his patient is, in fact, suffering from a chronic form of hypertension which calls for treatment, but which cannot be eliminated by surgical measures; 2. The G.P. is not in a position to carry out haemodynamic studies, or to determine renin activity; 3. There is apparently no difference between essential and renal hypertension as far as amenability to treatment with beta-blockers is concerned.

Let me state right away which forms of hypertension should not, in my opinion, be treated with beta-blockers as the drugs of first choice – namely, cases of very severe, and above all malignant, hypertension and hypertensive emergencies. For these indications, of course, we already possess drugs that exert a reliable effect. Naturally, this doesn't exclude the possibility of employing beta-blockers as *supplementary* medication in such cases. For mild, labile hypertension, on the other hand, I believe that today beta-blockers may well be the drugs of first choice. As regards the large group of cases falling between these two extremes, the beta-blockers – due allowance being made for their contra-indications – can be considered to be on a par with the classic antihypertensive agents.

L. Werkö: Much of what I wanted to say has just been said by Dr. Bock. Let me emphasise, though, that the main difficulty facing the G.P. is to decide whether or not the patient has hypertension. If the G.P. accidentally discovers that a patient has a blood pressure of, say, 165/110 mm. Hg, he should keep an eye on him, measure his blood pressure on several further occasions, and make sure that chronic hypertensive disease is present before he starts treatment.

P. Cottier: I merely wish to endorse what Dr. Bock has said. We must make it clear to the general practitioner that empirical tests with beta-blockers should of course only be carried out in cases of mild to moderate hypertension and not in severe cases; otherwise, we shall be providing the G.P. with a negative motivation, because monotherapy with beta-blockers will probably not yield satisfactory results in severe cases. This, however, also applies to the use of diuretics as monotherapy.

F. R. Bühler: There are three points that I would emphasise in connection with guidelines for the general practitioner: firstly, drugs should be used in a systematic fashion; secondly, due consideration should be given to the advantages of monotherapy; and, thirdly, since low-renin hypertension is more frequent in higher age groups, treatment should be started with a beta-blocker in young patients and perhaps with a diuretic in older ones.

J. G. Lewis: To revert to the question of contra-indications to the use of beta-blockers, I feel that heart failure should not necessarily be regarded as an absolute contra-indication. Provided the patient is first given treatment with digoxin and diuretics, beta-blockers can in fact quite safely be used in the presence of heart failure. I myself very often employ them in such cases because they possess certain advantages. Fibrillation, for example, can be brought under control without one having to resort to unduly high doses of digoxin. Moreover, elderly people *can* take beta-blockers together with digoxin and diuretics. In short, heart failure, provided it is controlled, constitutes no great danger – certainly not so far as treatment with, for instance, oxprenolol (®Trasicor) is concerned.

U. C. DUBACH: Concerning the types of case which are to be regarded as unsuitable for treatment with beta-blockers, I should like to revert to what has already been said about the use of these drugs in elderly subjects suffering from angina pectoris. In elderly and very old patients with hypertension, I myself would be extremely cautious about employing beta-blockers, and I should be interested to know whether anyone here has had experience with them in the treatment of patients belonging to this age group. In my view, combined therapy with a diuretic and reserpine has yielded sufficiently good results in such cases, and – in view of possible changes in the sinus node – I hardly think that in elderly subjects one would initiate treatment for hypertension with beta-blockers.

A. EISALO: In connection with the treatment of elderly hypertensives, I should like to refer to a trial we conducted with the beta-blocker alprenolol in a group of 43 such patients, whose ages ranged from 61 to 89 years. Of the five patients who dropped out of the trial, only one had congestive heart failure, and he was withdrawn not because of his heart failure but mainly because he refused to cooperate. With regard to previous drug therapy, the patients could be considered as being in a "steady state", inasmuch as the drugs they were taking had been commenced at least six months before the trial. It is interesting to note that the systolic blood pressure in these patients fell during the first week of treatment with alprenolol, administered in a dosage of 400 mg. daily. Although the diastolic pressure dropped more slowly than the systolic, it, too, showed a significant reduction. In many patients heart volume decreased during the trial even without supplementary treatment with diuretics or digitalis. After three months of treatment the patients' condition had improved considerably and they were completely free from subjective symptoms. Congestive heart failure developed in two cases – on both occasions during the first week of treatment with the beta-blocker. In other words, if this complication is going to occur, it is likely to do so within the first week of treatment and not later. In both our cases the heart failure could easily be compensated by administering low doses of digitalis.

F. GROSS: I take it then, Dr. EISALO, that you, too, feel that beta-blockers can be given to elderly patients, provided these patients are carefully followed up. This would answer Dr. DUBACH's query and would also be of interest to the general practitioner.

K. D. BOCK: I, too, would not hesitate to give an elderly patient with mild hypertension and angina pectoris a beta-blocking agent, because this treatment can achieve two objectives in such cases: it can reduce the blood pressure and it can also relieve the angina.

3.3. Systolic hypertension and labile hypertension

W. SOMERVILLE: I would remind you of one of the commonest manifestations of "high blood pressure" in the elderly, in which the systolic level is excessively raised and the diastolic pressure little if at all. In this situation, antihypertensives of the ganglion-blocker type are disastrous. Beta-blockers, on the other hand, can sometimes be of great help, but they should be used with care; much smaller doses than are customarily prescribed are often capable of combating the excessive rise in systolic pressure.

D. W. GAU: As one of the few general practitioners in this audience may I ask a question at this point? G.P.s are repeatedly told that, before considering the type of treatment to prescribe, they first have to decide whether the patient has chronic hypertension. But wasn't it demonstrated quite a long time ago that labile hypertension develops into chronic hypertension in a high percentage of cases? Specialists tell us that we must ignore labile hypertension, but, when such a case becomes chronic, they wonder why we haven't already been treating it. My own attitude now is, in fact, to treat labile hypertension and then slowly to wean the patients off the tablets if necessary.

W. SOMERVILLE: The question of labile hypertension raised by Dr. GAU should not be passed over lightly, since many people subject to phases of labile hypertension do eventually develop true, fixed, high-resistance hypertension. One particular group of subjects in whom labile hypertension is important is that of commercial air pilots. In the United Kingdom, these pilots are examined first by a general practitioner. If he encounters readings of 160 over 95–105, he has to say to the pilot: I cannot pass you for your licence, you will have to proceed to further examinations. With each subsequent examination, anxiety-induced tachycardia further escalates the blood pressure. A practice I often use when these pilots reach me is to repeat the blood pressure measurement one and a half hours after a dose of 20–40 mg. oxprenolol; the figures are then often found to be normal.

F. GROSS: What you just mentioned, Dr. SOMERVILLE, seems to confirm that, as has already been said before, beta-blockers and/or diuretics are indeed useful for treating patients with labile hypertension. I think this is something that can be recommended to the practitioner.

K. D. BOCK: The points raised by Dr. SOMERVILLE relate mainly to the diagnosis of hypertension. Firstly, there is the problem of the emotional stress to which the patient is exposed in presenting himself to the doctor; the only way of offsetting this factor is by repeated blood pressure measurements, which should preferably be taken by auxiliary personnel and not by the doctor himself. Secondly, there is the problem of systolic blood pressure elevation in elderly patients, which we believe is due in most of these cases to a loss of elasticity. Except in cases where the systolic pressure is extremely elevated, we wouldn't regard these as patients with chronic hypertension in the proper sense of the term; in my opinion, many of them don't require any medical treatment at all for this elevated systolic pressure, provided that their diastolic pressure is below 95 mm. Hg.

S. H. TAYLOR: The notion that the sequelae and complications of hypertension derive entirely from the diastolic pressure really doesn't stand up to scientific scrutiny. Epidemiological studies have, in fact, shown quite clearly that cardiovascular morbidity and mortality are as closely linked with the systolic pressure as with the diastolic. In the U.S.A., for example, GUBNER* has shown that at any level of diastolic pressure the cardiovascular morbidity and mortality increases in direct proportion to the associated systolic pressure. This may well mean that, in a patient with elevated diastolic pressure, we should pay at least as much attention to the accompanying increase in systolic pressure – though I admit that the elderly patient perhaps falls into a separate category. It also highlights one possibly important aspect of the effect of the beta-blockers that seems to have been largely overlooked.

K. D. BOCK: Though I am well aware that in recent years papers have been published on the positive statistical correlation existing between the systolic pressure and arteriosclerotic complications, these studies do not prove that there is any causal connection between the two. It is possible and indeed probable that, in elderly people in whom the rise in blood pressure is confined to the systolic readings, this rise is a sign of arteriosclerosis and of a correspondingly enhanced danger of arteriosclerotic complications. But this would mean that the increase in the systolic pressure is simply an *indicator* of the risk involved and not a *factor contributing to that risk*.

R. C. TARAZI: I'd like to make a few points concerning systolic hypertension and hypertension of stress in relation to treatment with beta-blockers. The first point arises

---

* GUBNER, R. S.: Systolic hypertension: a pathogenetic entity. Significance and therapeutic considerations. Amer. J. Cardiol. *9, 773* (1962)

from what Dr. TAYLOR has just said about the significance of the systolic pressure. Systolic hypertension in our view is not a homogeneous entity. In some patients it is related to an increased stroke volume and possibly an increased heart rate, as Dr. SOMERVILLE pointed out. In quite a large number, however, it is associated with a low stroke volume and diminished aortic distensibility; in such cases, a particularly cautious approach to therapy is indicated, because the drop in systolic pressure in response to treatment will be much more precipitous than whatever decrease occurs in diastolic pressure. But both forms of systolic hypertension – whether associated with a high stroke volume or with a low stroke volume – should in fact be treated.

As regards the episodic rises in pressure occurring in connection with the stress of a visit to the doctor, pain, or static exercise, beta-blockers blunt the increase in heart rate and cardiac output encountered in such cases, but not the increase in blood pressure; that is to say, they alter the mechanism of the episodic hypertension, but do not prevent its occurrence.

3.4. Tolerability and side effects of beta-blockers

B. GARNIER: Before we discuss the question of treatment schedules, allow me briefly to recapitulate certain facts about the beta-blockers. If we advise the general practitioner to use these drugs, it is because they possess certain major advantages as regards tolerability and side effects. The first advantage seems to me to be the absence of postural hypotension, an advantage which distinguishes the beta-blockers from virtually all other antihypertensive agents; I am thinking here not only of the alpha-blockers and the sympathicolytics, but also of methyldopa, which can provoke quite severe postural hypotension in certain cases. A second advantage of the beta-blockers – and one which is of particular importance for young patients – is that they do not influence sexual function. Thirdly, you will no doubt all agree with Dr. KIELHOLZ that central nervous effects are fairly uncommon. The beta-blockers seldom give rise to insomnia or nightmares, and they may even induce a feeling of well-being. Some patients have told me spontaneously that they felt better during beta-blocker therapy than they had ever felt before. This is an important factor, because it may well encourage the patient to continue treatment and to take his drug regularly. There is also the preventive aspect to be considered – i.e. the fact that beta-blockers can prevent arrhythmias in hypertensive patients, that they possess a cardioprotective action (the last word has obviously not yet been said on this subject), and that they exert an anti-stress effect which enables hypertensives to take part without risk in activities liable to give rise to emotion and nervous tension. Finally, as has also been pointed out in the course of our symposium, the beta-blockers display a wide therapeutic margin, with the result that the doctor is relatively free to select whichever dose he considers appropriate for a given situation.

3.5. Treatment schedules

B. GARNIER: Now let us pass on to the question of treatment schedules. Should the general practitioner be advised to use beta-blockers as monotherapy or in association with other drugs? And in what stages or forms of hypertension can they be regarded as the drugs of first choice? May I ask Dr. MÉNARD to say a few introductory words on this subject?

J. MÉNARD: To simplify matters, let us perhaps begin by considering the type of treatment schedule that would be applicable to patients with fixed essential hypertension who are found at two or more examinations to have a blood pressure in excess of 160/95 mm. Hg. In general practice, I feel that in such cases diuretic treatment should still be given first of all for a period of three weeks or a little longer. If, despite systematic diuretic therapy, the blood pressure levels remain above 140/95 mm. Hg, another antihypertensive agent will have to be added to the regimen. The beta-blockers certainly offer interesting possibilities here. Their effectiveness has been amply demonstrated during the present symposium, and their side effects are definitely of less consequence than those of the antihypertensives in current use. The dosage should be increased gradually to a level of 160–320 mg. daily – at least as far as propranolol or oxprenolol is concerned. In everyday practice there seems to be no justification for increasing the dosage to the daily amounts of 1 g., or even 4 g., which have occasionally been employed. If, after three or four months of treatment, the blood pressure has still not returned to normal levels, one will have to administer in addition a drug whose pharmacological action complements that of the beta-blocker. Hydralazine, or dihydralazine, in a maximum daily dose of 75–150 mg., has advantages from this point of view, particularly since its cardiac side effects (tachycardia) are inhibited by prior administration of the beta-blocker. Headache, another side effect of the hydralazines, is unfortunately not suppressed by beta-blockade. Provided the hydralazine compound is not given in a daily dosage exceeding 150 mg., the risk of its provoking rheumatoid arthritis or disseminated lupus erythematosus is minimal. Finally, if satisfactory control of the hypertension is still not achieved with this already formidable battery of drugs, other agents, such as guanethidine or betanidine, will have to be added to the regimen.

This, then, is one kind of treatment schedule that can be put up for discussion at the present time, i. e. in 1974. It is applicable to patients with fixed essential hypertension who are aged between 30 and 65 years, and who do not have a hyperkinetic syndrome, renal insufficiency, progressive arterial hypertension, or insulin-dependent diabetes.

B. Garnier: I would remind you all that two other, somewhat different treatment schedules were proposed in the papers by Dr. Sannerstedt and Dr. Lewis. Dr. Sannerstedt advocates that treatment should be built up in accordance with the W.H.O. stage of the hypertension: in Stage 1 a beta-blocker should be given alone, in Stage 2 it should be combined with a vasodilator, and in Stage 3 the beta-blocker should be used to supplement basic treatment with a diuretic. Dr. Lewis bases his approach on the severity of the hypertension: in mild cases, a diuretic alone (if its effect is insufficient, a beta-blocker may be added); in moderate hypertension, a diuretic plus beta-blocker and, if necessary, methyldopa; and in severe hypertension, basic treatment with guanethidine or betanidine; in the presence of renal insufficiency, recourse should be had to methyldopa.

P. Cottier: Particularly in view of the side effects of the diuretics, one could just as well start with the beta-blocker and then add a diuretic if required.

A. Zanchetti: Personally, I fully agree with Dr. Ménard when he says that diuretics should still constitute the back-bone of antihypertensive therapy. Dr. Bartorelli stressed this point too in his introduction to our session on hypertension. In most cases of hypertension, diuretics should in fact still be the first drugs to try. But what should one use as a second drug? A list of good reasons for choosing a beta-blocker has been given here, but at present – on the European continent at any rate – the second drug used is still generally reserpine. It therefore seems to me extremely important at this stage to examine very thoroughly the question as to what advantages the beta-blockers do indeed offer as compared with reserpine, which still has the advantage of a prolonged duration of action.

J. Brod: With regard to the question of using beta-blockers in combination with other antihypertensive agents, I'd like to point out once again that beta-blockers not only lower cardiac output but at the same time also eliminate the muscular vasodilatation which forms part of the haemodynamic pattern of early hypertension. I therefore believe that there may definitely be justification for a combination of beta-blockers not only with a diuretic agent – which can also be regarded as a mild vasodilator – but even with a stronger vasodilator.

A. Froment: I just wanted to point out that – in France at all events, and I think the same applies to many other countries at the present time – the main problem with hypertension is not so much the choice of treatment as the need to ensure that the hypertensive patient actually is treated. Out of every ten hypertensives there are eight or nine who are not receiving any treatment. Bearing this in mind, I believe that the diuretics must still be advocated as the first weapon to use, because general practitioners are now familiar with them.

B. Garnier: You have obviously raised a very important question there, Dr. Froment, but it strikes me that Dr. Cottier's suggestion is just as valid, especially if we remember that, as was discussed at length in the first part of our symposium, the diuretics stimulate the release of renin. On the other hand, we also have to consider that, if our aim is to ensure that large numbers of hypertensives receive the treatment they require, then the medication we recommend must be simple. Both reserpine and diuretics such as chlortalidone have a prolonged duration of action and in many cases need be given in a dose of only one tablet daily or even every other day.

C. Bartorelli: It is certainly important, as Dr. Froment said, to ensure that hypertensives are actually treated by general practitioners, but we must also bear in mind the fact that hypertensives, like diabetics, require treatment every day for the rest of their lives. Consequently, not only the doctors but also the patients have to be made to appreciate this requirement. I agree that the treatment must be very simple. You cannot go on adding one pill to another, because, if you do, the patient will stop taking his medication after two or three months. As regards the type of treatment to be advocated, I, too, feel that preference should be given to the diuretics, because they are simple to handle.

K. D. Bock: I don't think that, *vis-à-vis* general practice, we should lay down a firm rule that the drug of first choice has to be a diuretic. After all, long-term treatment with diuretics calls for biochemical determinations which are not always easy to carry out under the conditions of general practice. For example, hypopotassaemia induced by diuretics is apt to increase the toxicity of digitalis – to say nothing of the disturbances diuretics may cause in uric acid and glucose metabolism. I consider it to be a great advantage of the beta-blockers that, so far as we are aware today, their use in general practice does not necessitate biochemical monitoring.

A. Froment: In support of what Dr. Garnier and Dr. Bartorelli have just said, may I draw your attention to the results of a screening campaign we conducted in a population of 12,000 gainfully employed persons. In this population we found 825 people who at three separate examinations fulfilled the W.H.O.'s criteria for the diagnosis of hypertension (i.e. blood pressure levels of 160 and/or 95 mm. Hg or above). Their average blood pressure at the third examination was 165/100 mm. Hg. This underlines the fact that the overwhelming majority of hypertensives in the population at large have only slightly raised blood pressure levels, and that the general practitioner is thus confronted with a type of hypertension radically different from that encountered in a hospital setting. I say this despite the fact that in the general practice study referred to by Dr. Burley in his paper the mean blood pressure levels were in the neighbourhood of 200/115 mm. Hg, and therefore comparable to mean levels recorded in hypertensives treated in, for example, our hospital department.
May I also endorse Dr. Bartorelli's plea that the treatment should be simple. Finnerty et al.*, who used diagnostic criteria similar to those of the W.H.O., observed that hypertension could be controlled in 80% of cases with only one tablet of 50 mg. chlortalidone daily – which is a cogent argument for recommending the wide-scale use of diuretics as the initial weapon in the type of hypertension seen in general practice.

---

* Finnerty, F.A., Jr., Shaw, L.W., Himmelsbach, C.K.: Hypertension in the inner city. Detection and follow-up. Circulation *47*, 76 (1973)

W. SOMERVILLE: I was particularly impressed by Dr. MÉNARD's suggestions. However, for practitioners there is one stage – and to the best of my memory it is Stage 3 in the MÉNARD classification – at which they may run the risk of overlooking secondary hypertension. Among 100 patients whose blood pressure is elevated, there will be two to five who have secondary hypertension. Consequently, we should advise practitioners to take the following extra step at this stage: if the patient's blood pressure has still failed to respond, despite the measures suggested by Dr. MÉNARD, then the practitioner should refer him to hospital for a few simple screening tests which can be done in 24 hours.

M. E. CARRUTHERS: As one of the two other general practitioners present besides Dr. GAU, I'd like to raise two points. One is that in epidemiological studies it is usually the casual initial systolic blood pressure reading which is taken to be the important prognostic factor. The second point is that, if we recommend diuretics or perhaps Rauwolfia, we should bear in mind that these were in fact among the drugs used in the Veterans Administration study which failed to reduce the incidence of cardiac complications of hypertension.

H. J. BEIN: So far as I am aware, there is no evidence that Rauwolfia has really failed to reduce the incidence of cardiac complications in hypertensives. Combinations of Rauwolfia with a diuretic or some other antihypertensive constitute today probably the most widely used form of treatment for hypertension. Such combinations lead the field in the United States as well. If we consider the incidence of side effects encountered with the Rauwolfia preparations, including in particular depression, in relation to the frequency with which these preparations are prescribed, we have to acknowledge that they are in fact extremely well tolerated.

J. MÉNARD: The main question, then, is to decide whether we now have in 1974 sufficient documentation to be able to tell the general practitioner that the drug to use first of all in the treatment of moderate hypertension is a beta-blocker and not a diuretic. I don't think this question can really be answered until further clinical trials have been carried out. As for the immediate prospects, it may well be that monotherapy with beta-blockers possesses practical advantages, particularly since these drugs have few side effects. In the more distant future, they may also open up interesting possibilities as a means of preventing myocardial infarction and sudden death. Nevertheless, it still remains to be clearly demonstrated that these drugs are at least as effective and as easy to handle as the diuretics, and that they do not display any harmful side effects in the very long term.

B. GARNIER: I think the views expressed by Dr. MÉNARD correspond to those that are still fairly widely held today. The general consensus of opinion seems in fact to be that, for the time being and until further information becomes available, the diuretics should in principle continue to be regarded as the basic antihypertensives for use in general practice; on the other hand, it is acknowledged that beta-blockers alone or in combination with diuretics can also play a very valuable role in many cases.

F. R. BÜHLER: I feel that in this discussion we can agree on one common denominator, and that is that in a modern treatment schedule use should be made above all of beta-blockers, diuretics, and, if necessary, vasodilators. That would also fit in with the plan of treatment proposed in the paper I presented. Still left open is the question of the order in which the various drugs should be employed, and here I would not be too dogmatic.

B. GARNIER: Before we close this session, may I express my gratitude to Dr. GROSS and Dr. MÉNARD for the help and support they have given me. I should also like to thank you all for having taken such a lively part in this discussion of ours on the guidelines for the use of beta-blockers in general practice.

Closing addresses

W. Schweizer: This symposium, which now draws towards its conclusion, was intended to serve a dual purpose. Firstly, to provide practising physicians with a review of developments to date in the field of treatment with beta-blockers and, secondly, to suggest certain specific lines along which research workers might profitably undertake further studies with drugs of this category. I think we have come close to attaining both these objectives. Speaking for myself, I at all events now have a clearer impression as to what we know today and what we still need to know. For example, ignoring for a moment the anti-arrhythmic activity of the beta-blockers, I now have a better idea as to how much – or, rather, how little – significance can be ascribed to those differences between the various beta-blockers that have been demonstrated to date, and I am now also aware of the points on which it is absolutely essential that our knowledge should be extended, namely, points relating to the basic pharmacokinetic properties of each individual beta-blocker.

The present symposium has helped to define more precisely the role that beta-blockers can play in the treatment of arterial hypertension, and it has also shown that this role is undoubtedly of major interest; it will probably increase in scope and importance, depending on the number of doctors who can be prevailed upon to treat hypertension in its initial stages – in other words, especially in younger patients. In this connection, further symposia on hypertension are urgently required, particularly since specialists still differ considerably in their views. With regard to this problem, it would also be of great value if additional research teams were to re-examine the significance of renin activity.

The use of beta-blockers in myocardial ischaemia – together, of course, with nitroglycerin – is no longer in dispute and is now acknowledged to have marked a genuine advance. The circumstances under which one has to fall back upon surgical treatment have been clearly outlined by Dr. W. Somerville. We now have quite a good idea as to what can be achieved by reducing the oxygen requirement of the heart and as to when it becomes necessary to take measures to improve myocardial blood flow. That myocardial ischaemia may also occur in healthy subjects – i.e. in certain unpleasant situations associated with an excessive increase in the myocardial oxygen requirement – is in itself perfectly plausible, since there must presumably always be a limit to the extent to which blood flow can be stepped up. Physical activity as such is hardly likely to endanger us in this respect, because the Creator in his wisdom has so arranged matters that our legs give way before this limit is reached. It is quite conceivable, however, that we have not been provided with similar protection against the repercussions of psychic stress. But, before concluding that Western man requires the protection of beta-blockers, we would first of all have to prove that the psychic stress is pathogenic, i.e. that it really does constitute one of the causes of cardiovascular disease.

As regards the use of beta-blockers in disturbances of cardiac rhythm, here the situation is fairly clear: irrespective of whether or not Wolff-Parkinson-White syndrome is present, the beta-blockers prove extremely effective in preventing paroxysmal tachycardia. One point, on the other hand, which still urgently calls for clarification is the effect that beta-blockers – employed for pre-coronary and/or post-coronary care – can exert in ischaemic heart disease.

Another condition in which the value of treatment with beta-blockers is now undisputed is hypertrophic cardiomyopathy. Here I agree with Dr. Sowton that beta-blockers are invariably indicated, one reason being that they serve to prolong diastole and hence to improve ventricular filling, which in this disease is reduced as a result of abnormal "compliance". Incidentally, I also consider it necessary to make further efforts to ensure that doctors become more familiar with the diagnosis of this by no means rare disease.

The climax of our symposium has, I feel, been today's concluding discussion in which, under the joint chairmanship of Dr. Garnier, Dr. Gross, and Dr. Ménard, we have tried to work out guidelines for the use of beta-blockers in general practice. I regard it as extremely important that those of us who are specialists should provide doctors fighting in the front line, as it were, with a flow of reliable and practical information. We should, in fact, never forget that specialists and general practitioners must work together as a team.

This brings me to the end of the remarks I wished to make. It now only remains for me to perform the pleasurable duty of expressing, firstly, my personal thanks to all participants for their active and disciplined cooperation and, secondly, a general vote of thanks on behalf of you all to CIBA-GEIGY. We are grateful to the latter for having hit upon the excellent idea of holding a symposium on the present status and future prospects of treatment with beta-blockers, for having acted upon the idea so successfully, and for having so kindly invited us to discuss this highly topical theme at a meeting of eminent experts gathered together in the sunshine and gentle breezes of the Mediterranean. Last but not least, I think we owe a special word of thanks to the office staff of CIBA-GEIGY who have been toiling away in the Secretariat and particularly to the interpreters; their work has in my opinion been deserving of the highest praise.

R. Oberholzer: Having been accorded the privilege of also addressing a few words of thanks to you on behalf of the Management of the Pharmaceuticals Division of CIBA-GEIGY, I should like to begin – albeit, so to speak, after the event – by referring to the real purpose underlying symposia of this type organised by the pharmaceutical industry, at which usually a medley of information is presented.

Here in Juan-les-Pins, too, we have been discussing in varying depth questions relating not only to basic research but also to clinical diagnosis and to ways and means of making the best use of the therapeutic armamentaria at our disposal. In some instances, even the papers emanating from CIBA-GEIGY itself have been more concerned with the products of other manufacturers than with our own. I can therefore understand why, for example, one participant asked me in an almost pitying tone what use this symposium really was to our company.

The answer to this question is, I think, contained in the sentence with which Dr. Taggart concluded the paper he presented here: "Beta-blockers may well become unique in being the first group of drugs which, though developed in order to benefit the abnormal heart, can also be used to treat the normal person."

The prophecy implied in this statement may at the moment sound futuristic; perhaps it even smacks of utopianism. But it does impart the feeling that in the realm of cardiovascular diseases we are now about to witness the dawn of an era of prophylactic therapy. This is a feeling which we in the pharmaceutical industry also share, and it was for this reason that we considered it so important to invite experts from near and far to review the whole situation as it appears at present. The conclusions they have now reached will have a cardinal bearing on our future experimental and clinical research activites.

So much for the purpose of this symposium – a purpose which, I should like to emphasise, has in my view also been largely achieved. It will now be up to us in the pharmaceutical industry, as well as to those of you who are working in hospital or in private practice, to draw the correct inferences from the wealth of information given us here.

In conclusion, allow me to offer our warmest thanks to all participants for having undertaken the journey to Juan-les-Pins, for the painstaking preparatory work they have done prior to the symposium, and for the frankness and fairness with which they have presented and discussed their experience and findings. A very special vote of thanks – which I take the liberty of expressing on behalf of all of us here – is due to Dr. Schweizer for having chaired our symposium with such an excellent combination of strictness tempered with paternal indulgence; had it not been for him, I fear we might often have dallied upon trivialities or perhaps even wandered down the occasional blind alley.

I also wish to tender my own personal thanks to the accomplished team of interpreters who contributed so much to the success of our meeting. And, finally, I should like to thank Dr. ADAMS and his assistants. They have contrived to organise the symposium so well that we have been quite oblivious of the organisation – and that is no mean feat on their part.

Index

312

313